Data/Knowledge-Driven Behaviour Analysis for Maritime Autonomous Surface Ships

Data/Knowledge-Driven Behaviour Analysis for Maritime Autonomous Surface Ships

Editors

Yuanqiao Wen
Axel Hahn
Osiris Valdez Banda
Yamin Huang

MDPI • Basel • Beijing • Wuhan • Barcelona • Belgrade • Manchester • Tokyo • Cluj • Tianjin

Editors

Yuanqiao Wen
Intelligent Transportation Systems Research Center
Wuhan University of Technology
Wuhan
China

Axel Hahn
Department of Computing Science
Carl von Ossietzky Universität Oldenburg
Oldenburg
Germany

Osiris Valdez Banda
Department of Mechanical Engineering (Marine Technology)
Aalto University
Espoo
Finland

Yamin Huang
Intelligent Transportation Systems Research Center
Wuhan University of Technology
Wuhan
China

Editorial Office
MDPI
St. Alban-Anlage 66
4052 Basel, Switzerland

This is a reprint of articles from the Special Issue published online in the open access journal *Journal of Marine Science and Engineering* (ISSN 2077-1312) (available at: www.mdpi.com/journal/jmse/special_issues/maritime_autonomous_surface_ships).

For citation purposes, cite each article independently as indicated on the article page online and as indicated below:

LastName, A.A.; LastName, B.B.; LastName, C.C. Article Title. *Journal Name* **Year**, *Volume Number*, Page Range.

ISBN 978-3-0365-7443-1 (Hbk)
ISBN 978-3-0365-7442-4 (PDF)

Contents

About the Editors

Yuanqiao Wen

Dr. Yuanqiao Wen is a full professor in Wuhan University of Technology, Deputy Director of the National Water Transport Safety Engineering Technology Research Center, and Deputy Director of the Intelligent Transportation System Research Center of Wuhan University of Technology. He mainly engaged in research on water traffic safety and environment, waterborne unmanned systems (unmanned boats), ship intelligent navigation, and artificial intelligence (knowledge engineering) water traffic applications. He has successively led four projects in the National Natural Science Foundation of China and participated in two projects from the National Major Science and Technology Support Program (China).

Axel Hahn

Dr. Axel Hahn is a full professor and leading the Working Group System Analysis and Optimization at the Carl von Ossietzky University Oldenburg. He was a board member at the associated institute OFFIS—an institute for Computer Science and coordinated research on maritime transport systems. Since January 2022, he has been Director of the DLR Institute Systems Engineering for Future Mobility, which emerged from the OFFIS Transport R&D department. His research addresses design, simulation, and analysis of critical social technical systems in maritime transportation.

Osiris Valdez Banda

Dr. Osiris A. Valdez Banda is Assistant Professor of Marine Technology (Research Group on Safe and Efficient Marine and Ship Systems). Their research focuses on the analysis of marine risks and safety systems engineering with applications in the concepts of smart shipping and ship winter navigation. Their work focuses on the development of methods and processes for modeling and analyzing the management of the risks, safety, and efficiency of the entire maritime ecosystem with a focus on ship design and operation.

Yamin Huang

Dr. Yamin Huang is a full professor in the Intelligent Transport System Research Centre (ITS), Wuhan University of Technology. He has worked on enhancing Maritime Safety using emerging techniques. Specifically, he is interested in developing a Human–Machine Cooperative system supporting autonomous sailing, which aims to develop trustworthy Maritime Autonomous Surface Ships, and consequently develop state-of-the-art risk assessment tools for Waterborne Traffic Control.

Preface to "Data/Knowledge-Driven Behaviour Analysis for Maritime Autonomous Surface Ships"

With the development of artificial intelligence and ICT, ships are expected to be smarter than traditional ships in the future, as they can autonomously navigate from one point to another point in the waters. Several studies have developed various systems to achieve such goals. As we can see, the MASS can handle many tasks with explicit references, such as speed following, course keeping, path following, etc. However, the MASS does have some limitations in operating some complicated tasks that need the machine to make decisions and adjust its reference according to the recognized traffic scene, such as collision avoidance, emergent operations, etc. In this process, we found that the recognition and prediction of ship behavior are essential for the recognition of traffic scenes, which will influence the decision outcomes. For instance, when two ships encounter each other, the give-way ship's behavior will influence the decision of the stand-on ships. Thus, we believe the study on ship behavior would benefit the development of MASS.

Recently, the developments of equipment onboard ships enrich our data source to analyze ship behavior, such as radar, Automatic Identification Systems (AIS), CCTV, etc. These Maritime traffic data (e.g., radar data, AIS data, CCTV data) provide designers, officers on watch (OOW), and traffic operators with extensive information about the states of ships at present and in history, which are a treasure for behavior analysis. Additionally, the development of knowledge analysis tools, e.g., Fuzzy systems, knowledge graphs, etc., offer a new insight to analyze the ship's behavior based on human knowledge, e.g., navigation rules and regulations. Combining multisource heterogeneous big data and artificial intelligence techniques inspires innovative and important means for understanding ship behavior and developing MASS. Thus, under the support of the Key R&D Program of Zhejiang Province (China) through Grant No. 2021C01010, this reprint collects 12 papers working on data/knowledge-driven behavior analysis for MASS and its applications, including data-driven behavior modeling, knowledge-driven behavior modeling, multisource heterogeneous traffic data fusion, risk analysis and management of MASS, etc.

Yuanqiao Wen, Axel Hahn, Osiris Valdez Banda, and Yamin Huang
Editors

Journal of
Marine Science and Engineering

MDPI

Editorial

Data/Knowledge-Driven Behaviour Analysis for Maritime Autonomous Surface Ships

Yuanqiao Wen [1,2], Axel Hahn [3], Osiris Valdez Banda [4] and Yamin Huang [1,2,*]

[1] Intelligent Transportation Systems Research Center, Wuhan University of Technology, Wuhan 430063, China; yqwen@whut.edu.cn
[2] National Engineering Research Center Water Transport Safety, Wuhan University of Technology, Wuhan 430063, China
[3] Carl von Ossietzky Universität Oldenburg, German Aerospace Center, 26121 Oldenburg, Germany; axel.hahn@uol.de
[4] Marine Technology, Department of Mechanical Engineering, Aalto University, 02150 Espoo, Finland; osiris.valdez.banda@aalto.fi
* Correspondence: yaminhuang@whut.edu.cn

Citation: Wen, Y.; Hahn, A.; Valdez Banda, O.; Huang, Y. Data/Knowledge-Driven Behaviour Analysis for Maritime Autonomous Surface Ships. *J. Mar. Sci. Eng.* **2023**, *11*, 635. https://doi.org/10.3390/jmse11030635

Received: 7 March 2023
Accepted: 10 March 2023
Published: 17 March 2023

This Special Issue, "Data-/Knowledge-Driven Behavior Analysis of Maritime Autonomous Surface Ships", includes twelve contributions [1–12] published during 2021–2022. Maritime traffic data (e.g., radar data, AIS data, and CCTV data) provide designers, officers on watch, and traffic operators with extensive information about the states of ships at present and in history, representing a treasure trove for behavior analysis. Additionally, navigation rules and regulations (i.e., knowledge) offer valuable prior knowledge about ship manners at sea. Combining multisource heterogeneous big data and artificial intelligence techniques inspires innovative and important means for the development of MASS. Thus, this Special Issue aimed to collect studies that provide new views on data-/knowledge-driven analytical tools for maritime autonomous surface ships, including data-driven behavior modeling, knowledge-driven behavior modeling, multisource heterogeneous traffic data fusion, risk analysis and management of MASS, etc. A brief overview of all the contributions, emphasizing the main investigation topics and the outcomes of the analyses, follows below.

Data-driven behavior modelling methods are powerful tools that can be used to discover ship manners from large amounts of data. Guo et al. [5] developed a deep convolutional neural network (CNN) for ship trajectory classification. The improved QuickBundle clustering algorithm was used to preprocess the trajectory data, the trajectory data were further converted into image data, and then a deep CNN-based trajectory classification model was developed. Based on the proposed model, the manually annotated dataset was set as the input for model training. By comparison with the traditional connected neural network model and SVM model, the proposed method can effectively distinguish ship trajectories in different waterways. Xu et al. [9] developed a prediction model for ship traffic flow in wind farms area. Instead of using time series data, a spatiotemporal dependence feature matrix was developed to predict the ship traffic flow, and a Gated Recurrent Unit (GRU) of a Recurrent Neural Network (RNN) was used to identify multiple traffic flow sections from complex waters. By comparison with traditional methods using traffic data from wind farms in Yancheng City (China), e.g., the Autoregressive Integrated Moving Average (ARIMA), Support Vector Machine (SVM), and Long Short-Term Memory (LSTM), the proposed method, based on spatiotemporal dependence, performs better than the current traffic flow prediction methods.

Knowledge-driven methods offer tools that teach the machine to understand a ship's behaviors. Zhong et al. [12] proposed an ontological ship behavior model based on COLREGs, which is expected to automatically perform reasoning based on the knowledge derived from COLREGs. Knowledge graph techniques were employed. The ship behavior was viewed as the changes in temporal–spatial attributes of the ship, as described using the Resource Description Framework (RDF), function mapping, and set expression methods. Rule 9 (Narrow Channel Article) from COLREGs was inputted into the proposed method to demonstrate the feasibility of the proposed method. The results show its potential for the complete machine reasoning of ship behavior knowledge in the future. Song et al. [6] proposed a semantic model of ship behavior based on the ontology model, which aims to help the machine to understand ship behavior from ship trajectory data. Multi-scale features of ship behavior are observed, and the behaviors are divided into four sub-scales in cognitive space, namely, action, activity, process, and event. As demonstrated in case studies, some typical behaviors are deduced using a reasoner, such as Pellet, based on defined axioms and semantic web rule language (SWRL). The proposed model shows potential for smart maritime management.

Multisource heterogeneous traffic data fusion broadens the range of sources for the observation of ship behavior, and vision-based sensors have become an important means. Chen et al. [2] studied ship intention detection and prediction methods based on observed ship behaviors using radar, cameras, and Automatic Identification Systems and proposed a vision and Bayesian framework. Traditionally, radar and AIS data have been used for ship behavior analysis and intention detection, whereas it is still difficult to detect real-time ship intention due to low data frequency. Thus, the authors proposed the addition of a vision-based sensor for intention detection and prediction and argued that it could be used for real-time intention detection and prediction in intersection waters. Specifically, an algorithm based on the fusion of image sequences and radar information was proposed. The RANSAC method was used to fit radar and image detection information, and the YOLOv5 detector was used to track ship motions in the image sequence. Wu et al. [8] developed a multi-sensor hierarchical detection and tracking method for inland waterway ship chimneys which can be used to monitor the emission behavior of ships in inland waters. A convolutional neural network was developed to extract the ships from visible images. Then, the Ostu binarization algorithm and image morphology operation were employed to obtain the chimney target from the ship image, and an improved DeepSORT algorithm was developed for ship chimney tracking.

Safety is an important issue for the development of MASS and is also an ultimate goal of behavior analysis. Five contributions focused on the safety of MASS in the design phase and operation phase, and one contribution overviewed recent achievements regarding intelligent algorithms for MASS.

To investigate the safety of MASS in the design phase, Zhang et al. [10] proposed a hybrid causal logic method for the preliminary hazard analysis of maritime autonomous surface ships, which is expected to provide a reference for the MASS design and safety assessment process. Due to limited historical data, it is difficult to conduct comprehensive hazard analysis of MASS. To overcome this limitation, the authors developed a hybrid causal logic (HCL). Specifically, the event sequence diagram (ESD) was used for hazardous scenarios, the fault tree (FT) method was utilized to analyze mechanical events in ESD, the Bayesian Belief Network (BBN) was applied to analyze the human factors in MASS, and conventional ship operation data and MASS experiments data were used to determine the accident probability. As the authors demonstrated, the proposed method can be used to identify the key influential factors and accident-causing event chains for MASS in the case of autonomy level III.

To enhance the safety of MASS in the operation phase, Du et al. [3] developed the onboard available-maneuvering-margin (AMM)-based ship collision alert system (CAS) that supports the evasive behavior of ships. The AMM is an important factor for avoiding the types of collisions experienced by human navigators in ship real encounters, and it can reflect the risk perceived by the navigators. Thus, it can be used for ship collision alerts. Some typical encounter scenarios from historical AIS data were selected for the demonstration of the AMM-based CAS, and the results show that the proposed method can be used for two-ship and multi-ship encounters, providing timing alerts to autonomous systems or navigators onboard ships. Gu et al. [4] developed a motion-planning algorithm for unmanned surface vehicles that considers wind and currents and is based on regularization trajectory cells. A regularization trajectory cell library incorporating the influences of wind and current was developed, and the search cost was updated. Through simulation experiments, the authors showed that the proposed method can offer a trackable trajectory for a USV in some complex environments. Song [1] proposed a collision avoidance algorithm for USVs based on obstacle classification and fuzzy rules. Specifically, the time to the closest point of approach (TCPA) was used to determine the priorities of collision avoidance; the velocity obstacle algorithm was used to determine the safety avoidance strategy; fuzzy rules were designed to understand the multi-encounter scenario; and the particle swarm optimization (PSO) algorithm was introduced to identify the optimal solution. The simulation verified and validated the proposed method's effectiveness in complex scenarios. Zhang et al. [11] developed a novel decision support method for ship collision avoidance based on the deduction of the maneuvering process. A fuzzy-based collision risk indicator, modified velocity obstacle algorithm, and fuzzy adaptive PID method were proposed to determine the time required for collision avoidance, identify evasive decisions, execute the selected evasive decision, and resume sailing operations. The simulation results show that the proposed method can support ship collision avoidance in some complex encounter environments.

Tang et al. [7] analyzed and summarized the intelligent algorithms for MASS related to risk perception, decision making, and execution that have been published in the last five years. By reviewing the existing achievements, the authors concluded that the establishment of a risk perception system with digital and visual integration would improve the quality of risk identification. MASS strongly relies on intelligent algorithms to achieve both safe and efficient collision avoidance goals in a high-complexity manner, and the speed and accuracy of ship motion control still require improvement. Lastly, the authors also discussed the roles of humans and machines based on different autonomy levels.

Conflicts of Interest: The authors declare no conflict of interest.

References

1. Song, L.; Shi, X.; Sun, H.; Xu, K.; Huang, L. Collision Avoidance Algorithm for USV Based on Rolling Obstacle Classification and Fuzzy Rules. *J. Mar. Sci. Eng.* **2021**, *9*, 1321. [CrossRef]
2. Chen, Q.; Xiao, C.; Wen, Y.; Tao, M.; Zhan, W. Ship Intention Prediction at Intersections Based on Vision and Bayesian Framework. *J. Mar. Sci. Eng.* **2022**, *10*, 639. [CrossRef]
3. Du, L.; Banda, O.A.V.; Sui, Z. Available-Maneuvering-Margins-Based Ship Collision Alert System. *J. Mar. Sci. Eng.* **2022**, *10*, 1123. [CrossRef]
4. Gu, S.; Zhou, C.; Wen, Y.; Xiao, C.; Knoll, A. Motion Planning for an Unmanned Surface Vehicle with Wind and Current Effects. *J. Mar. Sci. Eng.* **2022**, *10*, 420. [CrossRef]
5. Guo, T.; Xie, L. Research on Ship Trajectory Classification Based on a Deep Convolutional Neural Network. *J. Mar. Sci. Eng.* **2022**, *10*, 568. [CrossRef]
6. Song, R.; Wen, Y.; Tao, W.; Zhang, Q.; Papadimitriou, E.; van Gelder, P. Semantic Modeling of Ship Behavior in Cognitive Space. *J. Mar. Sci. Eng.* **2022**, *10*, 1347. [CrossRef]
7. Tang, Y.; Mou, J.; Chen, L.; Zhou, Y. Review of Ship Behavior Characteristics in Mixed Waterborne Traffic. *J. Mar. Sci. Eng.* **2022**, *10*, 139. [CrossRef]
8. Wu, F.; Chen, Q.; Wen, Y.; Xiao, C.; Zeng, F. Multi-Sensor-Based Hierarchical Detection and Tracking Method for Inland Waterway Ship Chimneys. *J. Mar. Sci. Eng.* **2022**, *10*, 809. [CrossRef]
9. Xu, T.; Zhang, Q. Ship Traffic Flow Prediction in Wind Farms Water Area Based on Spatiotemporal Dependence. *J. Mar. Sci. Eng.* **2022**, *10*, 295. [CrossRef]
10. Zhang, D.; Han, Z.; Zhang, K.; Zhang, J.; Zhang, M.; Zhang, F. Use of Hybrid Causal Logic Method for Preliminary Hazard Analysis of Maritime Autonomous Surface Ships. *J. Mar. Sci. Eng.* **2022**, *10*, 725. [CrossRef]

11. Zhang, K.; Huang, L.; Liu, X.; Chen, J.; Zhao, X.; Huang, W.; He, Y. A Novel Decision Support Methodology for Autonomous Collision Avoidance Based on Deduction of Manoeuvring Process. *J. Mar. Sci. Eng.* **2022**, *10*, 765. [CrossRef]
12. Zhong, S.; Wen, Y.; Huang, Y.; Cheng, X.; Huang, L. Ontological Ship Behavior Modeling Based on COLREGs for Knowledge Reasoning. *J. Mar. Sci. Eng.* **2022**, *10*, 203. [CrossRef]

Journal of
Marine Science and Engineering

MDPI

Article

Ontological Ship Behavior Modeling Based on COLREGs for Knowledge Reasoning

Shubin Zhong [1], Yuanqiao Wen [2,3], Yamin Huang [2,3,*], Xiaodong Cheng [2,3] and Liang Huang [2,3]

1 School of Navigation, Wuhan University of Technology, Wuhan 430063, China; zhongzou@whut.edu.cn
2 Intelligent Transportation Systems Research Center, Wuhan University of Technology, Wuhan 430063, China; wenyqwhut@foxmail.com (Y.W.); cxd921121@whut.edu.cn (X.C.); leung.huang@whut.edu.cn (L.H.)
3 National Engineering Research Center for Water Transport Safety, Wuhan University of Technology, Wuhan 430063, China
* Correspondence: yaminhuang@whut.edu.cn

Abstract: Formal expression of ship behavior is the basis for developing autonomous navigation systems, which supports the scene recognition, the intention inference, and the rule-compliant actions of the systems. The Convention on the International Regulations for Preventing Collisions at Sea (COLREGs) offers experience-based expressions of ship behavior for human beings, helping the humans recognize the scene, infer the intention, and choose rule-compliant actions. However, it is still a challenge to teach a machine to interpret the COLREGs. This paper proposed an ontological ship behavior model based on the COLREGs using knowledge graph techniques, which aims at helping the machine interpret the COLREGs rules. In this paper, the ship is seen as a temporal-spatial object and its behavior is described as the change of object elements in time spatial scales by using Resource Description Framework (RDF), function mapping, and set expression methods. To demonstrate the proposed method, the Narrow Channel article (Rule 9) from COLREGs is introduced, and the ship objects and the ship behavior expression based on Rule 9 are shown. In brief, this paper lays a theoretical foundation for further constructing the ship behavior knowledge graph from COLREGs, which is helpful for the complete machine reasoning of ship behavior knowledge in the future.

Keywords: COLREGs; ship object; ship behavior; formal expression

Citation: Zhong, S.; Wen, Y.; Huang, Y.; Cheng, X.; Huang, L. Ontological Ship Behavior Modeling Based on COLREGs for Knowledge Reasoning. *J. Mar. Sci. Eng.* **2022**, *10*, 203. https://doi.org/10.3390/jmse10020203

Academic Editor: Dracos Vassalos

Received: 11 January 2022
Accepted: 30 January 2022
Published: 2 February 2022

1. Introduction

Ship behavior refers to the movement of the ship in response to the traffic situation, which usually reflects the intention of the officer on watch (OOW) at present and influences the trajectory of the ship in the future. Hence, the recognition of ship behavior is the key to judging the intention of the OOW and predicting the movement of ships in dangerous encounters, which benefits the safety and efficiency of autonomous navigation and traffic management [1]. From the perspective of traffic management, the vessel traffic service operators (VTSO) need to judge the development of the situation based on the analysis of the ship behavior and identify the near-miss as early as possible; from the perspective of ship navigation, the OOW or intelligent systems need to infer the intention of other ships and predict their trajectories based on the observed ship behavior before taking evasive actions [2]. In brief, to improve the intelligence level of VTS and ships, the study of ship behavior has become an essential topic.

In order to help the machine understands the behavior of the ship based on COLREGs, the techniques from the knowledge graph are introduced and the methodology of ontological ship behavior modeling is developed by using Resource Description Framework (RDF), function mapping, and set expression methods. The concept of ship object and ship behavior described in COLREGs rules are incorporated in the proposed method. The ship is seen as a temporal-spatial object containing attribute elements and relational elements;

the behavior, then, is described as the changes of the elements in time-spatial scales. Based on these techniques, the proposed method can be used to identify the intentions of the ships and their violation behavior, which has the potential of improving the autonomy level of the ships and decision support system in VTS.

In summary, the main contribution of this paper is developing a knowledge model of ship behavior according to the rules from COLREGs, which could be used to realize ship behavior knowledge expression in the machine. The rest of this paper is organized as follows: the studies on ship behavior modeling are overviewed in Section 2; Section 3 introduces the definitions of ship objects, attribute elements, and relational elements, followed by a conceptual model of ship behavior and the formal expression of ship behaviors according to the COLREGs in Section 4; case studies, discussion, and conclusions are addressed in Sections 5–7, respectively.

2. Literature Review

Studies on ship behavior modeling fall into the following two categories: data-driven behavior modeling and knowledge-driven behavior modeling. In addition, due to the recent focus on rule-compliant collision avoidance, many researchers studied ship behavior in encounters, which are also overviewed.

2.1. Data-Driven Behavior Modelling

Data-driven behavior modeling usually utilizes ship trajectory data to learn the ship's behavior. A group of researchers proposed to learn the characteristics of ship behavior from traffic data from a certain region and use the characteristics to predict the trajectory of the ship in the future [3]. Specifically, researchers obtained ship motion trajectories from AIS data [4], analyzed characteristics of trajectory [5], and concluded the distribution of ship state in history that reflects the characteristics of ship behavior [6]. The characteristics of ship behavior, then, are used to predict the trajectories of the ships. Some typical methods to predict the trajectory are Kalman filter [7], Long Short-Term Memory Neural Network (LSTM) [8], Bayesian networks [9], backpropagation neural network (BP) [10], etc.

Some researchers focus on the identification of abnormal behavior of ships by learning historical trajectory data. Patroumpas et al. [11] designed a method to identify the flow of ship events through AIS data, and on this basis, performed cognitive inferences on abnormal behavior of ships. Zouaoui et al. [12] introduced the Hidden Markov model and formal language for analyzing the ship movement data in the harbor to get the normal ship behavior and abnormal ship behavior. Lei et al. [13] proposed the MT-MAD framework, which can automatically detect abnormal behavior based on the evaluation of the ship's historical sub-trajectory data, and defines the ship's activity space, behavior sequence, and behavior characteristics.

Another group of researchers concentrates on ship behavior prediction. Zissis et al. [14] used machine learning, especially artificial neural networks, as a tool to increase the predictive ability of ship behavior. The developed systems can learn and accurately predict in real-time the future behavior of any ship, in a relatively low computing time, which can be used as the basis of prediction for various intelligent systems, e.g., ship collision prevention, ship route planning, ship operation, etc. Perera et al. [15] proposed a ship behavior recognition module in autonomous ships using historical ship trajectory data, which is also used to predict ship trajectory in the future.

In short, the data-driven ship behavior model is usually based on the observed traffic data, e.g., AIS data, etc., which are used to predict the trajectories of ships based on the characteristics of the majority and identify "abnormal behaviors" that are different from the majority. However, it is not easy for these models to infer the behavior of the ship that is rule-compliant or not (i.e., reasoning the knowledge of ship behavior). In particular, the machine lacks knowledge about rule-compliant behavior.

2.2. Knowledge-Driven Behavior Modeling

Knowledge-driven behavior modeling accepts that the ship cannot move freely but follows certain regulations/rules (i.e., prior knowledge). Thus, researchers intend to gain knowledge of ship behavior from semantic knowledge. Expert systems [16], expression logic [17], semantic network [18], the Resource Description Framework (RDF) [19], ontology [20], etc. are popular methods to construct knowledge and realize knowledge reasoning.

The semantic network is a popular tool to describe ship behavior in recent years. The information loss is inevitable when the researchers use trajectory data only for recognizing ship behavior [21]; thus, some researchers tried to enrich the semantic information of the trajectory. Parent et al. [22] proposed the semantic modeling method and defined the semantic model of the ship trajectory.

In addition, the ontology model of ship behavior becomes popular, which can realize knowledge expression for machines. Nogueira et al. [23] used ontology tools to combine the ship's trajectory motion characteristics, such as velocity and acceleration, to express the ship's trajectory. Lamprecht et al. [24] used the ontology's knowledge organization ability and reasoning function to realize the cognition of the conceptual modeling of ship behavior. Wen et al. [25] introduced a dynamic Bayesian network, combined with a semantic network to carry out dynamic uncertainty reasoning and knowledge expression of ship behavior in port waters. Huang et al. [26] combined machine learning and semantic behavior for pattern recognition. Adibi et al. [27] predicted ship behavior, analyzed and discovered ship behavior at the semantic level, and improved maritime supervisors' understanding of water traffic. However, these semantic models lack consideration of the influence of environmental disturbance and do not fully consider the constraints of COLREGs on ship behavior.

The knowledge-driven approach presents tools to model behaviors for behavior inference. The reasoning process uses techniques such as rule-based systems, case-based reasoning, and ontological reasoning to produce activity models. Knowledge-driven approaches can represent the context of the environments at multiple levels of abstraction to create generalized and personalized behavior modeling. In particular, ontologies have been widely used to represent semantic concepts and their relationships in a structural manner. Advantages of ontologies include the ability to express knowledge in a clearly organized and structured manner, machine-readable expression, and the expressive power to support the reasoning process.

2.3. Behavior Modeling of COLREGs

To our best knowledge, traditional methods basically considered some key rules from COLREGs and designed the rule-based expert system that helps the machine to recognize the traffic scene and apply certain reaction rules [28].

Some researchers use a question-and-answer method to construct an expert system of ship collision avoidance and give an avoidance plan in the form of question and answer. Others focus on quantifying the COLREGs rules. Many descriptions from COLREGs are ambiguous, vague, and unquantified, which made them difficult for the machine to use in practice. Thus, many researchers proposed quantification methods that quantified the conditions for each encounter [29] (e.g., heading, crossing, and overtaking) and addressed the link between encounters conditions and reaction rules with the help of captains and fuzzy theory [30]. Xu et al. [31] clarified the concepts of "head-on ship", "give-way ship", "overtaking", "crossing" and "heading" according to COLREGs, set up a corresponding reward function for each concept and designed the reward function. In the deep learning algorithm, the optimal collision-avoidance strategy is finally obtained. He et al. [32] put forward the COLREGs quantitative model by combining the ship domain model and the ship heading control system based on the four-stage theory of the ship encounter process. Eriksen H et al. [33] introduced a three-layer hybrid collision-avoidance (COLAV) system for surface unmanned boats, which complies with Articles 8 and 13 to 17 of the COLREGs.

The performance of the COLAV system is tested by numerical simulations of three different challenge scenarios (i.e., heading, crossing, and overtaking).

These studies can be used to develop the MASS that follows the rules inputted by developers, but it is challenging to enumerate all the possible scenarios and reaction rules. To develop a practical rule-compliant ship, the developers need to enumerate the scenes that one ship might encounter and design the reaction rules. However, it is almost impossible to address all the scenes one ship might encounter. Thus, adding additional reaction rules become necessary. For example, in a crossing encounter, one ship that is on the portside of another ship is usually seen as a "give-way" ship, whereas if the first ship is a fishing ship, the ship becomes the "stand-on" ship. To handle this exception, additional reaction rules would be needed, which address the special arrangements when the ship encounters fishing ships. However, it is hard to list the endless exceptions.

In this paper, we propose another way to handle this issue. Instead of humans adding patches for exceptions, we proposed the ontological knowledge model helping the machine to deconstruct the conditions and reactions, extract the common concepts, and define the relationships among concepts. With the help of the ontological model, the machine not only can perform the reactions based on the explicit rules but also can infer the implicit rules, i.e., interpretation of rules from COLREGs. It offers a new line of thought to develop a rule-complaint MASS.

3. Conceptual Modeling of Ship Object from COLREGs

The COLREGs, formulated by the International Maritime Organization (IMO), define different types of ships, different scenes one ship might encounter, and obligations of the ship in these scenes [34]. The ship is the core concept, and the formal expression (i.e., formulaic and structured expression) of the ship object introduced in this section is a prerequisite for the machine to understand the ship behavior described by COLREGs.

3.1. Conceptual Modeling of Ship Object

Ships usually have many spatiotemporal characteristics, e.g., velocity, course, position, etc., which implies that the ship is a spatiotemporal object. Thus, in this paper, the ship object is defined as Definition 1:

Definition 1. *Ship object is a spatiotemporal object with the characteristics in time and space scales, which can be expressed in the form of data, models, rules, logic, or knowledge by computers in cyberspace.*

In general, one ship has many characteristics helping us to distinguish one ship from another ship, and these characteristics are usually named as an "attribute" of the ship. By the type and values of the attributes, one can distinguish the ship from different objects.

Among these attributes, the attributes that describe the characteristics of the ship independent from the surrounding objects, e.g., the ship name, position, velocity, types, etc., are named as "attribute elements" in this paper, whereas other attributes rely on surrounding objects to express its characteristics, e.g., the bearing of objects, relative distance between objects and the relative speed, etc., are named as "relational elements". The formal definitions of attribute elements and relational elements are shown as Definition 2 and Definition 3:

Definition 2. *The ship's attribute elements are the expression of the specific characteristics of the ship object that are independent of other objects., e.g., ship name, velocity, course, flag state, etc.*

Definition 3. *Ship's relational elements are to describe the association relationship between objects (e.g., ship objects and environment objects), e.g., relative velocity, relative heading, relative location between the ship and the environment or between one ship and another ship, etc.*

In order to facilitate the understanding of the definitions in the paper, we have made Figure 1 to show that the entities (e.g., ships, channels, etc.) in the physical space are extracted and modeled in cyberspace, named as objects. Each object has attribute elements and relational elements that help us distinguish one from another. These elements might vary as time moves on, such as the course, the velocity of ship A, the relative distance, and relative bearing from $ti-1$ to $ti+1$.

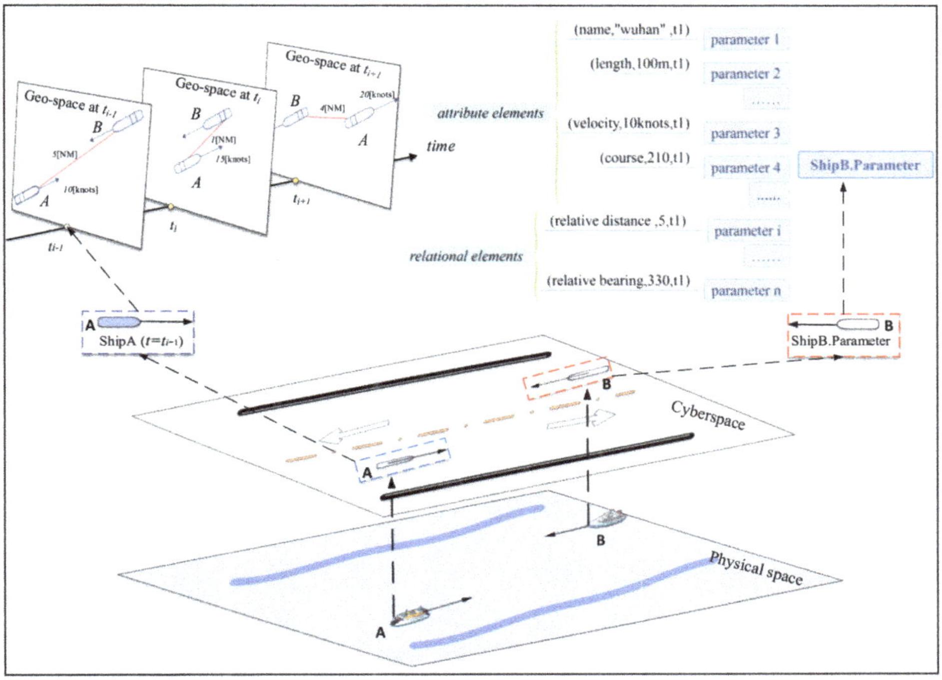

Figure 1. Abstract schematic diagram of Ship entity.

According to Definitions 1–3, the ship object has attribute elements and relational elements that might vary as time moves on or with the changes of positions. For instance, in an encounter scenario, relational elements (e.g., relative distance) of the ship would change as time moves on; in a curved channel, attribute elements (e.g., course) of the ship will be diverse according to the curvature of the channel. In brief, the values of attribute elements and relational elements have a time or spatial "stamp". Thus, each ship object can be expressed in the form of a triple-element model:

$$shipObject = \{Attribute\ elements, Relational\ elements, Time\_Space\} \tag{1}$$

where *shipObject* represents the ship object, *Attribute elements* represents the attribute elements of objects, *Relational elements* represent the relational elements of objects, and *Time_Space* represents the time and space scales.

Each element of the ship object can be formally expressed by a cell containing "*Type*", "*Value*", and "*t*", named as "*Object.parameter*" and defined as:

$$Object.parameter = (Type_{ti}, Value_{ti}, t_i), (i \in N^+) \tag{2}$$

where *Object.parameter* represents the smallest unit describing the elements of the specific ship object (say "*Object*"), "*Type*" represents the type of attribute elements or relational elements of the specific ship object; "*Value*" represents the value of the "*Type*", and *t* represents the moment when the "*Type*" has the "*Value*".

Based on these definitions, all characteristics of one object (with attribute elements and relational elements) can be collected in a set of the *Object.Parameter*, i.e.,

$$Object.Parameter = \left\{Object.parameter_1{}^{Type}, \ldots, Object.parameter_n{}^{Type}\right\} \quad (3)$$

where *Type* represents the type of *Object.parameters*, such as velocity, course, relative distance, relative bearing, etc. Additionally, the parameters relating to the attribute elements are collected in $Object.Parameter^{attribute}$, and the parameters relating to the relational elements between *Object* and $Object_2$ are collected in $[Object_1, Object_2]$. $Parameter^{relation}$. Thus, Equation (3) can be expressed as:

$$Object.Parameter = \left\{Object.Parameter^{Attribute}, \ldots, \left[Object_1, Object_2\right].Parameter^{Relation}\right\} \quad (4)$$

The Parameter of object can be expressed as Example 1:

Example 1. *Take the scene in Figure 1. as an example. The Parameter of ship A can be expressed as formula as:*

$$\begin{aligned} shipA.Parameter &= \left\{shipA.parameteter_3{}^{velocity}, [shipA, shipB].sparameter_3{}^{distance}\right\} \\ &= \left\{ \begin{array}{l} (velocity, 10, t_{i-1}), (velocity, 15, t_i), (velocity, 20, t_{i+1}), \\ (distance, 5, t_{i-1}), (distance, 1, t_i), (distance, 4, t_{i+1}) \end{array} \right\} \end{aligned} \quad (5)$$

3.2. Expression of Elements of Ship Object

3.2.1. Attribute Elements

According to COLREGs, the ship object has various attribute elements, and these attribute elements might influence the role of the ship and its obligations in a certain traffic scene. According to the feature of these elements, attribute elements can be categorized into two types, namely static attribute elements and dynamic attribute elements, see Figure 2.

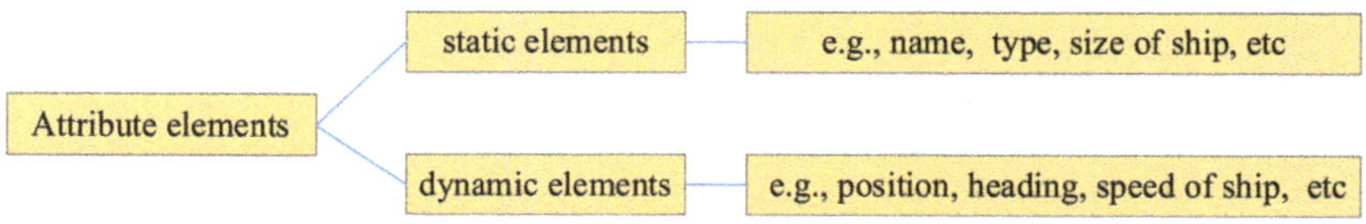

Figure 2. Attribute elements of ship entity.

The static attribute elements describe the attributes that are usually relatively invariant, such as ship name, ship type, ship size, etc., while the dynamic attribute elements are the attributes that might change over time, such as the ship's position, heading, velocity, ship's draft, etc.

3.2.2. Relational Elements

According to COLREGs and Definition 2, the ship also has many relational elements; some relational elements, such as the position and relative distance between two ships, can be used to determine the encounter scene of the two ships (overtaking, crossing, and heading scenes). Additionally, the obligation of one ship might change as the relational element changes. For example, when two ships are in a crossing scene, one of the ships has the obligation to give way to the other ship. When the two ships pass by, this obligation is relieved.

The relational elements between objects are categorized into three types, namely, spatial relations, temporal relations, and semantic relations.

(1) Spatial relational elements

The spatial relations among the objects in COLREGs include topological, bearing, and distance relations. The regional link calculus model [35] has been introduced to describe the topological relation between objects, e.g., ship object–ship object, ship object– area object, and area object–area object. The topological relation includes separation, inclusion, intersection, coincidence, inscribed, and circumscribed, which are shown in Figure 3a–f.

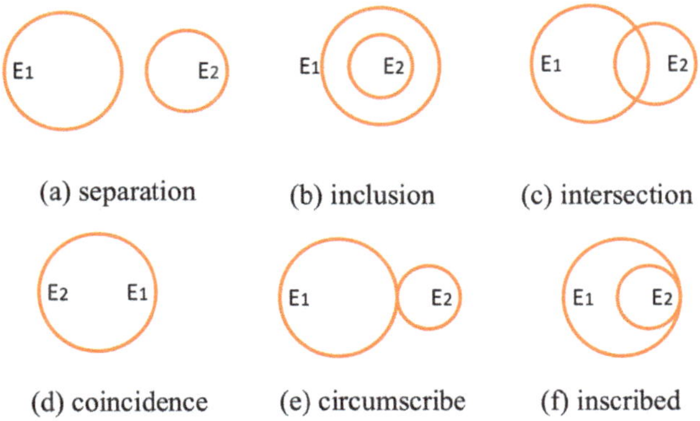

Figure 3. Topological relations of entity elements contains (**a**–**f**).

According to the statement from the COLREGs, the topological relation between two ship objects includes separation and circumscribe. The topological relation between one ship object and one area object includes the following four types: separation, inclusion, inscribed, circumscribe, and intersection. The topological relation between two area objects includes the following six types: separation, inclusion, inscribed, circumscribed, intersection and coincidence.

The bearing relation mainly describes the relative bearing between two ships. This paper constructs the ship coordinate system, which forms four directional regions by the intersection of the ship's headline and the ship's transverse line. For example, the coordinate system of ship A and ship B is shown in Figure 4. Ship B is in front of the starboard transverse 45° of ship A, while ship A is in front of the port side transverse 30° of ship B.

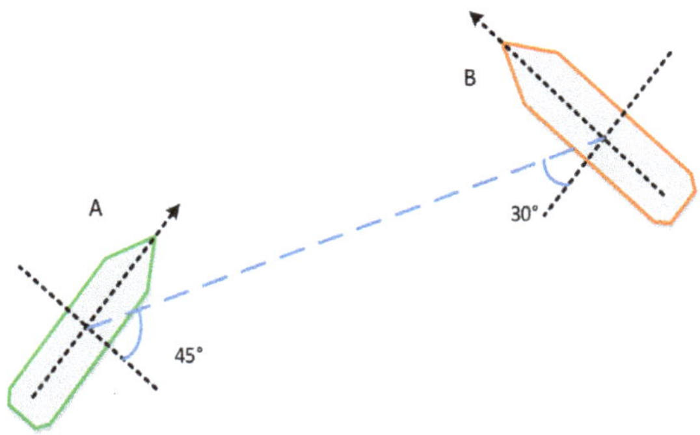

Figure 4. Bearing relational elements of ship objects.

The distance relation describes the distance between two ship objects, including quantitative expression and qualitative expression.

The quantitative expression refers to the Euclidean distance between two ship objects, as shown in Equation (6).

$$D = \sqrt{((x_A - x_B)^2 + (y_A - y_B)^2)} \tag{6}$$

where D represents the distance between ship A and ship B, (xA,yA), (xB,yB) represents the position coordinates of ship A and ship B.

According to COLREGs (Rule 7, Rule 8, Rule 13, Rule 15), the relative distance is divided into the following four stages: safety distance, urgent situation, risk of collision, and collision. The criteria for dividing these stages are depending on the encounter scenes. For the readers interested in the studies on the quantitative analysis of these criteria, the readers are encouraged to see the paper [36]. Although the quantitative analysis of the scenes is not the focus, the qualitative result, i.e., the stage of the encounter, is crucial for the subsequent deduction. Thus, a qualitative expression of the relative distance is introduced:

$$Dt = \begin{cases} \text{safety distance}, & Dn \leq D \\ \text{urgent situation}, & Dm \leq D \leq Dn \\ \text{risk of collision}, & Dl \leq D \leq Dm \\ \text{collision}, & D \leq Dm \end{cases} \tag{7}$$

where D represents the distance between ship objects, D_t is a qualitative expression of "D", and D_n, D_n, D_l are the threshold that defines the distance between ship objects.

(2) Time relational elements

The time relation is the expression of the ship's behavior and events in the time scales, which usually contain two forms, namely points and periods. The time point describes a specific moment. For instance, the time point when the ship performs a left turn, the time when two ships collide, etc. The time period is a range of time. For instance, when the ship is anchored at the anchorage, the ship passes through the narrow space, the time of the ship in the waterway, etc.

In Rule 13 of COLREGs, the definition of the two ships overtaking scene is given as follows: "A vessel shall be deemed to be overtaking when coming up with another vessel from a direction more than 22.5 degrees abaft her beam, that is, in such a position with reference to the vessel she is overtaking, that at night she would be able to see only the stern light of that vessel but neither of her sidelights." In this rule, there is actually a time relationship. For instance, the overtaking "begins at" the moment of catching up with the previous ship and "ends at" the time when the two ships pass by. In COLREGs, we can conclude the time-related concepts into five types, namely "earlier than", "later than", "between", "beginning at", and "ending at", which can be described by time points or time periods. The details see Table 1.

Table 1. Time relational elements of ship objects in COLREGs.

Time Relation Elements	Expression	Illustration
earlier than	*before* t_1 *before* (t_1, t_2)	
later than	*after* t_2 *after* (t_1, t_2)	
between	*Between* (t_1, t_2)	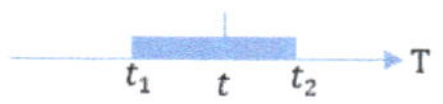
beginning at	*Begin with* t_1	
ending at	*End with* t_2	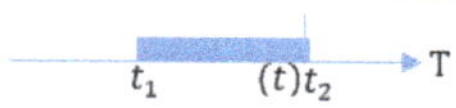

(3) Semantic relational elements

Semantic relational elements are used to describe the semantic relational elements between ship objects. For example, for the message that the name ship A is "007", there is a relationship ("hasName") between ship A and "007". We call "hasName" is a semantic relational element, ship A is the domain of the semantic relational element, and "007" is the value range of semantic relational elements. The semantic relationship is described as a triple structure <domain, relation, range> using the Resource Description Framework (RDF).

The COLREGs contain many semantic relations, and some typical semantic relations from COLREGs are concluded in Table 2.

Table 2. Time relational elements of ship objects in COLREGs.

Domain	Relation	Range	Expression
One ship object	*hasType*	attribute elements	The type of ship
	hasName	attribute elements	The name of a ship
	hasVelocity	attribute elements	The velocity of a ship
	hasCourse	attribute elements	The course of a ship
	hasSize	attribute elements	The size of a ship
Two ship objects	*hasRelative distance*	relational elements	The relative distance
	hasRelative bearing	relational elements	The relative bearing

4. Conceptual Modeling of Ship Behavior and Its Expression

Ship behavior is another important concept from the COLREGs. Specifically, COLREGs address the promoted and non-promoted behavior in different traffic scenes with different ship objects. According to Section 3, the ship entity in COLREGs is expressed as a ship object, and its element composition is expressed as attribute elements and relational elements for the machine. Based on that, ship behavior can be defined as the changes of elements in time and space scales, and the formal expression of ship behavior is presented in this section.

4.1. Conceptual Modeling of Ship Behavior

In general, "behavior" refers to the activities of spatiotemporal objects caused by external influences or internal action. In order to clearly classify and model the behavior of ship objects, and further express and reason about ship behavior, the definition of the ship behavior is introduced as Definition 4:

Definition 4. *Ship behavior refers to the change of the ship object's attribute elements and relational elements in time and space scales.*

Based on Definition 4, the ship behavior can be defined as ship behavior can be divided into attribute behavior and relational behavior, the definitions are introduced as Definition 5 and Definition 6 The ship behavior is formulated as:

$$Object.Behavior = \left\{Behavior^{Attribute}, Behavior^{Relation}\right\} \quad (8)$$

Definition 5. *Ship's attribute behavior refers to the change of ship object attribute information, e.g., ship's position, course, velocity and light type, denoted as* $Behavior^{Attribute}$.

Definition 6. *Ship's relational behavior refers to changes in ship relational elements over time, including spatial relationships, temporal relationships, semantic relationships, also including the generation, change, and demise of relationships, denoted as* $Behavior^{Relation}$.

Similarly to Equation (2), each characteristic of ship behavior (either attribute behavior or relational behavior) can be expressed by a cell, named as "*Behavior.parameter*":

$$Behavior.parameter = (dType, dValue, T) \quad (9)$$

where "*dType*" represents types of changes in specific object elements, "*dValue*" is the amount of change in the value of the same element at different times, the value of "*dValue*" can be calculated by $Value_{ti}$-$Value_{ti-1}$, T represents the period when the "*dType*" has the "*dValue*", T can be represented by $[t_{i-1}, t_i]$.

Similarly to Equation (3), *ObjectBehavior.Parameter* is a set of *Behavior.parameters* that change their values during T_s, which is formulated as:

$$\begin{aligned} ObjectBehavior.Parameter &= f(Object.Parameter) \\ &= \{(dType, dValue, T_s) | dValue \neq \phi\} \end{aligned} \tag{10}$$

where *ObjectBehavior.Parameter* represents a set of *Behavior.parameters*, $f(\cdot)$ is the function that finds the "*dType*" that "*dValue*" is non-empty from t_{i-1} to t_i. Then, the *Object.Behavior* can be expressed by the following formula:

$$\begin{aligned} Object.Behavior &= g(ObjectBehavior.Parameter) \\ &= (dType, BehaviorSemantic, T_s) \end{aligned} \tag{11}$$

where $g(\cdot)$ is the function that input the "*dType*" that has non-empty "*dValue*" and output the semantical meaning of the behavior (*BehaviorSemantic*), see Table 3.

Table 3. The semantics of behavior.

Elements Type		$Value_{ti}$-$Value_{ti-1}$	Behavior Type	
Attribute elements	velocity	>0 =0 <0	accelerate keep velocity decelerate	Attribute behavior
	course	>0 =0 <0	turn port keep course turn starboard	
Relational elements	relative distance	>0 =0 <0	far away keep distance near	Relational behavior
	relative bearing	>0 =0 <0	move to stern keep bearing move to bow	
	topology (R_{out} = −1, R_{in} = 1)	>0 =0 <0	sailing in keep topology sailing out	

The object.Parameter can be expressed as example 2:

Example 2. *Take the scene in Figure 1 as an example. The shipA.Parameter is expressed:*

$$shipA.Parameter = \{(velocity, 10, t_{i-1}), (velocity, 15, t_i)\} \tag{12}$$

according to Equation (11), the behavior of ship A can be expressed as:

$$\begin{aligned} shipA.Behavior &= g(f(shipA.Parameter)) \\ &= \{(dvelocity, \text{"}accelerate\text{"}, [t_{i-1}, t_i])\} \end{aligned} \tag{13}$$

Equation (13) means that the ship A is accelerated from the time t_{i-1} to t_i.

4.2. Formal Expression of Ship Behavior

Since the machines can only understand characterized, formulaic, and structured knowledge, it is necessary to express the knowledge of ship behavior in the way machines can "read", and such process is named as "formal expression". Thus, the definition of formal expression of ship behavior is shown as Definition 7.

Definition 7. *Formal expression of ship behavior is a formulaic and structured expression of ship behavior using methods, such as functions and sets.*

4.2.1. Attribute Behavior

According to Definition 5, attribute behavior is the change of the attribute elements, which include the changes of ship's position, velocity, course, and signal, etc. Some typical attribute behaviors are shown as follows:

- The change of velocity attribute implies the acceleration or deceleration attribute behavior;
- The change of course attribute can be divided into turning left and right steering attribute behavior;
- The change of signal attribute behavior refers to the signal number, color, and shape that will be changed in time scales

Based on Equation (11), the attribute behavior can be formulated as:

$$Object.Behavior^{Attribute} = g(f(Object.Parameter^{Attribute})) \tag{14}$$

Therefore, it is necessary to input multiple attribute element values at different times for the $f(\cdot)$ function, and $Object.Parameter^{attribute}$ can be formally expressed as:

$$Object.Parameter^{Attribute} = \left\{(dType_{t_j}, dValue_{t_j}, t_j)\right\}, (j \in N^{+}) \tag{15}$$

where $Object.Parameter^{attribute}$ represents the smallest unit describing the attribute elements of the specific ship object, "*Type*" represents the type of attribute elements or relational elements of the specific ship object; "*Value*" represents the value of the "*Type*", t represents the moment when the "*Type*" has the "*Value*".

4.2.2. Relational Behavior

In COLREGs, the relational behavior (e.g., variable relative distance and bearing) of ship objects are mainly used to determine the criteria of certain scenes and ships obligations. Some typical relational behaviors are shown as follows:

1. The change of relative distance relation implies the "near" or "far away" relation behavior;
2. The change of relative bearing relation can be divided into the angle of bearing turning smaller and the angle of bearing turning bigger;

Based on Equation (12), the relational behavior can be formulated as:

$$Object.Behavior^{Relation} = g(f([Object_1, Object_2].Parameter^{Relation})) \tag{16}$$

It is necessary to input multiple relation element values at different times for the $f(\cdot)$ function, the $Object.Parameter^{Relation}$ can be formally expressed as:

$$[Object_1, Object_2].Parameter^{Relation} = \{(dType_{t_k}, dValue_{t_k}, t_k)\}, (k \in N^{+}) \tag{17}$$

where $[Object_1, Object_2]$ $Parameter^{Relation}$ represents the smallest unit describing the relational elements of the specific ship object, "*Type*" represents the type of relational elements of the specific ship object; "*Value*" represents the value of the "*Type*", t represents the moment when the "*Type*" has the "*Value*".

5. Case Analysis

In order to demonstrate the proposed models, Rule 9 (the Narrow Channel clause) from COLREGs is introduced (the content of Rule 9 is shown in the Table A1), and the ontological behavior model based on Rule 9 is used. The Narrow Channel clause (Rule 9) addresses the promoting or non-promoting behavior when the ship object (e.g., O_{ship_in}) enters, leaves, and navigates in a narrow channel.

5.1. Ontological Expression of Ship Object Based on Rule 9

By analyzing the text information from Rule 9, there are two types of objects, namely the ship object and the waterway object, specifically, sailboats, ships less than 20 m in length, vessels engaged in fishing, narrow channel, etc., that are shown in Table 4.

Table 4. Objects under the Narrow Channel clause.

Object	Meaning	Object	Meaning
Ship	A set of ships	$Ship_{in}$	A set of ships in the narrow channel
NC	Narrow channel	$Ship \leq 20$ m	A set of ships which length less than 20 m
$Ship_{sailing}$	Sailboat	$Ship_{fishing}$	Engaged in fishing boats

For the ship object, the attribute elements contain static attributes and dynamic attributes, which are listed in Table 5.

Table 5. Attribute elements between water traffic objects in the Narrow Channel clause.

Object		Attribute Elements ($Object.parameter^{Attribute}$)	Meaning
Ship	Static attribute	$(Name\_Ship,h,t_i)$	"Ship's name is "h" at t_i"
		$(MMSI,i,t_i)$	"Ship call sign is "i" at t_i"
		$(Size,j,t_i)$	"The value of ship size is "j" at t_i"
		$(Type\_ship,k,t_i)$	"The value of ship type is "k" at t_i"
	Dynamic attribute	$(Location,a,t_i)$	"Ship's location is "a" at t_i"
		$(Velocity,b,t_i)$	"Ship's velocity is "b" at t_i"
		$(Course,c,t_i)$	"Ship's course is "c" at t_i"
		$(Draft,d,t_i)$	"Ship's draft is "d" at t_i"
		$(Sound,e,t_i)$	"Ship's sound is "e" at t_i"
Narrow Channel (NC)	Static attribute	$(Name\_NC,l,t_i)$	"The value of Narrow channel name is "l" at t_i"
		$(Boundry\_NC,m,t_i)$	"The value of boundary position of the narrow channel is "m" at t_i"
		$(Width\_NC,n,t_i)$	"The value of navigable water width of the narrow channel is "n" at t_i"
		$(Location\_NC,o,t_i)$	"The value of the center position of each water depth area of the narrow channel is "o" at t_i"
	Dynamic attribute	$(Visibility,f,t_i)$	"Visibility in narrow channel is "f" at t_i"
		$(Flow\ velocity,g,t_i)$	"Flow velocity in narrow channel is "g" at t_i"

1. The static attributes include ship's type, call sign, size, etc.
2. The dynamic attributes include some time-varying attributes, such as position, velocity, course, draft, sound signal, etc.

For the waterway object, the attribute elements also include static attributes and dynamic attributes, which are shown in Table 5.

1. The static attributes of narrow water channels are the name of the narrow water channel, the center position of each water depth; the width of the navigable water area; the boundary information of the narrow water channel.
2. The dynamic attributes of narrow water channels are the flow velocity, flow direction, and visibility of narrow water channels.

According to Section 3.2.2, the relational elements among these objects (ships and the waterway) can be analyzed from the following three aspects: time, space, and semantic. Table 6 lists different objects, the relationships between objects, and the semantic expressions of the relationships.

Table 6. Relational elements of objects in the Narrow Channel clause.

Objects		Relational Elements ($Object.parameter^{Relation}$)	Meaning
[Ship, NC]	Time relational	($Time.Before,1,t_i$)	"Before the ship enters the narrow channel"
		($Time.After,-1,t_i$)	"After the ship enters the narrow channel"
		($Time.Between,2,[t_i,t_{i+1}]$)	"The time period during which the ship is sailing in the narrow channel"
	Spatial topological relation	($Topology.Separation,-1,t_i$)	"The ship is outside the narrow channel"
		($Topology.Inclusion,1,t_i$)	"The ship is in the narrow channel"
		($Topology.Inclusion\_{starboard},12,t_i$)	"The ship is in the narrow channel on its starboard side"
		($Topology.Inclusion\_{elbow},13,t_i$)	"The ship is driving in the waters of the elbow of the narrow channel"
	Semantic relation	($Semantic.Avoid\_{anchoring},1,t_i$)	"Ships avoid anchoring in the narrow channel"
		($Semantic.Avoid\_{crossring},1,t_i$)	"Ships avoid crossing narrow channel"
[ShipA, ShipB]	Spatial relation	($Relative\ bearing,a,t_i$)	"The bearing relation between ship A and ship B"
		($Relative\ distance,b,t_i$)	"The distance relation between ship A and ship B"
	Semantic relation	($Semantic.Overtaking\_{Port},1,t_i$)	"Ship A attempts to overtake the port side of Ship B"
		($Semantic.Overtaking\_{Starboard},2,t_i$)	"Ship A attempts to overtake from the starboard side of Ship B"
		($Semantic.Agree\_{Overtaking},3,t_i$)	"Ship B agrees to ship A overtaking"
[$Ship_{sailing}$, $Ship_{in}$]	Semantic relation	($Semantic.Avoid\_{impede},1,t_i$)	"Sailing boats should not impede ships that can only navigate safely in the narrow channel"
[$Shipl \leq 20$ m, $Ship_{in}$]		($Semantic.Avoid\_{impede},1,t_i$)	"Ships less than 20 m in length should not impede ships that can only navigate safely in narrow channels"
[$Ship_{fishing}$, $Ship_{in}$]		($Semantic.Avoid\_{impede},1,t_i$)	"Vessels engaged in fishing shall not impede any vessel navigating safely in the narrow channel"

1. The time relations between the ship and the narrow water channel include the time before the ship enters the narrow water channel, after entering the narrow water channel, and when the ship moves in the narrow water channel.
2. The spatial topological relationship includes the ship outside the narrow water channel and the ship in the narrow water channel. Ships are in narrow channel elbow waters or boundary waters, etc.
3. The semantic relations include ships avoiding anchoring and crossing in narrow channels. Specific numerical values express the spatial position relationship and spatial distance relationship between ships and ships; semantic relations include the ship attempting to overtake another ship, the other ship agrees or suspects overtaking, and sailboats and ships less than 20 m in length should not interfere with ships that can only navigate safely in narrow channels. Vessels engaged in fishing shall not hinder any ships that navigate safely in narrow channels, etc.

5.2. Formal Expression of Ship Behavior Based on Rule 9

The text information of Rule 9 addresses the attribute and relational elements of objects. Table 7 lists the attribute elements of one ship at different moments in time. By comparing the attribute elements at different moments, the ship's attribute behavior is inferred, and the attribute behavior is concluded in the last column of the table. Based on Table 7, the machine can reason about the behavior of the ship by analyzing or comparing the values of position, velocity, heading, and other ship attributes in a narrow channel at different moments. Specifically, the machine can judge whether the ship has moved, accelerated, decelerated, and turned in the period between the two moments.

Table 7. Relational elements of objects in the Narrow Channel clause.

Object	Attribute Elements ($Object.Parameter^{Attribute}$)	Attribute Behavior ($Object.Behavior^{Attribute}$)
$Ship_{in}$	$\{(Location, a, t_i),(Location, b, t_{i+1})\}$, $a \neq b$ $\{(Velocity, n, t_i),(Velocity, m, t_{i+1})\}$, $m < n$ $\{(Velocity, n, t_i),(Velocity, m, t_{i+1})\}$, $m > n$ $\{(Course, c, t_i),(Course, d, t_{i+1})\}$, $c \neq d$	{*dLocation*, "move", [t_i, t_{i+1}]} {*dVelocity*, "decelerate", [t_i, t_{i+1}]} {*dVelocity*, "accelerate", [t_i, t_{i+1}]} {*dCourse*, "turn course", [t_i, t_{i+1}]}

Table 8 lists the relational elements of one ship w.r.t. other objects (i.e., the ship and the waterway). By comparing the relational elements at different moments, the ship's relational behavior is inferred, and the relational behavior is expressed semantically. Based on Table 8, the machine can reason about the behavior of the ship by analyzing the topological relationship, and the spatial topological behaviors including sailing in, sailing out, and crossing can be inferred. By analyzing the spatial bearing relationship and spatial distance relationship between ships, the pursuit and crossing behavior between ships in the narrow channel can be inferred.

Table 8. Ship relational behaviors in the Narrow Channel clause.

Objects	Relational Elements	Relation Behavior
[*Ship*, *NC*]	$\{(Topology,-1, t_i), (Topology,1, t_{i+1})\}$ $\{(Topology,1,t_i), (Topology,-1,t_{i+1})\}$ $\{(Course_{\_ship},a,t_i), (Course_{\_NC},a,t_i), (Course_{\_ship},a-90,t_{i+1}), (Course_{\_NC},a,t_{i+1})\}$	{*dTopology*, "sailing in", [t_i, t_{i+1}]} {*dTopology*, "sailing out", [t_i, t_{i+1}]} {*dCourse*,"crossing narrow channel", [t_i, t_{i+1}]}
[*Ship*A, *Ship*B]	$\{(Bearing,M,t_i), (Bearing,M,t_{i+1}), (Distance,n,t_i), (Distance,m,t_{i+1})\}$,	If $n > m$ and {*dBearing*, "keep bearing", [t_i, t_{i+1}]}∩{d*Distance*, "near", [t_i, t_{i+1}]}, then {d*Semantic*, "Ship A overtaking Ship B", [t_i, t_{i+1}]}.

5.3. Reasoning Based on the Proposed Method

Based on the above formal expression of the behavior of ships in the narrow channel terms of COLREGs, a formal expression of ship behavior can be applied in conjunction with AIS data and nautical chart data.

In Figure 5, we introduce a scene where two ships encountered in a narrow channel. Ship B is navigating in the starboard channel and move towards the north; Ship A is navigating in the port channel and move towards the south.

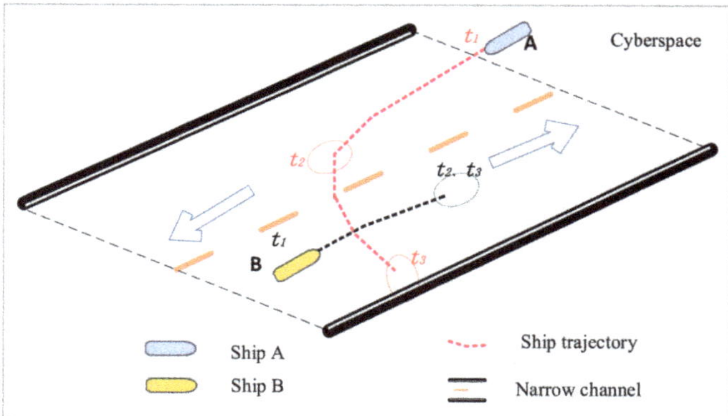

Figure 5. Application of formal expression of ship behavior in narrow channel scenarios.

By analyzing the changes of the attribute elements and relational elements of ship A and ship B at the moments of time t_1,t_2, and t_3, and expressing the attribute behavior and relational behavior of the ships formally in this way, the machine can finally judge whether the ship behavior complies with the COLREGs.

According to the above research on the expression of ship objects and ship behavior, the attribute elements, relational elements of ship objects, and the ship's attribute behavior and relational behavior can be expressed as follows:

(a) The expression of the attribute elements of the *ship A*

$$\begin{aligned} shipA.parameter^{Attribute} &= \left\{ shipA.parameter^{velocity}, shipA.parameter^{course} \right\} \\ &= \left\{ \begin{array}{c} (velocity, 10, t_1), (velocity, 15, t_2), (velocity, 20, t_3), \\ (course, 220, t_1), (course, 220, t_2), (course, 150, t_3) \end{array} \right\} \end{aligned} \tag{18}$$

(b) The expression of the attribute elements of the *ship B*

$$\begin{aligned} shipB.parameter^{Attribute} &= \left\{ shipB.parameter^{velocity}, shipB.parameter^{course} \right\} \\ &= \left\{ \begin{array}{c} (velocity, 18, t_1), (velocity, 0, t_2), (velocity, 0, t_3), \\ (course, 60, t_1), (course, 60, t_2), (course, 60, t_3) \end{array} \right\} \end{aligned} \tag{19}$$

(c) The expression of the relational elements between *ship A* and *ship B*

$$\begin{aligned} [shipA, shipB].Parameter^{Relation} &= \left\{ \begin{array}{c} [shipA, shipB].Parameter^{distance} \\ [shipA, shipB].Parameter^{bearing} \end{array} \right\} \\ &= \left\{ \begin{array}{c} (distance, 18, t_1), (distance, 10, t_2), (distance, 3, t_3), \\ (bearing, 050, t_1), (bearing, 030, t_2), (bearing, 230, t_3) \end{array} \right\} \end{aligned} \tag{20}$$

(d) The expression of the relational elements between *ship A* and Narrow channel

$$\begin{aligned} [shipA, O_{NC}].Parameter^{Relation} &= \left\{ [shipA, O_{NC}].Parameter^{Topolopy} \right\} \\ &= \{ (Topolopy, -1, t_1), (Topolopy, 1, t_2), (Topolopy, 1, t_3) \} \end{aligned} \tag{21}$$

(e) The expression of the relational elements between *ship B* and Narrow channel

$$\begin{aligned} [shipB, O_{NC}].Parameter^{Relation} &= \left\{ [shipB, O_{NC}].Parameter^{Topolopy} \right\} \\ &= \{ (Topolopy, 1, t_1), (Topolopy, 1, t_2), (Topolopy, 1, t_3) \} \end{aligned} \tag{22}$$

(f) The expression of the attribute behavior of *ship A*

$$\begin{aligned} shipA.Behavior^{Attribute} &= g(f(shipA.Parameter^{Attribute})) \\ &= \left\{ \begin{array}{l} (dvelocity, \text{"}accelerate\text{"}, [t_1, t_2]), \\ (dvelocity, \text{"}decelerate\text{"}, [t_2, t_3]), \\ (dcourse, \text{"}keep\ course\text{"}, [t_1, t_2]), \\ (dcourse, \text{"}turn\ starboard\text{"}, [t_2, t_3]) \end{array} \right\} \end{aligned} \tag{23}$$

According to the changes of the velocity and course of *ship A*, the semantics of the ship behaviors are expressed as "*accelerate*" and "*keep course*" from the time t_1 to t_2, "dec*elerate*" and "*turn starboard*" from the time t_2 to t_3. From time t_2 to t_3, the course of *ship A* is perpendicular to the total flow direction of the narrow channel, which means a spatial topological behavior of "*crossing*" between *ship A* and the narrow channel. Therefore, it violates the COLREGs rule that "Ships should avoid crossing narrow channel".

(g) The expression of the attribute behavior of *ship B*

$$\begin{aligned} shipB.Behavior^{Attribute} &= g(f(shipB.Parameter^{Attribute})) \\ &= \left\{ \begin{array}{l} (dvelocity, \text{"}decelerate\text{"}, [t_1, t_2]), \\ (dvelocity, \text{"}keep\ velocity\text{"}, [t_2, t_3]), \\ (dcourse, \text{"}keep\ course\text{"}, [t_1, t_2]), \\ (dcourse, \text{"}keep\ course\text{"}, [t_2, t_3]) \end{array} \right\} \end{aligned} \tag{24}$$

According to the changes of the velocity and course of *ship B*, the semantics of the ship behaviors are expressed as *"decelerate"* and *"keep course"* from the time t_1 to t_2, *"keep velocity"* and *"keep course"* from the time t_2 to t_3. *Ship B* is *"anchored"* in the narrow channel from the time t_2 to t_3. Therefore, it violated the COLREGs stipulation that "ships should avoid anchoring in the narrow channel".

(h) The expression of the relational behavior of *ship A* and *ship B*

$$[shipA, shipB].Behavior^{Relation} = g(f([shipA, shipB].Parameter^{Relation})) = \left\{ \begin{array}{l} (ddistance, \text{"near"}, [t_1, t_2]), \\ (ddistance, \text{"near"}, [t_2, t_3]), \\ (dbearing, \text{"move to bow"}, [t_1, t_2]), \\ (dbearing, \text{"move to bow"}, [t_2, t_3]) \end{array} \right\} \quad (25)$$

According to the changes of the relative distance and relative bearing between *ship A* and *ship B*, the semantics of the ship behaviors are expressed as *"near"* and *"move to bow"* from the time t_1 to t_2, *"far away"* and *"move to stern"* from the time t_2 to t_3.

(i) The expression of the relational behavior of *ship A* and Narrow channel

$$[shipA, O_{NC}].Behavior^{Relation} = g(f([shipA, O_{NC}].Parameter^{topology})) = \left\{ \begin{array}{l} (dtopology, \text{"sailing in"}, [t_1, t_2]), \\ (dtopology, \text{"keep topology"}, [t_2, t_3]) \end{array} \right\} \quad (26)$$

According to the changes of the topology relation between *ship A* and the narrow channel, the semantics of the ship behaviors are expressed as *"sailing in"* from the time t_1 to t_2, *"keep topology"* in the narrow channel from the time t_2 to t_3.

(j) The expression of the relational behavior of *ship B* and Narrow channel

$$[shipB, O_{NC}].Behavior^{Relation} = g(f([shipB, O_{NC}].Parameter^{topology})) = \left\{ \begin{array}{l} (dtopology, \text{"keep topology"}, [t_1, t_2]), \\ (dtopology, \text{"keep topology"}, [t_2, t_3]) \end{array} \right\} \quad (27)$$

According to the changes of the topology relation between *ship B* and the narrow channel, the semantics of the ship behaviors are expressed as *"keep topology"* in the narrow channel from the time t_1 to t_2, *"keep topology"* in the narrow channel from the time t_2 to t_3.

According to the above-mentioned expression of the attribute behavior of *ship A* and *ship B*, and the relational behavior between *ship A* and *ship B*, *ship A* and the narrow channel, and *ship B* and the narrow channel at the time from t_1 to t_3. Based on these expressions, we can clearly judge whether the ship behavior complies with COLREGs, see Table 9.

Table 9. Behavior of objects in the narrow channel.

Object	Time	Attribute Behavior	Relational Behavior	COLREGs-Compliant (Yes/No)
ship A	[t1,t2]	*"accelerate"* *"keep course"*		Yes
	[t2,t3]	*"decelerate"* *"turn starboard"*		No
ship B	[t1,t2]	*"decelerate"* *"keep course"*		Yes
	[t2,t3]	*"keep velocity = 0"* *"keep course"*		No
[ship A, ship B]	[t1,t2]		*"near"* *"move to bow"*	Yes
	[t2,t3]		*"far away"* *"move to stern"*	Yes

Table 9. *Cont.*

Object	Time	Attribute Behavior	Relational Behavior	COLREGs-Compliant (Yes/No)
[ship A, O_{NC}]	[*t*1,*t*2] [*t*2,*t*3]		*"sailing in"* *"keep topology = 1"*	Yes Yes
[ship B, O_{NC}]	[*t*1,*t*2] [*t*2,*t*3]		*"keep topology = 1"* *"keep topology = 1"*	Yes Yes

6. Discussion

With the development of knowledge engineering, knowledge expression has been widely explored and utilized in multiple knowledge-driven tasks, which significantly improves their performance. In this section, we first give a summary of this research then summarize the advantages and disadvantages of the method of this research.

6.1. Discussion on Case Study

In this paper, we provide a broad overview of currently available techniques, including RDF, function mapping, and set expression methods. The proposed method imitates human understanding ability, which makes it possible to incorporate prior knowledge to assist machine recognizing.

In Section 3, we abstractly express the ship objects in COLREGs as attribute elements and relationship elements, and in Section 4, we express the dynamic changes of the ship object's attribute elements and relationship elements over time as ship behavior. The expression method through RDF, function and collection is similar to human thinking, which is more in line with our COLREGs ship behavior ontology knowledge modeling. Based on the ship behavior ontology method in Sections 3 and 4, we use COLREGs (Rule 9) for example verification in Section 5, and the results show that our method can formally express the ship behavior of COLREGs.

However, this research is only the initial work for realizing ship behavior knowledge reasoning to the machine. Based on this research, in the future, the ship behavior knowledge graph, COLREGs knowledge graph, and the knowledge graph of water traffic scene can be further constructed to realize the autonomous recognition of water traffic scenes, judge water Traffic situation, reason about the violations of COLREGs by ships, and support decision making of MASS.

6.2. Advantages and Disadvantages of the Proposed Method

(1) Advantages of this method

In this paper, the ship behavior, based on COLREGs, is modeled as the change of entity elements in time and space scales by using RDF, function mapping, and set expression methods. The advantages of this method are as follows: first, it can capture hidden semantic information in COLREGs; second, it can improve the accuracy of knowledge recognizing significantly; finally, it can simulate human recognizing ability, which makes it possible to incorporate prior knowledge to assist in recognizing.

(2) Disadvantages of this method

On the basis of Sections 3–5, we realize the formal expression of the ship behavior ontology model in COLREGs, but the ontology model still has some deficiencies. The knowledge model of ship behavior established in this paper is still in the enlightenment stage in the maritime industry, which has not yet formed a unified industry standard. Its disadvantage is that it has not solved a series of problems such as dependence on domain experts and poor generalization ability. On the one hand, this method requires manual modeling of ship behavior knowledge, and its modeling efficiency is low. On the other hand, semantic calculation and reasoning methods are still missing.

6.3. Future Work

The formal expression of ship behavior is the basis for developing autonomous navigation systems that support the scene cognition, the intention inference, and the rule-compliant actions of the systems. This paper studies the formal expression of ship behavior based on COLREGs. However, there is still a certain distance for the machine to truly realize the autonomous recognition of the navigation scene, the autonomous reasoning of the ship's intention, and the autonomous judgment of the ship's behavior in compliance with the COLREGs rules. Based on the research in this paper, we give several directions for future research, as follows:

(1) Constructing the ontology of ship behavior

Ontology plays an important role in enriching the semantic information of things and realizing knowledge sharing. Based on the formal expression of ship objects and ship behavior in this paper, the ship behavior ontology is further constructed to form a knowledge base with semantic information, and the custom SWRL rules are input into the ontology inference engine to realize the machine's autonomous cognition of ship behavior.

(2) Constructing the ontology of traffic scene

COLREGs are the norms of ship behavior in different traffic scenarios. According to different traffic scenarios, ships should take corresponding behaviors, the traffic scene ontology is constructed based on COLREGs. The custom SWRL rules are input into the ontology inference engine to realize the machine's autonomous cognition of traffic scenarios.

(3) Constructing the knowledge graph of ship behavior

Based on the formal expression of ship behavior in this article, and the ship behavior ontology and traffic scene ontology constructed in future research, the knowledge graph of ship behavior can be further constructed in the future. Then, the machine can be queried, and it can be inferred that the actions whether the acctions are COLREGs-compliant or not in different scenarios.

7. Conclusions

For developing rule-compliant maritime autonomous surface ships (MASS), understanding the Convention on the International Regulation for the preventing Collision at Sea (COLREGs) is the foundation for the machine. The existing expert systems for MASS did not teach the machine to understand the COLREGs rules but list condition-and-reaction rules for endless exceptions. To handle this issue, this paper proposed an ontological method to model the ship behavior and try to build the first step to help the machine to interpret the COLREGs in a manner of humans.

The attributes of the ship are categorized into "attribute elements" and "relational elements", and the ship behaviors then are defined as the changes on "attribute elements" (i.e., attribute behavior) and "relational elements" (i.e., relational behavior). Based on these definitions, the attribute elements, relational elements, attribute behavior, and relational behavior are formally expressed by using the Resource Description Framework (RDF), function mapping, and set expression methods. By introducing Rule 9 from COLREGs, this paper demonstrates the performance of the proposed method, which has laid a theoretical foundation for structural modeling and semantic understanding of ship behavior.

The proposed method addressed a novel way to develop the rule-compliant machine, which is promising in the development of MASS. This paper is the first step for the rule-compliant MASS, and the proposed model is still at the conceptual and logical levels. Thus, it is necessary to construct the ship behavior ontology further, construct the knowledge model driven by the ship behavior, and use it in actual cases in the future.

Author Contributions: Conceptualization, S.Z. and Y.W.; methodology, Y.H.; software, X.C.; validation, S.Z., Y.H. and L.H.; formal analysis, S.Z.; resources, Y.H.; writing—original draft preparation, S.Z. and Y.H.; writing—review and editing, S.Z. and Y.H.; funding acquisition, Y.H. All authors have read and agreed to the published version of the manuscript.

Funding: This work was supported by the Key R&D Program of Zhejiang Province China through Grant No. 2021C01010, the National Natural Science Foundation of China (NSFC) through Grant No. 52072287 and 52001241, and "the Fundamental Research Funds for the Central Universities (WUT: 2021IVA106)".

Institutional Review Board Statement: No applicable.

Informed Consent Statement: No applicable.

Data Availability Statement: No applicable.

Conflicts of Interest: The authors declare no conflict of interest and the funders had no role in the design of the study, in the collection, analyses, or interpretation of data, in the writing of the manuscript, or in the decision to publish the results.

Appendix A

Table A1. Narrow Channel Provisions Text Information.

Rule 9 **Narrow Channel**
(a) A vessel proceeding along the course of a narrow channel or fairway shall keep as near to the outer limit of the channel or fairway which lies on her starboard side as is safe and practicable.
(b) A vessel of less than 20 m in length or a sailing vessel shall not impede the passage of a vessel which can safely navigate only within a narrow channel or fairway.
(c) A vessel engaged in fishing shall not impede the passage of any other vessel navigating within a narrow channel or fairway.
(d) A vessel shall not cross a narrow channel or fairway if such crossing impedes the passage of a vessel which can safely navigate only within such channel or fairway. The latter vessel may use the sound signal prescribed in Rule 34(d) if in doubt as to the intention of the crossing vessel.
(e) (i) In a narrow channel or fairway when overtaking can take place only if the vessel to be overtaken has to take action to permit safe passing, the vessel intending to overtake shall indicate her intention by sounding the appropriate signal prescribed in Rule 34(c)(i). The vessel to be overtaken shall, if in agreement, sound the appropriate signal prescribed in Rule 34(c)(ii) and take steps to permit safe passing. If in doubt she may sound the signals prescribed in Rule 34(d). (ii) This Rule does not relieve the overtaking vessel of her obligation under Rule 13.
(f) A vessel nearing a bend or an area of a narrow channel or fairway where other vessels may be obscured by an intervening obstruction shall navigate with particular alertness and caution and shall sound the appropriate signal prescribed in Rule 34(e).
(g) Any vessel shall, if the circumstances of the case admit, avoid anchoring in a narrow channel.

References

1. Sui, Z.; Wen, Y.; Huang, Y.; Zhou, C.; Xiao, C.; Chen, H. Empirical analysis of complex network for marine traffic situation. *Ocean Eng.* **2020**, *214*, 107848. [CrossRef]
2. Zhou, Y.; Daamen, W.; Vellinga, T.; Hoogendoorn, S.P. Impacts of wind and current on ship behavior in ports and waterways: A quantitative analysis based on AIS data. *Ocean Eng.* **2020**, *213*, 107774. [CrossRef]
3. Zhou, Y.; Daamen, W.; Vellinga, T.; Hoogendoorn, S.P. Ship classification based on ship behavior clustering from AIS data. *Ocean Eng.* **2019**, *175*, 176–187. [CrossRef]
4. Pallotta, G.; Vespe, M.; Bryan, K. Vessel pattern knowledge discovery from AIS data: A framework for anomaly detection and route prediction. *Entropy* **2013**, *15*, 2218–2245. [CrossRef]
5. Zhu, F.; Zhang, Y. Research on Marine Traffic Data Mining System Based on AIS. *Energy Procedia* **2011**, *13*, 8254–8259.
6. Zhu, F. Mining ship spatial trajectory patterns from AIS database for maritime surveillance. In Proceedings of the 2nd IEEE International Conference on Emergency Management and Management Sciences (ICEMMS), Beijing, China, 8–10 August 2011; pp. 772–775.
7. Gao, M.; Shi, G.; Li, S. Online prediction of ship behavior with automatic identification system sensor data using bidirectional long short-term memory recurrent neural network. *Sensors* **2018**, *18*, 4211. [CrossRef] [PubMed]
8. Sun, Y.; Peng, X.; Ding, Z.; Zhao, J. An approach to ship behavior prediction based on AIS and RNN optimization model. *Int. J. Transp. Eng. Technol* **2020**, *6*, 16–21.

9. Mascaro, S.; Korb, K.B.; Nicholson, A.E. Learning abnormal vessel behaviour from ais data with bayesian networks at two time scales. *Tracks A J. Artist. Writ.* **2010**, *4*, 1–34.
10. Xu, T.; Liu, X.; Yang, X. Ship Trajectory online prediction based on BP neural network algorithm. In Proceedings of the 2011 International Conference of Information Technology, Computer Engineering and Management Sciences, Nanjing, China, 24–25 September 2011; pp. 103–106.
11. Patroumpas, K.; Alevizos, E.; Artikis, A.; Vodas, M.; Pelekis, N.; Theodoridis, Y. Online event recognition from moving vessel trajectories. *GeoInformatica* **2017**, *21*, 389–427. [CrossRef]
12. Zouaoui-Elloumi, S.; Maïzi, N. Securing harbor by combining probabilistic approach with event-based approach. *Appl. Ocean Res.* **2014**, *47*, 98–109. [CrossRef]
13. Lei, P.-R. A framework for anomaly detection in maritime trajectory behavior. *Knowl. Inf. Syst.* **2016**, *47*, 189–214. [CrossRef]
14. Zissis, D.; Xidias, E.K.; Lekkas, D. Real-time vessel behavior prediction. *Evol. Syst.* **2016**, *7*, 29–40. [CrossRef]
15. Perera, L.P.; Oliveira, P.; Soares, C.G. Maritime traffic monitoring based on vessel detection, tracking, state estimation, and trajectory prediction. *IEEE Trans. Intell. Transp. Syst.* **2012**, *13*, 1188–1200. [CrossRef]
16. Cowan, R. Expert systems: Aspects of and limitations to the codifiability of knowledge. *Res. Policy* **2001**, *30*, 1355–1372. [CrossRef]
17. Sarker, K.U.; Deraman, A.B.; Hasan, R. Descriptive Logic for Software Engineering Ontology: Aspect Software Quality Control. In Proceedings of the 2018 4th International Conference on Computer and Information Sciences (ICCOINS), Kuala Lumpur, Malaysia, 13–14 August 2018; pp. 1–5.
18. Mullen, J.; Cockell, S.J.; Tipney, H.; Woollard, P.M.; Wipat, A. Mining integrated semantic networks for drug repositioning opportunities. *PeerJ* **2016**, *4*, e1558. [CrossRef] [PubMed]
19. Fan, T.; Yan, L.; Ma, Z. Mapping fuzzy RDF(S) into fuzzy object-oriented databases. *Int. J. Intell. Syst.* **2019**, *34*, 751–780. [CrossRef]
20. Smith, B. Ontology. In *The Furniture of the World*; Brill: Leiden, The Netherlands, 2012; pp. 47–68.
21. Zhang, Z.; Suo, Y.; Yang, S.; Zhao, Z. Detection of Complex Abnormal Ship Behavior Based on Event Stream. In Proceedings of the 2020 Chinese Automation Congress (CAC), Shanghai, China, 6–8 November 2020; pp. 5730–5735.
22. Yan, Z.; Chakraborty, D.; Parent, C.; Spaccapietra, S.; Aberer, K. Semantic trajectories: Mobility data computation and annotation. *ACM Trans. Intell. Syst. Technol.* **2013**, *4*, 1–38. [CrossRef]
23. Nogueira, T.P.; Braga, R.B.; Martin, H. An ontology-based approach to represent trajectory characteristics. In Proceedings of the 2014 Fifth International Conference on Computing for Geospatial Research and Application, Washington, DC, USA, 4–6 August 2014; pp. 102–107.
24. Lamprecht, D.; Strohmaier, M.; Helic, D.; Nyulas, C.; Tudorache, T.; Noy, N.; Musen, M.A. Using ontologies to model human navigation behavior in information networks: A study based on wikipedia. *Semant. Web* **2015**, *6*, 403–422. [CrossRef]
25. Wen, Y.; Zhang, Y.; Huang, L.; Zhou, C.; Xiao, C.; Zhang, F.; Peng, X.; Zhan, W.; Sui, Z. Semantic modelling of ship behavior in harbor based on ontology and dynamic bayesian network. *ISPRS Int. J. Geo-Inf.* **2019**, *8*, 107. [CrossRef]
26. Huang, L.; Wen, Y.; Guo, W.; Zhu, X.; Zhou, C.; Zhang, F.; Zhu, M. Mobility pattern analysis of ship trajectories based on semantic transformation and topic model. *Ocean Eng.* **2020**, *201*, 107092. [CrossRef]
27. Adibi, P.; Pranovi, F.; Raffaetà, A.; Russo, E.; Silvestri, C.; Simeoni, M.; Soares, A.; Matwin, S. Predicting fishing effort and catch using semantic trajectories and machine learning. In Proceedings of the International Workshop on Multiple-Aspect Analysis of Semantic Trajectories, Würzburg, Germany, 16 September 2019; Springer: Cham, Switzerland, 2019; pp. 83–99.
28. Kose, K.; Yang, C.; Ishioka, Y.; Kato, Y.; Nagasawa, A.; Hara, K. A collision avoidance expert system for integrated navigation system and its brush-up. *J. Soc. Nav. Archit. Jpn.* **1995**, *1995*, 399–407. [CrossRef]
29. He, Y.; Jin, Y.; Huang, L.; Xiong, Y.; Chen, P.; Mou, J. Quantitative analysis of COLREG rules and seamanship for autonomous collision avoidance at open sea. *Ocean Eng.* **2017**, *140*, 281–291. [CrossRef]
30. Hwang, C.N. The integrated design of fuzzy collision-avoidance and H∞-autopilots on ships. *J. Navig.* **2002**, *55*, 117–136. [CrossRef]
31. Xu, X.; Lu, Y.; Liu, X.; Zhang, W. Intelligent collision avoidance algorithms for USVs via deep reinforcement learning under COLREGs. *Ocean Eng.* **2020**, *217*, 107704. [CrossRef]
32. He, Y.; Huang, L.; Xiong, Y.; Hu, W. The Research of Ship ACA Actions at Different Stages on Head-On Situation Based on CRI and COLREGS. *J. Coast. Res.* **2015**, *73*, 735–740. [CrossRef]
33. Eriksen, B.-O.H.; Bitar, G.; Breivik, M.; Lekkas, A. Hybrid Collision Avoidance for ASVs Compliant With COLREGs Rules 8 and 13–17. *Front. Robot. AI* **2020**, *7*, 11. [CrossRef]
34. Weng, J.; Liu, M.; Zhou, Y. *Watch and Collision Avoidance of Ship*; Wuhan University of Technology Press: Wuhan, China, 2021. (In Chinese)
35. Hoang, V.-N.; Nguyen-Xuan, H. Extruded-geometric-component-based 3D topology optimization. *Comput. Methods Appl. Mech. Eng.* **2020**, *371*, 113293. [CrossRef]
36. He, Z.; Deng, M.; Cai, J.; Xie, Z.; Guan, Q.; Yang, C. Mining spatiotemporal association patterns from complex geographic phenomena. *Int. J. Geogr. Inf. Sci.* **2020**, *34*, 1162–1187. [CrossRef]

Journal of
Marine Science and Engineering

MDPI

Article

Semantic Modeling of Ship Behavior in Cognitive Space

Rongxin Song [1,2,*], Yuanqiao Wen [3,4,*], Wei Tao [4], Qi Zhang [5], Eleonora Papadimitriou [1] and Pieter van Gelder [1]

1 Safety and Security Science Group, Faculty of Technology, Policy and Management, Delft University of Technology, 2628 BX Delft, The Netherlands
2 School of Navigation, Wuhan University of Technology, Wuhan 430063, China
3 National Engineering Research Center for Water Transport Safety, Wuhan University of Technology, Wuhan 430070, China
4 Intelligent Transportation Systems Research Center, Wuhan University of Technology, Wuhan 430062, China
5 Department of Civil and Architectural Engineering, KTH Royal Institute of Technology, 10044 Stockholm, Sweden
* Correspondence: r.song@tudelft.nl (R.S.); yqwen@whut.edu.cn (Y.W.); Tel.: +31-626-31-0261 (R.S.)

Abstract: Ship behavior is the semantic expression of corresponding trajectory in spatial-temporal space. The intelligent identification of ship behavior is critical for safety supervision in the waterborne transport. In particular, the complicated behavior reflects the long-term intentions of a ship, but it is challenging to recognize it automatically for computers without a proper understanding. For this purpose, this study provides a method to model the behavior for computers from the perspective of knowledge modeling that is explainable. Based on our previous work, a semantic model for ship behavior representation is given considering the multi-scale features of ship behavior in cognitive space. Firstly, the multi-scale features of ship behavior are analyzed in spatial-temporal dimension and semantic dimension individually. Then, a method for multi-scale behaviors modeling from the perspective of semantics is determined, which divides the behavior scale into four sub-scales in cognitive space, considering spatial and temporal dimensions: action, activity, process, and event. Furthermore, an ontology model is introduced to construct the multi-scale semantic model for ship behavior, where behaviors with different semantic scales are expressed using the functions of ontology from a microscopic perspective to a macroscopic perspective consecutively. To validate the model, a case study is conducted in which ship behavior with different scales occurred in port water areas. Typical behaviors, which include leveraging the axioms expression and semantic web rule language (SWRL) of the ontology, are then deduced using a reasoner, such as Pellet. The results show that the model is reasonable and feasible to represent multi-scale ship behavior in various scenarios and provides the potential to construct a smart supervision network for maritime authorities.

Keywords: semantic modeling; ship behavior; cognitive space; multi-scale analysis; ontology

Citation: Song, R.; Wen, Y.; Tao, W.; Zhang, Q.; Papadimitriou, E.; van Gelder, P. Semantic Modeling of Ship Behavior in Cognitive Space. *J. Mar. Sci. Eng.* **2022**, *10*, 1347. https://doi.org/10.3390/jmse10101347

Academic Editor: Apostolos Papanikolaou

Received: 18 August 2022
Accepted: 17 September 2022
Published: 22 September 2022

Publisher's Note: MDPI stays neutral with regard to jurisdictional claims in published maps and institutional affiliations.

1. Introduction

There is a high traffic density in some busy waterways, especially in port areas, where some severe situations have occurred. It increases the supervision difficulty to vessels for maritime authorities, such as the Maritime Safety Committee (MSC) and services. Specifically, the supervision to vessels includes static information inquiry, tracking of one or more vessels, ship behavior recognition, etc. The fact facilitates the autonomous supervision to vessels, especially whose behaviors in congestion areas are riskier than in normal areas. The rapid development of Maritime Autonomous Ships (MASS) in recent years has also placed a demand on the autonomous recognition and semantic transformation of ship behavior, which MASS should ideally satisfy to improve the perception of surrounding ship behavior. As a result, more and more researchers are paying attention to the automatic recognition and semantic enrichment of ship behavior.

Ship behavior is the representation of the trajectories of ships enriched with various types of semantic attributes. It is challenging to recognize ship behavior without any instructions from the human aspect for computers. In particular, complicated behaviors comprise a set of simple behaviors enriched with geographic attributes, temporal features, motion characteristics, etc. For example, the behavior of anchoring implies the place where the behavior occurred (anchorage), the time duration of the behavior (long-term), and the motion state (stationary). These behaviors are commonly used in realistic scenarios currently and require a deeper study. It is challenging to clarify all of the behaviors by computers without a proper model in which the semantic features of behaviors can be considered in depth. By contrast, a human expert can quickly and precisely understand exactly ship behavior. This is due to the excellent capability of processing information collected from multiple sources in a cognitive space for humans. Such a capability is what is required for the intelligent computers of MSC or MASS.

The semantic modeling of behaviors in cognitive space is a process of semantic reflection of the movement of physical objects, which enables computers to understand behaviors in the same way that humans think. Hence, it is a feasible way to empower a computer to be capable of recognizing behaviors enriched with rich semantics. However, there are a wide variety of behaviors with different semantics, as stated above. It is impractical to program each behavior manually. Thus, a model to extract and collate the semantic characteristics of ship behavior is desired to be provided to reach the final goal of semantic modeling.

This work is based on the previous work [1,2], focusing on modeling and reasoning of semantic ship behavior with different scales in multiple dimensions. We propose a semantic model to extract and recognize multi-scale behaviors automatically in cognitive space based on historical automatic identification system (AIS) data. In this study, the features of ship behavior that represent corresponding trajectories are analyzed from the perspective of spatial-temporal and semantic, respectively. Furthermore, a multi-scale semantic model is given to depict ship behavior in cognitive space, in which behaviors with different spatial scales are sorted out and a formalized cognitive model of behavior is presented. Moreover, by means of the ontology modeling method, multi-scale behaviors are explored and expressed further. Behaviors with different semantic scales are presented, leveraging the functions of ontology. Finally, a case study of a ship approaching into and leaving a port is given to show how the model works.

The remainder of the paper is structured as follows. In Section 2, an overview of related work is given. The analysis of the multi-dimensional feature is in Section 3. The model of multi-scale behaviors is proposed in Section 4. Section 5 constructs an ontology model for multi-scale semantic behaviors. Section 6 presents a case study to validate the feasibility of the semantic model. In Section 7, the results and discussion of the experiment are presented. Finally, the conclusion is given in Section 8.

2. Related Work

2.1. Cognitive Modeling

There is extensive literature on the topic of cognition modeling for human behaviors, which are influential in ship behavior modeling in cognitive space. A number of studies have examined the construction method of knowledge base [3–6], and knowledge reasoning [7] with ontology [8–10], which discusses cognitive modeling and knowledge reasoning for human activities. [4,6,8] constructed the ontology based on the relationships between humans and then environment to transform human behaviors from the data layer to the semantic level, which realized the recognition of human behavior intelligently. These studies are beneficial to provide some thoughts about how to construct the cognitive framework of ship behavior.

Currently, there are few works that focus on the research of semantic recognition of ship behavior. [9] designed a method to identify the ship events using AIS data that records relevant information about ship movement, such as position, speed, course, etc. [10] tries to deduce the complicated behaviors based on the method proposed in [9], providing the foundation of cognition modeling of complicated ship behavior.

2.2. Semantic Expressions of Trajectory

To address the problem of semantic behaviors recognition for computers, from the perspective of humans, some recent studies centered on semantic modeling from a human perspective have proposed semantic computational processing methods [11], using the same approach that humans use to perform semantic recognition of behaviors. Refs. [12,13] proposed a semantic computing frame to compute the trajectory generated by moving objects, such as vehicles, humans, and animals. Based on these efforts, some studies on ship semantic behaviors have also been carried out. A semantic model of ship behavior was proposed in [14], which takes into account the uncertainty of the occurring behavior; [15] mined the pattern of ship trajectories by means of semantic annotation and possibility modeling; and [1] constructed the ontology model of ship behavior, considering the temporal relationships between each other. In addition, some projects work on the semantic computing of trajectories in the maritime domain. An example of this is datAcron [10,16,17], a project focusing on the representation of semantic trajectories of aviation and maritime conceptualizations.

2.3. Ship Behavior Modeling

In order to model ship behavior explicitly, many studies have been focusing on behavior modeling from trajectory to behavior. There are two kinds of methods used to analyze it: probabilistic statistics and motion characteristics extracting and modeling. The former refers to the pattern mining of ship behavior by means of statistics analysis [18–20]. Another approach based on motion characteristics analysis accomplishes this by analyzing the relationship between the characteristics and behaviors and then modeling. Ref. [21] considered the motion characteristics of ship trajectories to construct the model. Ref. [14] proposed a model for ship behavior based the ship basic behaviors, such as turning to port side, turning to starboard side, and some semantic behavior occurred in the environment.

Few studies on ship behavior undertake basic behavior modeling and prediction, considering the structural and temporal features of complicated behaviors, which are necessary for computers to satisfy the requirements to ascertain a desirable understanding of behaviors. Ref. [2] proposed a framework for ship behavior from a cognitive and semantic modeling perspective and constructed a semantic model to represent behavior from data to trajectory to complex behavior, considering its motion data and environmental attributes.

2.4. Multi-Scale Modeling of Trajectory

There are extensive studies focusing on the topic of multi-scale characteristics analysis in geography [22,23], which have explored in detail about the multi-scale characteristics of spatial-temporal objects [24–26]. The trajectory, as the representation of the spatial-temporal characteristics of physical objects, exhibits multi-scale characteristics. Previous studies provide a benchmark of multi-scale feature analysis for spatial objects and a solid basis for building a cognitive framework for modeling multi-scale ship behavior. Ref. [27] discussed the multi-scale representation of battlefield situation. Ref. [28] proposed a multi-level model to explore the spatial-temporal patterns of crime in different spatial scales of area. They provide guidance for the construction of cognitive models of ship behavior.

It is necessary to propose a systematic approach to analyze complicated behaviors by comprehensively considering its various characteristics, such as motion characteristics, topological relationships with environmental entities, etc. In general, the modeling of complicated behaviors needs to be considered in different dimensions.

As the semantic representation of ship trajectory, the multi-scale features of ship trajectory can form the multi-scale features of behaviors in three dimensions, such as time, space, and semantics. However, few studies have considered the semantic multi-scale features of ship behavior that are crucial for behavior recognition. The relationship of ship behavior between different levels and between different scales of the same level has not been constructed properly, which limits the development of the modular computing capability of the autonomous system for the safety supervision of behavior.

To address the problem, Firstly, we characterize ship behavior from the scale of spatial-temporal and analyze the shortcomings of modeling ship behavior in this dimension.

Secondly, we analyze the way ship managers with different cognitive mindsets perceive ship behavior and propose a cognitive model for ship behavior from the semantic dimension, dividing ship behavior into four layers of action, activity, process, and event to describe ship behavior at different spatial-temporal and semantic scales. Finally, the cognitive ontology of ship behavior is constructed, taking the typical behavior of ships in port areas as an example for ontological modeling and expression, and exploring the mechanism of multi-scale semantic expression and reasoning of ship behavior in port waters.

3. Multi-Dimensional Characterization of Ship Behavior in Cognitive Space

A ship generates a series of trajectory segments driven by the intention of the seafarer. That means that the semantics implied by the trajectory reflects the seafarer intention to navigate. From simple behaviors, such as accelerating and going straight, to advanced behaviors are the semantics implied by a ship's trajectory, such as sailing along the fairway, berthing, etc. In other words, the behavior can be represented as the semantic reflection implied by the trajectories produced by physical objects in cognitive space where human operators process information on their own temporal and logical terms. That is, ship behavior has additional semantic features in addition to the spatial-temporal motion characteristics of ship trajectories. The semantics implied by trajectories are described differently within different spatial-temporal dimensions.

3.1. Previous Work for Semantic Modeling of Ship Behavior

For semantic modeling of ship behavior, we have explored in our previous studies [2], where a framework of semantic behavior generation process from trajectories enriched with motion semantics and topological environment semantics was given. In this paper, we proposed several concepts, such as atomic trajectory, atomic behavior, topological behavior, as well as traffic behavior, representing the semantic behavior with corresponding semantic features.

Specifically, we first divide the trajectory, generated from AIS data, into atomic trajectories, as trajectory units on the basis of our classification of atomic behavior. Atomic behavior represents the behavior of maintaining a constant motion state of both speed and course simultaneously, as shown in Figure 1. That means the trajectory was segmented according to its motion status instead of sample frequency or spatial grid division with same size, which is beneficial to reduce its computation complexity. Topological semantic enrichment is based on the atomic trajectory.

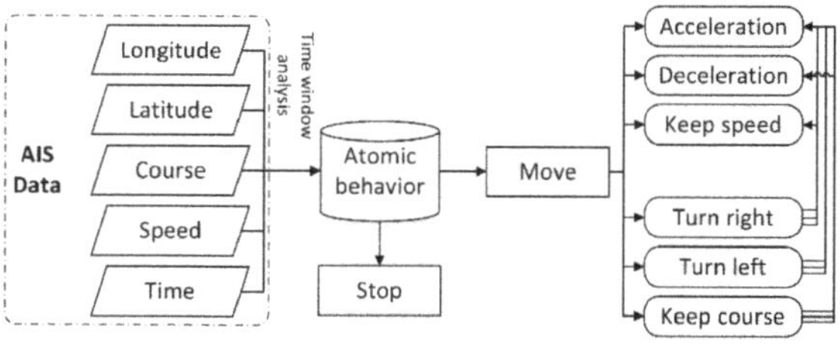

Figure 1. Classification of ship atomic behaviors adapted with permission from Ref. [2].

Following this, in order to enrich the semantics of geographical properties for trajectory unit, we introduced 15 spatial relations for the calculation of two objects involving point, line, and surface in maritime domain by adapting Dimensionally Extended 9-Intersection Model (DE-9IM) that is proposed for describing spatial relations of two regions.

Finally, we presented how traffic behavior, as semantic behavior unit of ships, including motion status and topological semantics, are formed through atomic trajectory, atomic behavior, topological behavior, and environment. Traffic behavior T_l^{r} can be represented as Formula (1) performing as a sentence, where $\underline{\underline{T_i^{\mathrm{a}}}}$ represents atomic trajectory as the subject, $\underline{T_j^{\mathrm{o}}}$ represents topological behavior as the predicate, $\underset{\sim\sim\sim}{r_k}$ represents the environment as the object, and $[B_i^{\mathrm{a}}]$ refers to the atomic behavior as the gerund of the sentence, respectively.

$$T_l^{\mathrm{r}} = \left\{ \underline{\underline{T_i^{\mathrm{a}}}} \cap \underline{T_j^{\mathrm{o}}} \cap \underset{\sim\sim\sim}{r_k} \cap [B_i^{\mathrm{a}}] \right\} \tag{1}$$

This model provides a way to reach the goal of semantic unit formation, supporting further high-level semantic modeling for complicated behaviors, which can be represented with a set of traffic behaviors. In addition, we explored the temporal relations preliminarily within complicated behaviors in [1], where we expected to depict complicated behaviors through combining simple behaviors.

Previous works present how to enrich semantics from different respects to trajectories, but there is a lack of extensive analysis on complicated behaviors, especially the relationships between different dimensions. On the basis of these work, we try to propose an extended semantic model for complicated behaviors combining human cognitive habits.

3.2. Multi-Dimensional Feature Analysis

Based on previous research, we expect to investigate how complicated behaviors can be represented in terms of basic semantic behaviors. Considering the intrinsic spatial-temporal and semantic scale features [29] of complex behaviors, we wish to propose a framework for the analysis of complex behaviors that considers spatial, temporal, and semantic dimensions. Thus, we analyze behavior in three dimensions.

In terms of the spatial-temporal dimension, ship trajectories as a form of spatial-temporal representation generated by physical objects, the determination of the spatial-temporal scale depends on the frequency with which the trajectories are sampled [30]. Therefore, the sampling frequency and granularity of ship trajectories must be determined when analyzing and modeling ship trajectories at multiple scales purely from the spatial-temporal dimension.

However, it is challenging to provide a standard method to determine the scale of the spatial-temporal dimension. Because people with different roles have different concerns about ship behavior, that is not appropriate. Therefore, the modeling of multi-scale features of the track also needs to be reworked around different needs for attention, which presents a higher standard and challenge for the accurate sampling of ship tracks. For example, mariners are more concerned with short-term vessel behavior, such as analyzing whether the target vessel around her is performing the maneuvers specified in COLREG. In contrast, VTS officers are more inclined to obtain a longer range or time interval of behavior, such as analyzing whether vessels within their jurisdiction are engaging in illegal activities. In other words, different people have different scales of attention to the behavior of vessels, involving differences in scale not only in the spatial-temporal dimension but also in the semantic dimension.

Therefore, the analysis of ship behavior should combine the spatial-temporal dimension with the semantic dimension. From a semantic point of view, when modeling ship behavior at multiple scales, we need to describe the behavior semantically in the spatial-temporal dimension at the same time. They need to obtain a good understanding of behavior by dividing the semantic space into several appropriate semantic scales, which are closer to the human habit of perceiving behavior.

4. Multi-Scale Cognitive Modeling of Ship Behavior from Semantic Dimension

Spatial-temporal data are prevalent with multi-level, multi-grain, and multi-resolution characteristics, and the analysis and extraction of these features is a prerequisite for their awareness and modeling. In addition, the model construction based on these features is also in line with the human cognitive habits of multi-dimensional and multi-features of spatial-temporal data. Therefore, for the spatial-temporal trajectories corresponding to ship behavior, we need to consider these multidimensional features mentioned above and consider the intrinsic relationship of each dimension and the relationship between them. In view of the cognitive habits of people with different roles in the maritime domain for ship behavior, ship behavior can be analyzed and modeled from microscopic scale to macroscopic scale.

4.1. Formalized Cognitive Expression of Ship Behavior

Behavioral cognition is the result of multifaceted description and expression of ship trajectory. Based on the analysis of cognitive elements, the cognitive expression of ship behavior, *Cog*, should be considered as a cognitive set, including four elements: who, what, when, and where, which can be expressed as Equation (2).

$$Cog = \{\mathbf{o},\mathbf{b},\mathbf{t},\mathbf{p}\} \tag{2}$$

where **o** denotes the object where the behavior occurs; **b** is the behavior that occurs at the object; **t** represents the time, including instant and interval; and **p** is the place where the behavior occurs.

Considering the multi-scale characteristics of spatial-temporal trajectories, this paper divides the cognition of ship behavior into four layers in the cognitive space: action, activity, process, and event, according to the expression habit of ship behavior in the semantic dimension. The division of behavior cognition is based on two aspects, including motion features and the topological features.

4.2. Multi-Scale Division of Ship Behavior in Cognitive Space

4.2.1. Action

Considering the practical needs of users for ship behavior, when describing and calculating the microscopic behavior of a ship, this paper avoids the situation that causes the redundancy of successive division of equal time interval or equal distance interval trajectories and the complexity of calculating topological relations. In this paper, from the perspective of behavior semantics, the concept of action is introduced to represent the cognition results of the micro-semantic behavior features, which is to represent the behavior that the ship motion characteristic, involving both speed and course, remains unchanged during the current behavior stage, such as keep course and deceleration (KC_DE), turn left and deceleration (TL_DE), turn right and deceleration (TR_DE), etc. Action behavior is a behavior to characterize the basic motion characteristics of the trajectory without additional semantic information related to environment. The behavior enriched with rich semantic can be formed based the action behavior.

4.2.2. Activity

Activity is the cognition results of the behavior represented by the trajectory of action behavior, enriched with topological and geographical semantics, which represents the behavior based on the topological interaction and geographical semantic enrichment. The behavior of the activity occurs on the trajectory of action, which interacts with the entities in the environment, such as anchorage, berth, etc., which is the basic semantic unit of ship behavior, and complex semantic behavior can be expressed by the combination of a set of consecutive activity behaviors.

The behavior difference between the activity and the action is that the action only reflects the semantic of motion characteristics of the trajectory and do not include the semantics of the interaction between the trajectory and its surroundings. In contrast, the activity has more semantics than the action but also involves the semantics of spatial topological calculations and geographic semantic enrichment performed by the action trajectory with environmental entities.

4.2.3. Process

A process represents the behavior of a ship in which the spatial topological relationship between its trajectory and environmental entities remains constant while keeping its speed or course unchanged. That is, when the characteristics of the speed or course of ships change or the spatial topological relationship between its trajectory and environmental entities changes, the current process behavior turns to the next process behavior.

In contrast to activity behaviors, process behaviors are the extension of activity behaviors, which describe the interaction behaviors that occur between trajectories with constant speed and constant course and entities in geographic space, respectively. For example, the behavior of anchoring preparing can be regarded as a process behavior, which is usually accompanied by a series of action behaviors of deceleration, while the trajectory of this behavior keeps the same spatial topology relationship with the geographic entity of anchorage during this process until the behavior of deceleration is stopped, at which point the action behavior of the ship changes, which also means that the activity of the ship changes. Therefore, the process behavior of anchoring preparing indicates that a process behavior occurs for the ship, and its connection with the next process behavior is made through the activity behavior of beginning to anchor, and the moment when the act of beginning to anchor begins is the moment when the behavior of anchoring preparing ends and the moment when the next process behavior begins.

Similarly, when the spatial topological relationship between the ship trajectory and the geographic entity changes, it means that the ship experiences an activity behavior, which indicates the beginning of the next process behavior.

A process behavior tends to have a larger temporal and spatial scope than an activity behavior, and it represents that the ship is executing a certain task, such as the process behavior of anchoring represents that a ship is anchoring at anchor, and the activity of anchoring does not change during this period. In contrast to the activity behavior, the process behavior does not consider the change or not of the combined motion characteristics of speed and course, and its focus is on the change of spatial and topological state between the trajectory that remains constant either on the speed or the course and the surrounding environment.

The behavior of activity, which can be considered as one of the components of the behavior of process, is a trigger element condition between different processes and serves to trigger the end of the previous process and the beginning of the next one.

4.2.4. Event

Event behaviors, which represent the overall behavior of the vessel occurring in the current observation view or macroscopic behavior relative to the current reference target, describe the logical and temporal relationships between the behaviors. For example, the entire behavior of berthing and unberthing can be considered as one event, including the three process behaviors of decelerating for preparing to berth, berthing, and accelerating for unberthing. Event behavior can also denote the behavior that occurs in a larger temporal and spatial context, which is extensible. For example, the behavior of a ship sailing from one port to another can be regarded as a whole event containing several sub-events, including the departure event from port A, several subsequent events, and the arrival event at port B.

The event behavior corresponding to the macro behavior is a semantic aggregation of activities and processes or, alternatively, can represent a semantic aggregation of multiple events with related relationships. As can be seen from Figure 2, the goal of transforming from trajectories to multi-scale spatial-temporal semantic behaviors can be achieved and multi-dimensional modeling and representation of behaviors is realized.

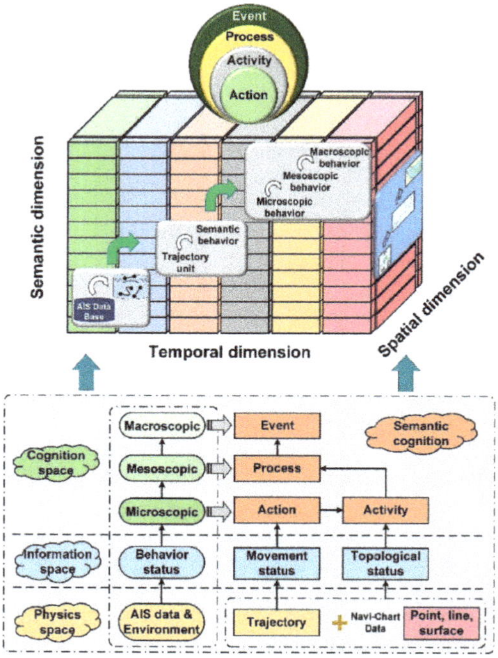

Figure 2. Representation process and construction model of ship multi-dimensional behavior.

5. Ontology Modeling for Ship Behavior

Based on the analysis of multi-scale behavior, this paper proposes an ontology model for ship semantic behavior modeling in a cognitive manner. The model introduces several concepts in cognition to model the cognitive framework of ship behavior, involving ship, behavior, time, and place, which supports a systematic interpretation of ship behavior by a computer, as shown in Figure 3. Figure 4 presents the ontology of ship behavior created according to our cognitive framework of ship behavior. Figures 4a, 4b and 4c show the class, object properties, and data properties interfaces in the ontology software Protégé, respectively.

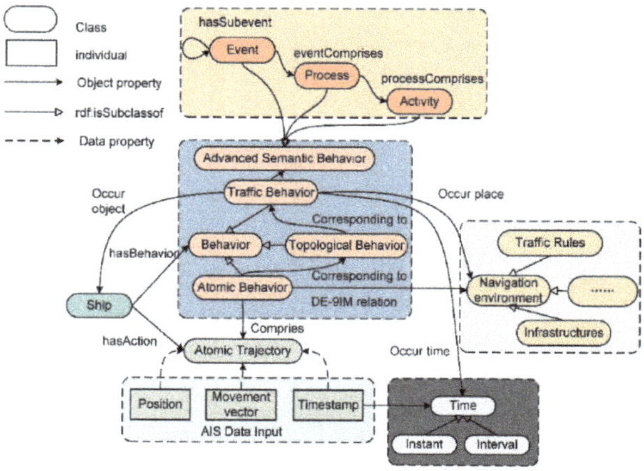

Figure 3. Ontology model of cognition for ship behavior.

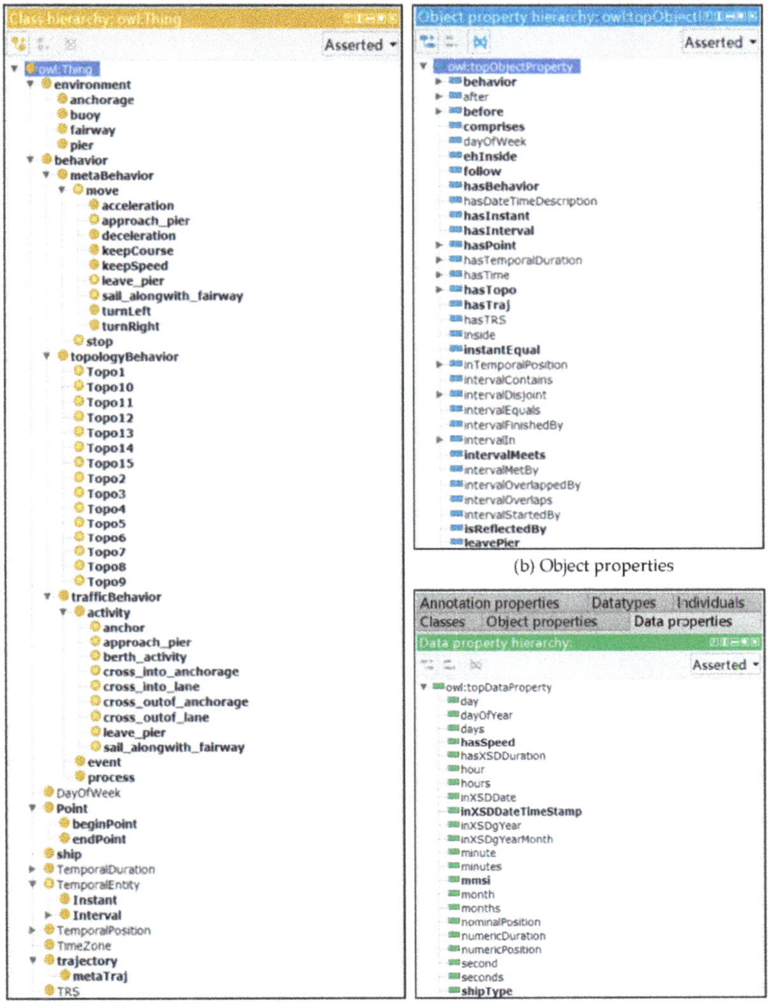

(a) Classes (b) Object properties (c) Data properties

Figure 4. Display of cognition ontology of ship behavior in port. (**a**) Classes; (**b**) Object properties; (**c**) Data properties.

5.1. Classes

- Ship

The ship is the generator of trajectory and the occurrence object of behavior. Therefore, it is necessary to identify the object of trajectory in behavior cognition. As a unique code for the ship, the number of the Maritime Mobile Service Identity (MMSI) can be used to determine the object that generates the behavior.

- Behavior

The class of behavior is one of core classes of the cognitive ontology. Based on previous work [2], ship semantic behaviors are divided into four categories, including atomic behaviors, topological behaviors, traffic behaviors, and advanced semantic behaviors. For a better understanding, The class of atomic behavior proposed in [2] are extended to refined divisions, including first-order and second-order atomic behaviors. The second-order atomic behavior corresponds to action, including 10 categories, while the first-order atomic behavior corresponds to the process where either the speed or the course keeps ships maintaining constant, as can be seen in Figure 2.

As the unit of semantic behavior, the traffic behavior is the basic element to describe advanced semantic behaviors, corresponding to the behavior of activity.

The advanced semantic behavior refers to high-level behaviors, such as the behavior of process and event, that can be formed combining multiple consecutive sort of traffic behaviors in specific application scenarios.

- Environment

The class of environment represents the set of surrounding spatial objects existing in the form of physical or virtual entities, such as anchorages, channels, control areas, and infrastructure.

- Time

In order to represent the temporal relationship of behaviors, we introduce an existed time ontology to our work, which is available online: http://www.w3.org/2006/time# (accessed on 27 September 2020), that includes both instant and interval that can fulfill the temporal functions, which is beneficial to describe complex behaviors that are temporal and logical. Specifically, the class of instant is to describe the transient state of behavior, and time interval represents the time quantum, which lasts for a certain period, including start time and end time to express the duration of behaviors.

According to the theory of interval algebra proposed by Allen [31], basic time relationship includes before, after, and equal. Furthermore, 10 types of relationships between instant and interval can be depicted with 3 kinds of basic relationships. It is the temporal and logic features of ship behavior that can be depicted in this way.

Besides the abovementioned, we connect them with their relationships to link this network. There are three kinds of arrows for that, including solid arrow, dashed arrow, and solid arrow with empty end. As for solid arrow, it refers to the relationship between two classes either same classes or different classes. For example, there is the relationship of "occur object" that has the domain—"traffic behavior" and the range—"ship". For the second one, it means what the own data properties the entity have. An example can be taken to illustrate that the dashed arrow pointing to "Atomic Trajectory" from "Position" represents that the former one has the data of the latter one. The final one means there is the relationship of parent–subclass relationship, such as the arrows of the three behaviors in the top yellow round box pointing to "Advanced Semantic Behavior".

Now that we have extracted the different classes, the next step is to connect them to form knowledge graph. For that reason, we then need to add the relationship properties with each other to them.

5.2. Property-Constrained Axiom

- The class of Atomic Behavior

Atomic behavior can be recognized and annotated by pre-processing and calculation of trajectories. Therefore, we import atomic behavior as instances of ship behavior directly into the ontology via its interface. On this basis, first-order atomic behavior can be expressed by second-order atomic behavior. For example, several instances can be illustrated using the property-constrained axiom as follows.

acceleration = (KC_AC or TL_AC or TR_AC)

keepSpeed = (KC_KS or TL_KS or TR_KS)

stop= stay

Furthermore, since the move behavior is one of the general behaviors, including all the first-order atomic behaviors, we likewise use the axiomatic expression of the property constraint to define the move behavior, which indicates that the ship is in the move state, including all the second-order atomic behaviors, which can be expressed as follows.

move = (reflects value KC_AC) or (reflects value KC_DE) or (reflects value KC_KS) or (reflects value TL_AC) or (reflects value TL_DE) or (reflects value TL_KS) or (reflects value TR_AC) or (reflects value TR_DE) or (reflects value TR_KS)

- The class of Topological Behavior

According to definition and semantic computing results for topological behavior, each topological behavior can be expressed with ontology via axiomatic expression method of attribute constraint. For example, Topo1 represents the topological relationships between trajectory and navigation environment, which can be expressed with spatial topological relationships as follows:

Topo1 = PL1 some trafficRule

- The class of Traffic behavior and Advanced Semantic behavior

The traffic behavior class corresponds to the activity behavior. The traffic behavior in port water areas can be divided into nine types of activity behavior, and the advanced semantic behavior can be divided into five types of process behavior and three types of event behavior, which can be seen in Figure 5.

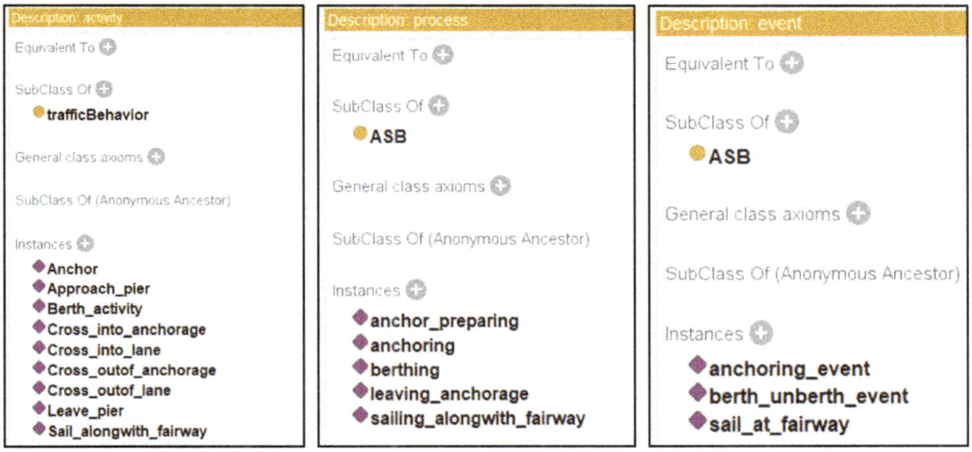

Figure 5. Results of ship semantic behavior in port water traffic areas.

Representation of behaviors with different semantic scales in the port water traffic areas can be done in different ways using ontology. Simple behaviors, such as active behaviors, can be expressed using property-constrained axioms based on atomic and topological behaviors.

Taking the behavior of Entering the Fairway as an example, the sufficient and necessary conditions for cross_into_lane should be as follows:

Trajectory T intersects with the line of fairway or the line between end points of the former, resulting in an intersection point located on an atomic trajectory AT which belongs to T. The beginning point of AT is located on the inside of the fairway and the endpoint is located on the outside of the fairway.. Likewise, the behavior of approach_pier can also be presented in a same way. The activity behavior of cross_into_lane and approach_pier can be represented as Figure 6.

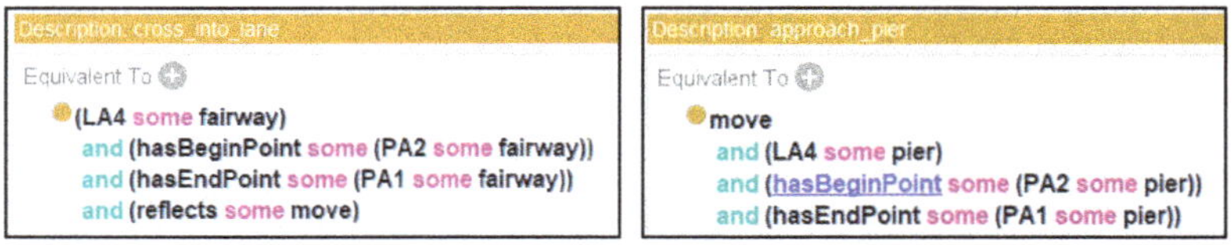

Figure 6. Knowledge representation of activity behaviors.

5.3. Complicated Behavior Expressions Using SWRL

Complicated behaviors, such as process behaviors and event behaviors, are difficult to express directly with property-constrained axioms due to complex intrinsic behavioral logic. For this reason, we introduce Allen's algebraic theory to model the temporal relationship of behaviors and express their complex behavioral intrinsic logical relations leveraging SWRL. Specifically, advanced semantic behaviors, such as event behaviors, consist of ordered specific activities and process behaviors, and as these behaviors occur, it can be triggered and inferred whether the advanced semantic behavior occurs or not. As shown in Figure 7, the event behavior of Anchor is explicitly temporal and logical, in which behaviors of the blue rectangular box and the gray arrow box make up the Anchor event.

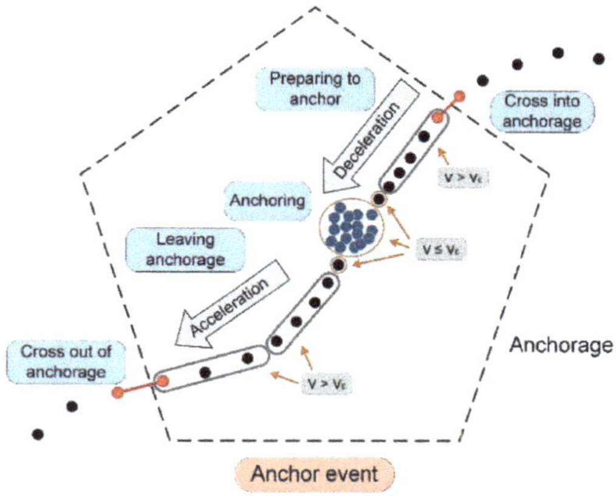

Figure 7. Anchor event of ship occurred in anchorage.

For an explicit explanation, the process of the Anchor preparing process is selected to illustrate how to formalize the behavior, which is described below.

Anchor preparing process: The behavior from the instant the ship enters the anchorage until the start of anchoring, it consists of a series of activities in *preparing to anchor*. It is based on the existing knowledge to infer the advanced behavior, but the activity of preparing to anchor is not easily identified. Therefore, we do not use it to deduce the process. However, it is also worth noting that the trajectory corresponding to the behavior of the process contains a series of trajectories of the activity behavior, i.e., from the trajectory reflecting the *cross_into_anchorage* behavior to the trajectory reflecting the activity of the first anchor activity. Since these two behaviors can be obtained computationally from the AIS-based preprocessing module, it is possible to determine whether the process occurs by judging whether the two behaviors occur sequentially, which can be expressed in SWRL as:

Anchor preparing process = cognition2:ship (?s) ˆ cognition2:trajectory (?t) ˆ cognition2:hasTraj (?s, ?t) ˆ cognition2:metaTraj (?stra) ˆ cognition2:comprises (?t, ?stra) ˆ cognition2:Point (?p1) ˆ cognition2:Point (?p2) ˆ cognition2:hasBeginPoint (?stra, ?p1) ˆ cognition2:hasEndPoint (?stra, ?p2) ˆ cognition2:LA5 (?stra, ?p) ˆ cognition2:anchorage (?p) ˆ cognition2:hasSpeed (?p2, ?x) ˆ swrlb:lessThanOrEqual (?x, "0.5" ˆˆ xsd:float) → cognition2:hasBehavior (?s, cognition2:anchor_preparing)

Similarly, other behaviors occurring in port areas can be stated in the same way, as shown in Table 1.

Table 1. Selected SWRL rules for reasoning about advanced behavior.

Number	SWRL Rules	Description
1	hasTraj(?s, ?t) ^ ship(?s) ^ reflects(?t, ?b) ^ behavior(?b) → hasBehavior(?s, ?b)	HasBehavior
2	ship(?s) ^ trajectory(?t) ^ metaTraj(?stra) ^ comprises(?t, ?stra) ^ hasTraj(?s, ?t) ^ approach_pier(?stra) → hasBehavior(?s, Approach_pier)	Behavior of approaching pier
3	trajectory(?t) ^ hasBeginPoint(?t, ?p1) ^ hasEndPoint(?t, ?p2) ^ Point(?p1) ^ Point(?p2) ^ Instant(?t1) ^ Instant(?t2) ^ hasInstant(?p1, ?t1) ^ hasInstant(?p2, ?t2) → occursEnd(?t, ?t2)	The occur begin time of trajectory
4	ship(?s) ^ trajectory(?t) ^ hasTraj(?s, ?t) ^ comprises(?t, ?straj) ^ metaTraj(?straj) ^ leave_pier(?straj) ^ LA4(?straj, ?p) ^ pier(?p) → leavePier(?s, ?p)	Behavior of leaving pier
5	ship(?s) ^ trajectory(?t) ^ hasTraj(?s, ?t) ^ berth_activity(?t) → hasBehavior(?s, berthing)	Berthing
6	ship(?s) ^ hasTraj(?s, ?t) ^ trajectory(?t) ^ cross_outof_anchorage(?subtra) ^ comprises(?t, ?subtra) ^ metaTraj(?subtra) → hasBehavior(?s, Cross_outof_anchorage)	Behavior of crossing out of anchorage
7	ship(?s) ^ trajectory(?t) ^ hasTraj(?s, ?t) ^ comprises(?t, ?straj) ^ metaTraj(?straj) ^ cross_outof_anchorage(?straj) ^ anchorage(?p) ^ LA4(?straj, ?p) → crossOutofAnchorage(?s, ?p)	Object properties for cross out of anchorage
8	ship(?s) ^ trajectory(?t) ^ hasTraj(?s, ?t) ^ metaTraj(?strai) ^ metaTraj(?straj) ^ comprises(?t, ?straj) ^ comprises(?t, ?strai) ^ LA4(?strai, ?p) ^ LA4(?straj, ?p) ^ fairway(?p) ^ cross_into_lane(?strai) ^ cross_outof_lane(?straj) → hasBehavior(?s, sailing_alongwith_fairway)	Behavior of sailing along with fairway
9	Instant(?i1) ^ Instant(?i2) ^ inXSDDateTimeStamp(?i1, ?it1) ^ inXSDDateTimeStamp(?i2, ?it2) ^ swrlb:lessThan(?it1, ?it2) → before(?i1, ?i2)	Time order
10	ship(?s) ^ trajectory(?t) ^ hasTraj(?s, ?t) ^ metaTraj(?stra) ^ comprises(?t, ?stra) ^ Point(?p1) ^ Point(?p2) ^ hasBeginPoint(?stra, ?p1) ^ hasEndPoint(?stra, ?p2) ^ LA5(?stra, ?p) ^ anchorage(?p) ^ hasSpeed(?p2, ?x) ^ swrlb:lessThanOrEqual(?x, 0.5) ^ deceleration(?a) ^ reflects(?stra, ?a) → hasBehavior(?s, anchor_preparing)	Behavior of preparing to anchor

6. Case Study

In order to investigate the feasibility of the cognitive model, we take the scenario as the experiment case where ship behavior, such as *arrival* and *departure* events, occurred in a port to show how complicated behaviors can be deduced in a cognitive way.

Firstly, based on the navigational experience of seafarers in port traffic areas, the most common ship behaviors occurring in order in these areas can be divided into three layers in which they occur, as shown in Figure 8. The overall behavior can be considered as an event of ship arrival-departure, in which the event of anchoring, entering fairway, berthing and unberthing, and departure are most commonly occur in an orderly manner. Likewise, the process layer and the activity layer can be extracted and depicted as follows.

6.1. Data Processing

The paper uses the AIS data and geographic data from Xiamen port for March and April 2016, including ship trajectory, fairways, anchorages, and piers. First, we pre-process ship trajectories, including data sorting and interpolation. Then, the dynamic AIS data are matched with the ship name, MMSI, and ship type in the static database to achieve the acquisition of ship attributes. Furthermore, the name, functional attributes, and location information of geographical objects can be obtained from www.chinaports.com (accessed on 18 September 2020).

Protégé is an ontology modeling tool [32] and is used here to construct an ontology model of ship behavior perception. We use version 5.5.0 of the software, version 2.2.0 of the Pellet reasoner and version 2.0.9 of SWRL. In addition, the model imports the time ontology abovementioned to support reasoning about complex behavioral temporal relationships.

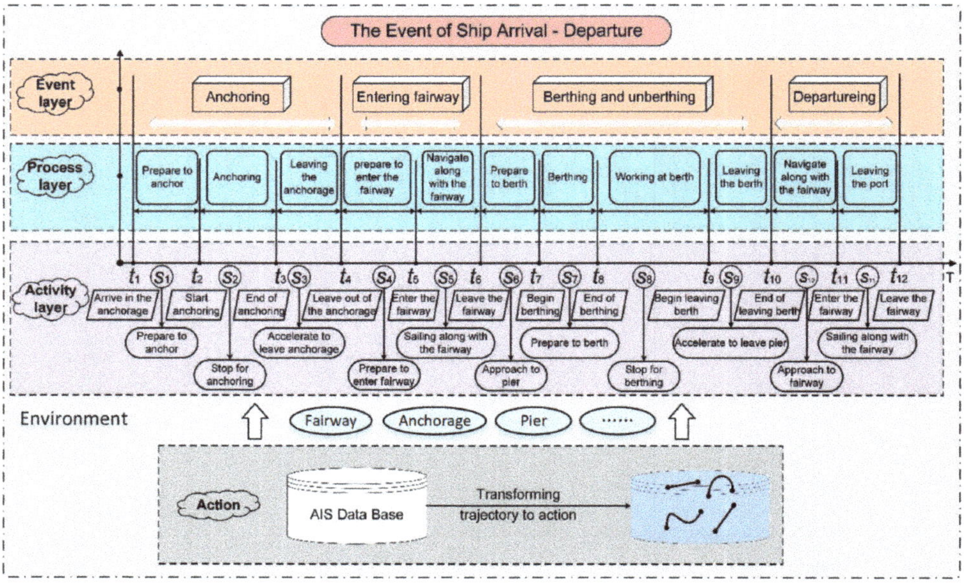

Figure 8. Cognition graph of the behavior of ship arriving and leaving the port.

6.2. Trajectory Segments and Semantic Annotation

In order to reduce the computational complexity and improve the ontological reasoning efficiency, 20,000 AIS data of ships are extracted for validation in this experiment. Firstly, the trajectories of different ships are sorted out in order to obtain the trajectories of each ship. Secondly, the trajectories are divided into "moving-stop" segments based on the recognition of stopping points to realize the labeling of moving trajectories. On this basis, we separate the moving trajectory from the stop trajectory to complete the annotation of atomic trajectory and the further recognition of atomic behavior. Finally, the start and end points of the trajectory are marked according to the stop, start, and end points of the atomic trajectory.

In order to calculate the spatial topological relationship between trajectories and the environment, the paper introduces a library for topology calculation based on Python programming language—Shapely. Firstly, various geographical objects and ship trajectories are converted into the format of spatial data. Then, the topological relationships of these converted objects are calculated to obtain the DE-9IM metrics of the relationships between trajectories and geographic objects. Finally, the computed results are mapped with the corresponding trajectories to prepare for the semantic annotation of ship behavior.

After the data level preparation is completed, the semantic information needs to be added to the ontology. In order to realize the combination of data and semantic information in the Python environment, the paper introduces Owlready2, a python-oriented ontology programming module that adds the already computed semantic information to the data layer and can load and save ontology files for modification and inference.

Figure 9 shows the overall process of behavioral cognitive computation, semantic reasoning and querying, which can support knowledge queries of behaviors with different semantic scales.

6.3. Semanticization of Ship Behavior

After importing the data related to ship behavior cognition into the ontology, including the ship, its trajectory segments, and the relationship between them, ship behavior can be clearly depicted. Figure 10 shows the importing results of ship RENLONG and the details of its trajectory segments, such as the place and time of occurrence.

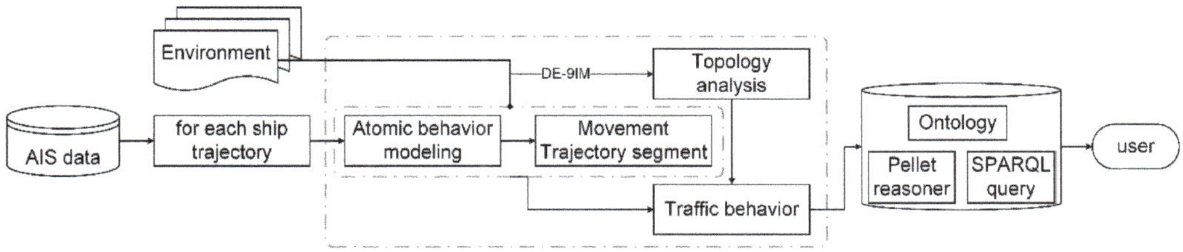

Figure 9. The process of cognitive computation, semantic reasoning, and querying of ship behavior.

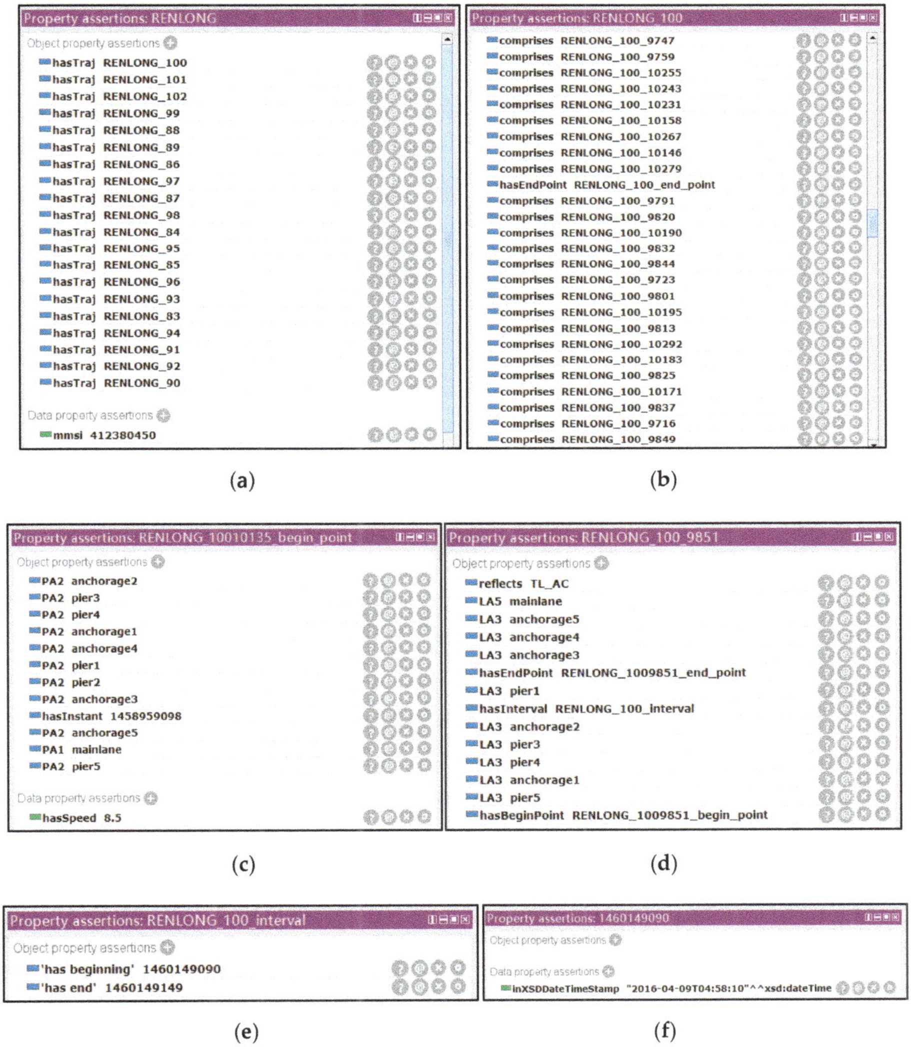

Figure 10. The importing result of ship behavior. (**a**) The trajectories occurred by RENLONG; (**b**) The sub-trajectories and key points of trajectory RENLONG_100; (**c**) The spatial topological relationships with surroundings for RENLONG_10010135_begin_point; (**d**) The spatial topological relationships with surroundings for RENLONG_100_9851; (**e**) The instant properties of RENLONG_100_interval; (**f**) The XSD date timestamp property corresponding to the moment 1460149090.

Specifically, Figure 10a shows the movement and stationary trajectory segmentation identified for the ship RENLONG based on the "move-stop" trajectory segmentation method. Figure 10b shows the different atomic trajectory segments contained in the motion trajectory segmentation RENLONG_100 and the end point of this trajectory. Figure 10c shows the topological and temporal properties of RENLONG_ 10010135_begin_point as the starting point of RENLONG_10010135, in relation to the geographic region. Figure 10d represents the topological properties of the RENLONG_100_9851 atomic trajectory segment with respect to the geographic region around it, as well as its beginning and end points. Figure 10e shows the temporal properties of the time period in which the RENLONG_100 trajectory segment occurs, where the property of "has beginning" indicates that the beginning point of the trajectory segment occurred at the moment 1460149090. Figure 10f shows the XSD date timestamp property corresponding to the moment 1460149090.

As mentioned in the previous section, the atomic and topological behaviors can be stated based on the property-constrained axioms. The first-order atomic and topological behaviors are defined in terms of sufficient and necessary conditional constraint axioms. When the second-order behavior or topological features satisfy the definition of the class of the corresponding behavior, they will be automatically derived and classified to the corresponding first-order atomic behavior. As shown in Figure 11, the trajectory segment XINHAIXIU_49_3699 is classified as the class of cross_into_lane. On the basis of simple semantic behaviors, complicated behaviors, such as behaviors of process and event, can be further deduced based on the rules stated using SWRL, as described in Section 5.3.

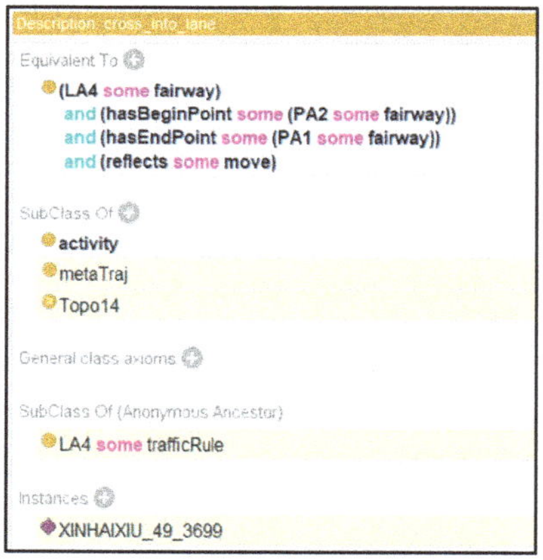

Figure 11. The reasoning results of activity behavior cross_into_lane.

For an intuitive comprehension of deduced behaviors shown in Figure 11, the trajectory marked with corresponding behavior is visualized in Figure 12 that shows ship semantic behaviors after trajectory segmentation, spatial topology calculation, and semantic annotation. Figure 12a,c,d show the ship's semantic behavior at anchorage, fairway, and berth, respectively, while Figure 12b is a zoomed-in view of the behavior in Figure 12a. The segmented ship trajectory segments can be clearly identified in these images, as well as the annotated advanced ship semantic behavior, such as the ship's approach to the anchorage, the ship's exit from the anchorage, the ship's approach to the channel, etc.

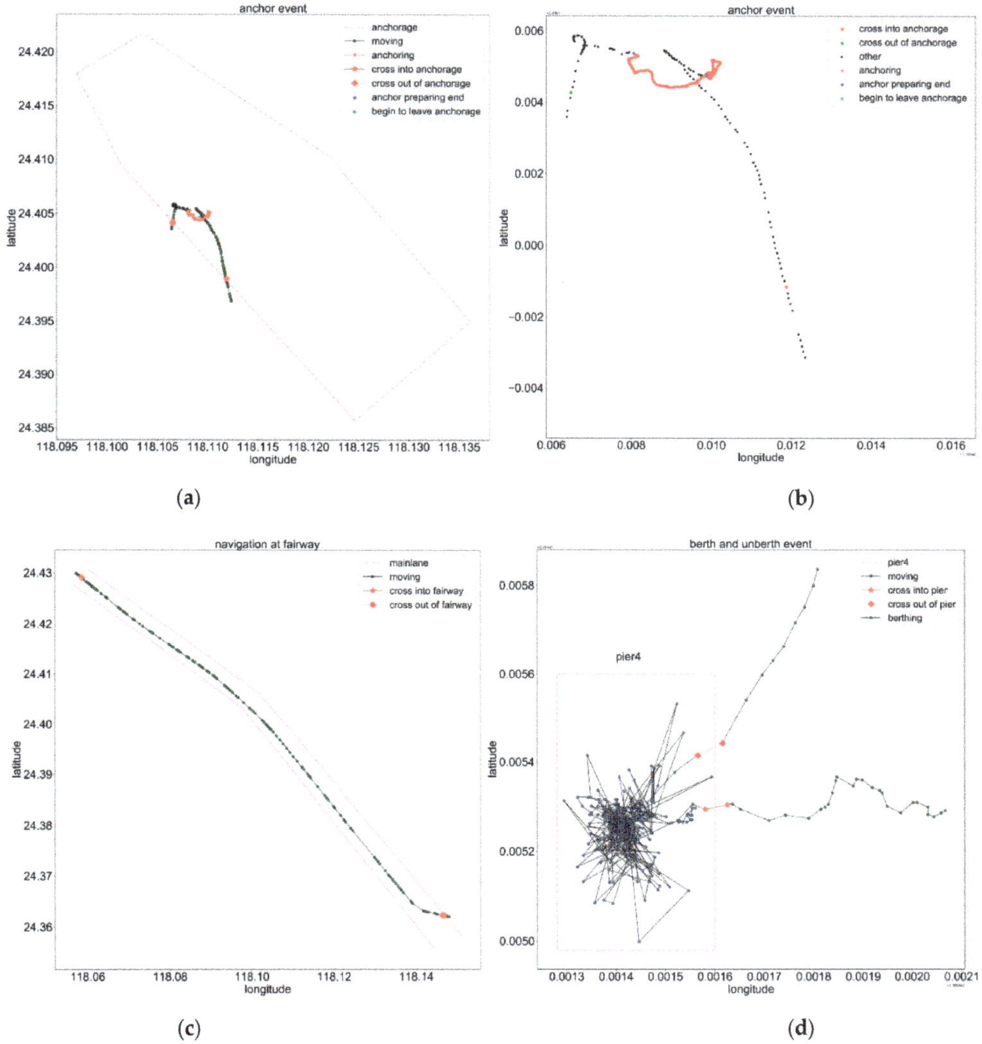

Figure 12. The visualization of the ship trajectories marked with corresponding behaviors. (Note: in order to facilitate the identification of ship behavior at the pier, a rectangular area of the pier in the port was extracted roughly from the ship *berthing* behaviors.). (**a**) Ship's semantic behavior occurring at anchorage; (**b**) Zoomed-in view ship's semantic behavior occurring at anchorage in (**a**); (**c**) Ship's semantic behavior occurring at fairway; (**d**) Ship's semantic behavior occurring at pier.

7. Results

7.1. Semantic Query

Based on this ontology model, users can execute semantic queries on behavior cognition, such as ship trajectory, behavior, occurrence time, and occurrence place. In addition, the behavior of changing speed, changing course, stopping, and so on can be obtained based on the query. The SPARQL language of the query is shown below, and the results of the query can be seen in Figure 13.

PREFIX rdf: <http://www.w3.org/1999/02/22-rdf-syntax-ns#>
PREFIX owl: <http://www.w3.org/2002/07/owl#>
PREFIX rdfs: <http://www.w3.org/2000/01/rdf-schema#>

PREFIX xsd: <http://www.w3.org/2001/XMLSchema#>
PREFIX cog: <http://www.semanticweb.org/song/cognition2#>
SELECT ?ship ?behavior ?trajectory ?metatraj WHERE { ?ship cog:hasTraj ?trajectory. optional {?trajectory cog:reflects ?behavior. ?trajectory cog:comprises ?metatraj.}}

ship	behavior	trajectory	metatraj
XINHAIXIU	stay	XINHAIXIU_49	XINHAIXIU_49_3809
XINHAIXIU	stay	XINHAIXIU_49	XINHAIXIU_49_3808
XINHAIXIU	stay	XINHAIXIU_49	XINHAIXIU_49_3699

Figure 13. The SPARQL query results of ship behavior.

7.2. Semantic Reasoning

Semantic reasoning is the process of acquiring implicit knowledge by driving the reasoning function of an ontology. Figure 14 shows an example of semantic reasoning about the behavior of a ship at anchorage. The attributes with yellow background of RENLONG are based on the inference results, including the implicit behaviors generated by RENLONG, its trajectory segments, and the place where and the time when these behaviors took place.

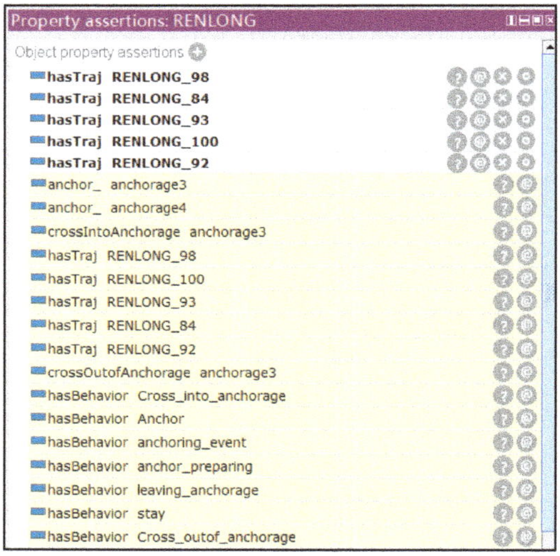

(**a**)

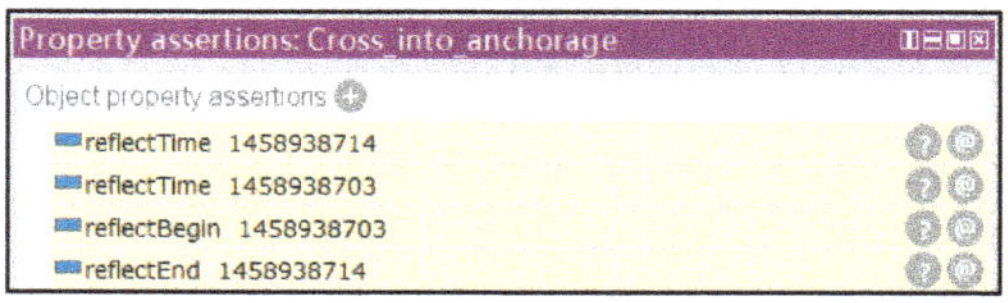

(**b**)

Figure 14. Reasoning results of behavior occurred in anchorage. (**a**) Deduced behaviors occurred by RENLONG; (**b**) The time of occurrence of the behavior Cross_into_anchorage, involving begin and end moments.

The result of the inference shows that RENLONG has the behavior of an anchor event. The start time is the beginning moment of the cross_into_anchorage behavior, and the end time is the end moment of the cross_out_of_anchorage behavior. Likewise, the behaviors occurring in the fairway and the pier of the ship can be reasoned out like the reasoning process of ship behavior in the anchorage.

As can be seen in Figure 13, the value of the object property of RENLONG *has behavior* is cross_into_anchorage, Anchor, and anchoring_event, but it cannot be deduced to the scale to which the behavior specifically belongs, such as activity and event. For this reason, the ontology sets different scales of behavior for object properties describing the scale of ship behavior, such as the properties of leave_pier and crossIntoAnchorage, which can provide a computational basis for reasoning about complicated behaviors.

7.3. Discussion

The behaviors of ships navigating around anchorages, fairways, and piers are selected for property-constrained axiom-based reasoning with SWRL for complicated behavioral reasoning, respectively. The results show that desired semantic behaviors can be recognized, leveraging the inference mechanism of behavior ontology, including from simple semantic behaviors, such as atomic behavior to large scale ship behavior, such as event in port waters. Key information of ship behavior cognition can be characterized, proving the advantage and effectiveness of the model in recognizing ship semantic behaviors, especially complicated temporal behaviors.

In addition, the behavior of a ship can be expressed by the object property 'has behavior', and the behavior with different scales can be characterized by setting the corresponding object property to achieve the multi-scale behavior of the ship. On this basis, the SWRL rule can be used to achieve the progressive reasoning of behavior between different scales, which is in line with the human habit of behavior cognition. However, such an approach is too cumbersome and all the rules need to be added manually by the people with expert knowledge, which consumes a lot of resources.

8. Conclusions

To enable autonomous objects in waterborne transport systems to have the capability of reasoning about and recognize historical complicated ship behavior semantically based on the historical AIS trajectory data, this paper proposes a framework for constructing semantic models of multi-scale ship behavior in cognitive space to achieve automatic extraction of semantic behavior of ships from the data layer to the semantic layer. On the basis of multi-scale characteristics of ship behavior reflected in ship trajectories, combined with the logical way humans perceive complicated behaviors, the cognition of ship behavior by an intelligent supervision system can be seen as an all-encompassing cognition involving the object, time, place, and behavior of the occurrence of ship behavior. Therefore, based on our previous work, this paper introduces a multi-scale behavioral semantic representation model to support the intelligent supervisory system's cognition of ship behavior in a multi-dimensional and multi-scale space. Using the logical reasoning capabilities of the ontology and the temporal ontology's modelling basis for time, ship behavior, including both simple and complex behaviors, can be accessed driven by the knowledge representation and logical reasoning capabilities of ontology. This suggests that it is feasible and reasonable to model the behavior of ships at multiple scales in a human cognitive manner.

However, there are some points that need further improvement. First, the model relies heavily on domain knowledge and needs to be constructed by domain experts, leading to inefficient application in practical scenarios. In addition, the paper does not consider the probability of ship behavior, especially in the continuous process, which limits the effectiveness of behavior implementation. In addition, there are various navigation scenarios where infrastructure exists that needs to be identified by the autonomous objects themselves or be considered as variables for human operator input for further analysis, which also needs to be addressed or clarified in the future. What needs to be done in the future is how to quickly extract and transform the textual information obtained from the website for various navigation scenarios, such as navigation modes, mooring information, etc., into knowledge that can be processed and understood by the autonomous system and expand it into a knowledge base with some scenario migration capability to make it highly reusable in different scenarios.

Future work can focus on the following points: firstly, online modelling, and identification of ship semantic behavior based on ship trajectory data; secondly, based on the semantic annotation results of historical ship trajectory data, combined with data mining algorithms, further mining of ship behavior at different semantic scales in port waters from the semantic layer to obtain implicit knowledge of high level ship behavior semantics. Finally, extending the individual semantic behavior model to interactive behaviors between two or more vessels can support the safety supervision of the waterborne transport system.

Author Contributions: R.S. and Y.W. carried out the experiment. R.S. wrote the manuscript with support from Y.W., R.S., W.T. and Q.Z. planned and carried out the experiment. E.P. helped supervise the project. R.S. and Y.W. conceived the original idea. Y.W. and P.v.G. supervised the project. All authors have read and agreed to the published version of the manuscript.

Funding: The study would express great gratitude to the support of China Scholarship Council under Grant 202106950002 and the National Natural Science Foundation of China (NSFC) through Grant No. 52072287.

Institutional Review Board Statement: Not applicable.

Informed Consent Statement: Not applicable.

Data Availability Statement: Not applicable.

Conflicts of Interest: The authors declare no conflict of interest and the funders had no role in the design of the study; in the collection, analyses, or interpretation of data; in the writing of the manuscript; or in the decision to publish the results.

References

1. Song, R.; Wen, Y.; Huang, L.; Zhang, F.; Zhou, C. Data-driven cognitive modeling and semantic reasoning of ship behavior. In *Developments in Maritime Technology and Engineering*; CRC Press: Boca Raton, FL, USA, 2021; pp. 269–276.
2. Yuanqiao, W.; Rongxin, S.; Liang, H.; Yamin, H.; Zhongyi, S.; Man, Z. Semantic modeling and expression of ship behavior. *Harbin Gongye Daxue Xuebao/J. Harbin Inst. Technol.* **2021**, *53*, 109–115.
3. Ding, Y.; Wu, R.; Zhang, X. Ontology-based knowledge representation for malware individuals and families. *Comput. Secur.* **2019**, *87*, 101574. [CrossRef]
4. Lamprecht, D.; Strohmaier, M.; Helic, D.; Nyulas, C.; Tudorache, T.; Noy, N.F.; Musen, M.A. Using ontologies to model human navigation behavior in information networks: A study based on Wikipedia. *Semant. Web* **2015**, *6*, 403–422. [CrossRef]
5. Roy, P.C.; Abidi, S.R.; Abidi, S.S.R. Possibilistic activity recognition with uncertain observations to support medication adherence in an assisted ambient living setting. *Knowl.-Based Syst.* **2017**, *133*, 156–173. [CrossRef]
6. Noor, M.H.M.; Salcic, Z.; Wang, K.K. Ontology-based sensor fusion activity recognition. *J. Ambient. Intell. Humaniz. Comput.* **2020**, *11*, 3073–3087. [CrossRef]
7. Giatrakos, N.; Alevizos, E.; Artikis, A.; Deligiannakis, A.; Garofalakis, M. Complex event recognition in the Big Data era: A survey. *VLDB J.* **2020**, *29*, 313–352. [CrossRef]
8. Rodríguez, N.D.; Cuéllar, M.P.; Lilius, J.; Calvo-Flores, M.D. A survey on ontologies for human behavior recognition. *ACM Comput. Surv.* **2014**, *46*, 43. [CrossRef]
9. Santipantakis, G.; Kotis, K.; Vouros, G. Ontology-Based Data Integration for Event Recognition in the Maritime Domain. In Proceedings of the 5th International Conference on Web Intelligence, Mining and Semantics, Larnaca Cyprus, 13–15 July 2015; pp. 1–11.
10. Santipantakis, G.M.; Vouros, G.A.; Doulkeridis, C.; Vlachou, A.; Andrienko, G.; Andrienko, N.; Fuchs, G.; Garcia, J.M.; Martinez, M.G. Specification of semantic trajectories supporting data transformations for analytics: The datAcron ontology. In Proceedings of the 13th International Conference on Semantic Systems, New York, NY, USA, 11–14 September 2017; pp. 17–24.
11. Signorelli, C.M. Can Computers Become Conscious and Overcome Humans? *Front. Robot. AI* **2018**, *5*, 121. [CrossRef]
12. Yan, Z.; Chakraborty, D.; Parent, C.; Spaccapietra, S.; Aberer, K. Semantic trajectories: Mobility data computation and annotation. *ACM Trans. Intell. Syst. Technol. (TIST)* **2013**, *4*, 1–38. [CrossRef]
13. Parent, C.; Spaccapietra, S.; Renso, C.; Andrienko, G.; Andrienko, N.; Bogorny, V.; Damiani, M.L.; Gkoulalas-Divanis, A.; Macedo, J.; Pelekis, N.; et al. Semantic trajectories modeling and analysis. *ACM Comput. Surv.* **2013**, *45*, 1–32. [CrossRef]
14. Wen, Y.; Zhang, Y.; Huang, L.; Zhou, C.; Xiao, C.; Zhang, F.; Peng, X.; Zhan, W.; Sui, Z. Semantic modelling of ship behavior in harbor based on ontology and dynamic bayesian network. *ISPRS Int. J. Geo-Inf.* **2019**, *8*, 107. [CrossRef]

15. Huang, L.; Wen, Y.; Guo, W.; Zhu, X.; Zhou, C.; Zhang, F.; Zhu, M. Mobility pattern analysis of ship trajectories based on semantic transformation and topic model. *Ocean Eng.* **2020**, *201*, 107092. [CrossRef]
16. Santipantakis, G.M.; Vouros, G.A.; Glenis, A.; Doulkeridis, C.; Vlachou, A. The datacron ontology for semantic trajectories. In *The Semantic Web: ESWC 2017 Satellite Events, Proceedings of the ESWC 2017 Satellite Events, Portorož, Slovenia, 28 May–1 June 2017*; Springer: Cham, Switzerland, 2017; pp. 26–30.
17. Vouros, G.A.; Santipantakis, G.M.; Doulkeridis, C.; Vlachou, A.; Andrienko, G.; Andrienko, N.; Fuchs, G.; Cordero Garcia, J.M.; Martinez , M.G. The datAcron Ontology for the Specification of Semantic Trajectories: Specification of Semantic Trajectories for Data Transformations Supporting Visual Analytics. *J. Data Semant.* **2019**, *8*, 235–262. [CrossRef]
18. Zhen, R.; Jin, Y.; Hu, Q.; Shao, Z.; Nikitakos, N. Maritime anomaly detection within coastal waters based on vessel trajectory clustering and Naïve Bayes Classifier. *J. Navig.* **2017**, *70*, 648. [CrossRef]
19. Rong, H.; Teixeira, A.P.; Soares, C.G. Ship trajectory uncertainty prediction based on a Gaussian Process model. *Ocean Eng.* **2019**, *182*, 499–511. [CrossRef]
20. Cao, J.; Liang, M.; Li, Y.; Chen, J.; Li, H.; Liu, R.W.; Liu, J. PCA-based hierarchical clustering of AIS trajectories with automatic extraction of clusters. In Proceedings of the 2018 IEEE 3rd International Conference on Big Data Analysis (ICBDA), Shanghai, China, 9–12 March 2018; pp. 448–452.
21. van Hage, W.; Malaisé, V.; de Vries, G.; Schreiber, G.; van Someren, M. Combining ship trajectories and semantics with the simple event model (SEM). In Proceedings of the 1st ACM International Workshop on Events in Multimedia, New York, NY, USA, 23 October 2009; pp. 73–80.
22. Ai, B.; Sun, D.; Liu, Y.; Li, C.; Yang, F.; Yin, Y.; Tian, H. Multi-Scale Representation of Ocean Flow Fields Based on Feature Analysis. *ISPRS Int. J. Geo-Inf.* **2020**, *9*, 307. [CrossRef]
23. Xiao, Y.; Ai, T.; Yang, M.; Zhang, X. A Multi-Scale Representation of Point-of-Interest (POI) Features in Indoor Map Visualization. *ISPRS Int. J. Geo-Inf.* **2020**, *9*, 239. [CrossRef]
24. Dong, Z.; Lin, B. BMF-CNN: An object detection method based on multi-scale feature fusion in VHR remote sensing images. *Remote Sens. Lett.* **2020**, *11*, 215–224.
25. Yang, F.; Li, W.; Hu, H.; Li, W.; Wang, P. Multi-Scale Feature Integrated Attention-Based Rotation Network for Object Detection in VHR Aerial Images. *Sensors* **2020**, *20*, 1686. [CrossRef]
26. Zheng, H.; Chen, J.; Chen, L.; Li, Y.; Yan, Z. Feature Enhancement for Multi-scale Object Detection. *Neural Process. Lett.* **2020**, *51*, 1907–1919. [CrossRef]
27. Yafeng, H.; Xudong, L.; Hangfeng, Z. Multi-scale Representation of Battlefield Situation. *J. Syst. Simul.* **2018**, *30*, 452.
28. Quick, M. Multiscale spatiotemporal patterns of crime: A Bayesian cross-classified multilevel modelling approach. *J. Geogr. Syst.* **2019**, *21*, 339–365. [CrossRef]
29. Yang, L.; Tao, L.; Chen, X.; Gu, X. Multi-scale semantic feature fusion and data augmentation for acoustic scene classification. *Appl. Acoust.* **2020**, *163*, 107238. [CrossRef]
30. Reis, M.S. Multiscale and Multi-Granularity Process Analytics: A Review. *Processes* **2019**, *7*, 61. [CrossRef]
31. Grüninger, M.; Li, Z. The time ontology of Allen's interval algebra. In Proceedings of the 24th International Symposium on Temporal Representation and Reasoning (TIME 2017), Schloss Dagstuhl-Leibniz-Zentrum fuer Informatik, Dagstuhl, Germany, 25 September 2017.
32. Zeng, Y.; Zhuang, J.; Su, Z. Construction of Domain Ontology for Engineering Equipment Maintenance Support. In *Knowledge Graph and Semantic Computing: Semantic, Knowledge, and Linked Big Data*; Chen, H., Ji, H., Sun, L., Wang, H., Qian, T., Ruan, T., Eds.; Springer: Singapore, 2016; pp. 33–38.

Journal of
Marine Science and Engineering

MDPI

Article

Research on Ship Trajectory Classification Based on a Deep Convolutional Neural Network

Tao Guo [1,2] and Lei Xie [2,3,*]

1 School of Transportation and Logistics Engineering, Wuhan University of Technology, Wuhan 430063, China; taoguo@whut.edu.cn
2 National Engineering Research Centre for Water Transport Safety (WTSC), Wuhan 430063, China
3 Intelligent Transportion Systems Research Centre (ITSC), Wuhan University of Technology, Wuhan 430063, China
* Correspondence: xielei@whut.edu.cn

Abstract: With the aim of solving the problems of ship trajectory classification and channel identification, a ship trajectory classification method based on deep a convolutional neural network is proposed. First, the ship trajectory data are preprocessed using the improved QuickBundle clustering algorithm. Then, data are converted into ship trajectory image data, a dataset is established, a deep convolutional neural network-based ship trajectory classification model is constructed, and the manually annotated dataset is used for training. The fully connected neural network model and SVM model with latitude and longitude data as input are selected for comparative analysis. The results show that the ship trajectory classification model based on a deep convolutional neural network can effectively distinguish ship trajectories in different waterways, and the proposed method is an effective ship trajectory classification method.

Keywords: inland waterway transportation; AIS data; trajectory classification; clustering; deep convolutional neural network

Citation: Guo, T.; Xie, L. Research on Ship Trajectory Classification Based on a Deep Convolutional Neural Network. *J. Mar. Sci. Eng.* **2022**, *10*, 568. https://doi.org/10.3390/jmse10050568

Academic Editors: Kostas Belibassakis and Fabio Bruno

Received: 28 March 2022
Accepted: 19 April 2022
Published: 22 April 2022

1. Introduction

A ship automatic identification system (AIS) is an open data transmission system widely used in the fields of ship traffic information collection and analysis, ship navigation monitoring, and water traffic planning. The ship trajectory data collected by AIS has the advantages of massiveness and large geographical scope, but the data time interval is too large, and the quality is not high, which introduces challenges to the classification of ship trajectory.

At present, the specific application scenarios of ship trajectory classification methods at home and abroad mainly include the identification of ship types and the classification of ship motion patterns. The realization process is divided into three parts: feature extraction, transformation of ship trajectory data, and modeling of classification models. Chen et al. [1] realized the classification of AIS ship trajectory based on the sparse representation classification algorithm and conducted experiments in the waters of the Yangtze River. The cubic spline method is used to approximate the trajectory of a ship, which may destroy the characteristics of the trajectory of the ship. Kraus et al. [2] used the random forest algorithm to classify ship type by extracting geographic features (navigation route, stay area, etc.) and behavior features (heading, speed, etc.) of the ship's trajectory and achieved 97.51% recognition accuracy. Based on the AIS ship trajectory, Sánchez et al. [3–7] used SVM and a decision tree to achieve binary classification of fishing boats and preprocessed the trajectory by data cleaning, data filtering, trajectory segmentation, feature extraction and other methods to improve the accuracy of classification. Liu et al. [8] used a semi-supervised deep learning model (SCEDN) for classification in the case of ship encounters, which used an encoder–decoder convolutional structure with four channels (distance, speed to approach point) for each segment time (TCPA) and distance to approach point (DCPA)). Sheng et al. [9–15] divided the ship's trajectory into three motion modes: anchoring, going straight, and turning. According to factors such as speed and heading, the behavior characteristics of the three modes were extracted, and the ship trajectory feature classification model was established by logistic regression. Cui Tong et al. [16–22] combined LSTM and CNN to establish a hybrid classification model, which is characterized by speed, acceleration, heading and curvature, with feature vectors as inputs and ship shape as output. In this method, CNN is used to extract the spatial features of the trajectory data, and LSTM is used to extract the temporal features of the trajectory data. Because ship trajectory data belong to spatial data, in this paper, we refer to some methods for trajectory image classification.

However, with respect to the relevant research results at home and abroad, the following research trends and directions are observed. Research on ship trajectory clustering is gradually developing towards efficient execution and extraction of diversified trajectory data features, and research on trajectory classification is gradually developing towards accurate feature extraction and the establishment of mathematical models based on deep learning. Combined with the main research objects of this paper, the current research has the following shortcomings:

1. Most of the current ship trajectory clustering methods are based on the density clustering algorithm of DBSCAN. Although the algorithm complexity is high, there is room for improvement in execution efficiency, and it is difficult to select the dual parameters of DBSCAN.
2. When domestic and foreign scholars use supervised algorithms for ship trajectory classification, there is still room for improvement in the use of ship trajectory spatial feature information and the process of extracting features, such as ship trajectory heading and speed.

The main work of this paper:

In this paper, we take ship trajectory data as the research object and investigate a fast, efficient and accurate ship trajectory clustering method for waters with dense and complex traffic flow that obtains the ship trajectory data of various clusters in the water area. In this paper, we use the clustered ship trajectory data as the basis to study ship trajectory anomaly detection a channel classification so as to provide decision support for intelligent risk management and control of ship traffic control departments. Specifically, the main research work of this paper is as follows:

The main task of ship trajectory preprocessing is to eliminate interference trajectories by eliminating ship trajectories that are concentrated in a small area of water with little movement or ship trajectories with a sampling interval that is too long to characterize continuous motion characteristics, eliminating the interference of ship anchor points in trajectory analysis of moving ships, and reducing the complexity of ship trajectories. Under the premise of ship trajectory preprocessing, in this paper, we use the QuickBundles algorithm as a basic method to carry out ship trajectory clustering research. First, we analyze the performance of three trajectory similarity measurement methods, MDF [23], DTW [24], and Hausdorff [25]. Then, aiming at the problem of insufficient sampling of local features of ship trajectory by the QuickBundles algorithm, a sampling method based on heading is used to improve it, and an improved QuickBundles ship trajectory clustering algorithm is proposed. We use the improved QuickBundles algorithm [26] to establish a clustering model of ship trajectories, determine appropriate thresholds according to a variety of evaluation indicators, complete the task of ship trajectory clustering, and conduct comparative experiments with the improved QuickBundles algorithm and the traditional DBSCAN [27] algorithm.

In view of the problem of ship trajectory classification based on latitude and longitude data, the spatial characteristics of the data are not obvious, and the classification effect is not ideal. In this paper, we propose a ship trajectory classification method based on a deep convolutional neural network to classify the channel to which a ship trajectory belongs, achieving the recognition of ship trajectories and waterways. Based on the clustering results, the latitude and longitude coordinates are mapped to the image pixel coordinates according to the scale, the spatial characteristics of the ship trajectory data are extracted, and the ship trajectory image dataset is established. The ship trajectory classification model based on a deep convolutional neural network is established according to on the ResNet50 [28] model, using the training set to train the model. On the test set, the fully connected neural network and multi-class SVM classifier [29] with latitude and longitude data as input are used for comparison with the deep convolution model with trajectory image data as input.

The main contributions include:

- An improved QuickBundles ship trajectory clustering algorithm is proposed.
- A method of ship trajectory classification based on a deep convolutional neural network is proposed that realizes the classification and identification of the waterway to which a ship trajectory belongs.

The contents of this paper are organized as follows: Section 2 provides details of the proposed scheme, the result analysis is shown in Section 3, and conclusions are presented in Section 4.

2. Methods

The working process of the proposed methodology is shown in Figure 1. This method takes specific ship trajectory AIS data as the research object and focuses on ship trajectory clustering, ship trajectory anomaly detection, and channel identification of ship trajectories in dense-traffic waters. Through the identification of abnormal trajectories and the classification of the channel to which a trajectory belongs, the ship supervision department provides technical support for targeted ship trajectory data analysis. Ship trajectory clustering research is carried out based on the QuickBundles clustering algorithm. The sampling method of QuickBundles is improved according to the local heading changes of the ship trajectory, and a fast, accurate, and efficient ship trajectory clustering method is proposed. Ship trajectory clustering research also provides cluster quantity parameters for anomaly detection models and data support for ship trajectory classification.

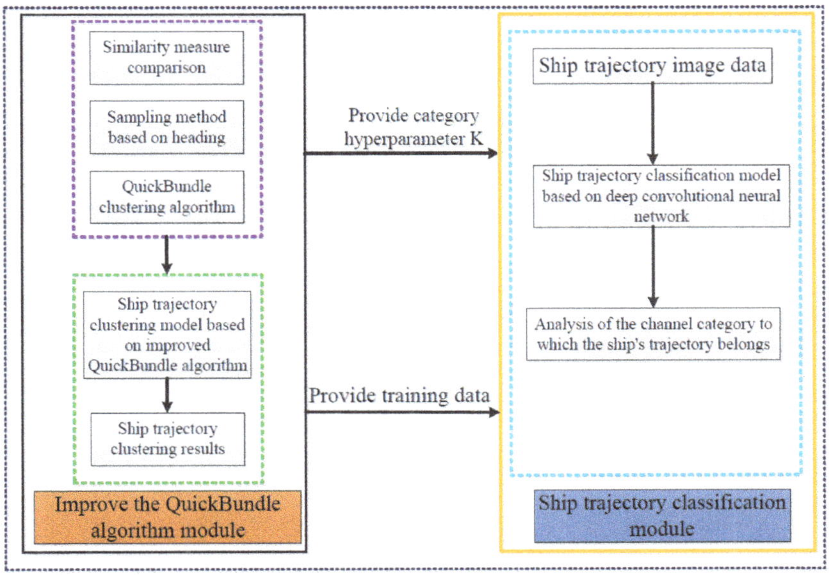

Figure 1. The working process of the proposed methodology.

2.1. The Improved QuickBundles Algorithm Module

The trajectory of a ship can be of any length. Before the task of clustering the trajectory of the ship, data need to be divided and filtered so that the subtrajectory segments with similar motion characteristics can be retained and some important information can be obtained; therefore, it is very important to properly divide the original trajectory. Commonly used methods of ship trajectory division are based on time interval and speed changes.

The data used in this paper come from the US Coastal AIS Vessel Traffic Data (https://marinecadastre.gov/ais/, accessed on 1 March 2022), which are collected by the US Coast Guard through on-board navigation and positioning equipment to monitor the location of large ships in the United States, as well as characteristics of coastal waters. In this section, we take the AIS dataset from January to March 2019 as the experimental object and use two methods to process ship trajectory data. The specific parameter settings are shown in Table 1, and the processing results are shown in Table 2.

Table 1. Division and filtering threshold settings.

Threshold Type	Default Value
Time threshold	600
Speed threshold	1
Track-point capacity threshold	20

Table 2. Division and filtering results.

	Ship MMSI Number	Total Tracks	Total Track Points
Before dividing and screening	871	/	263,149
After dividing and filtering	868	3120	144,438

The QuickBundles algorithm was originally designed for use with nerve bundles in the medical field. The local changes of nerve bundles are not complicated. Therefore, the QuickBundles algorithm uses only simple linear interpolation as the sampling method. However, if the clustering object is a ship trajectory with moving characteristics and the local heading changes are more complicated, then the characteristic changes of these local headings cannot be ignored, e.g., the 20 ship trajectory points shown in Figure 2a,b. In the original trajectory, the ship's course changes considerably due to reasons such as avoidance, and the changed trajectory is curved and smooth. After sampling by the QuickBundles algorithm, the local features of this heading change are replaced by simple polylines; the ship in the original trajectory in Figure 2c has a short, straight line at the turn. After being sampled by the QuickBundles algorithm, this short straight line is ignored.

In order to overcome the above shortcomings, we improve the sampling method of the QuickBundles algorithm. First, the ship's trajectory is compressed, with the heading as a factor, and the key position points of the ship's trajectory are extracted. Then, the ship trajectory is interpolated based on the distance between the trajectory points.

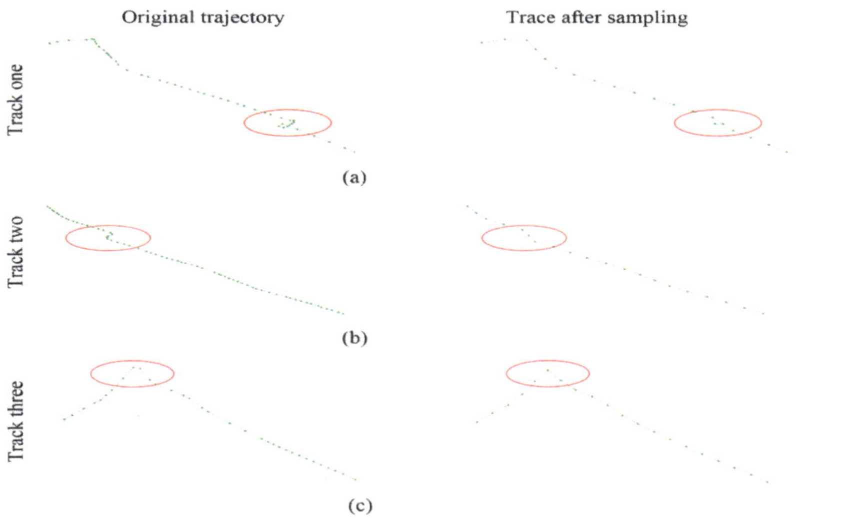

Figure 2. Three different types of original and sampled trajectories. (**a**,**b**) the 20 ship trajectory points; (**c**) the ship in the original trajectory.

- Ship trajectory compression considering heading

There are two purposes for ship trajectory compression in this paper: one is to reduce the number of trajectory points of all ship trajectories so as to more conveniently achieve the unification of the number of trajectory points in the future; the other is to reduce the number of trajectory points to improve the similarity between trajectories and calculate efficiency.

The course can indicate the direction of a ship's trajectory and the trend of a ship's movement. Figure 3 shows the difference in heading angle. The heading angle difference (AD) represents the difference in the direction angle of the adjacent ship trajectory segment, which can more clearly illustrate the change in the current trajectory segment compared to the previous trajectory segment. Through the calculation of the heading angle difference, the key position points in the trajectory of a ship can be accurately obtained, and the compression of the trajectory of the ship can be determined. The detailed calculation process is shown in Figure 4. The input is the angle threshold and the ship trajectory. The heading angle difference between the current trajectory point and the previous trajectory point is calculated. If the heading angle difference is greater than the threshold, the current trajectory point is retained; otherwise, the current trajectory point is deleted.

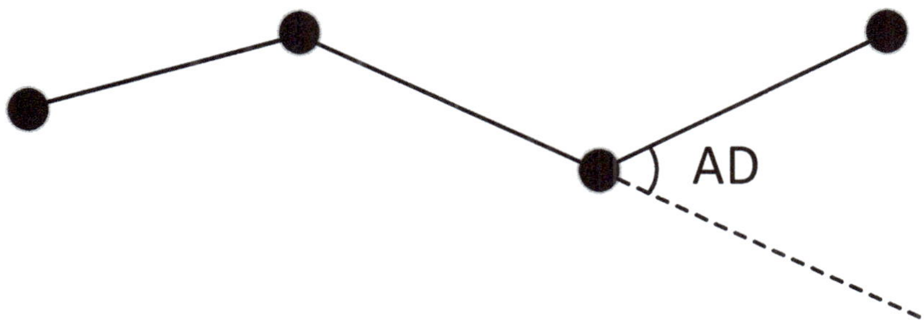

Figure 3. Heading angle difference.

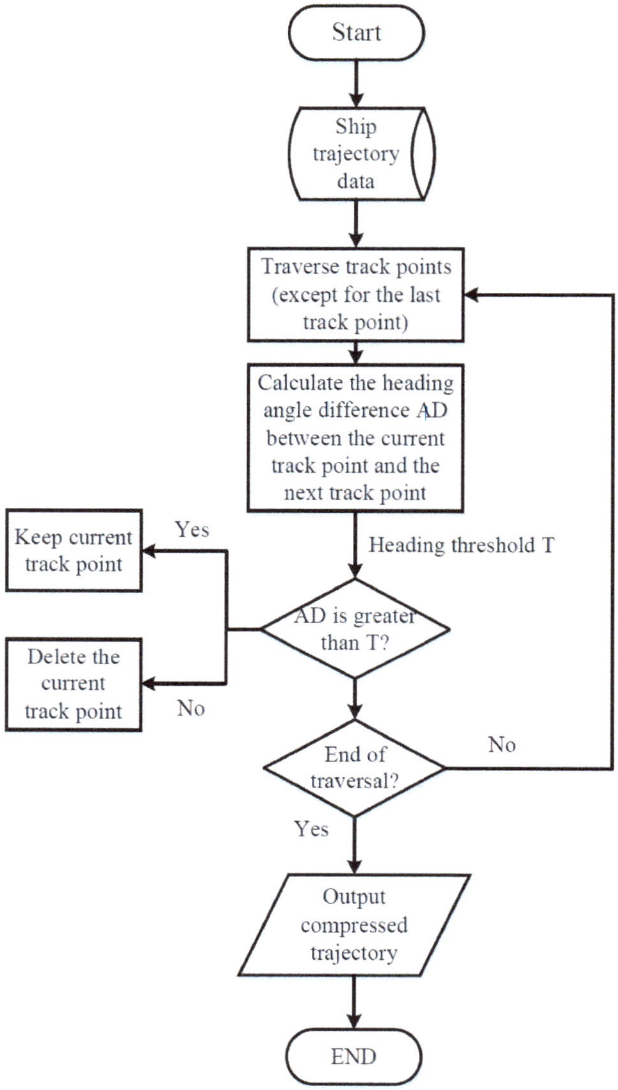

Figure 4. Ship trajectory compression.

- Subsection interpolation based on the distance between track points

The QuickBundles clustering algorithm requires that the trajectories to be clustered have the same number of trajectory points. After compressing the ship's trajectory, in order to meet this requirement, in this section, we adopt the segmented interpolation method based on the distance between the trajectory points to unify the number of ship trajectory points. The specific process is shown in Figure 5. First, the number of track points to be inserted is obtained, and then the distance between each adjacent track point is calculated. According to the ratio between the distances, the number of inserted track points to the track to be inserted in each segment is allocated a corresponding number of points.

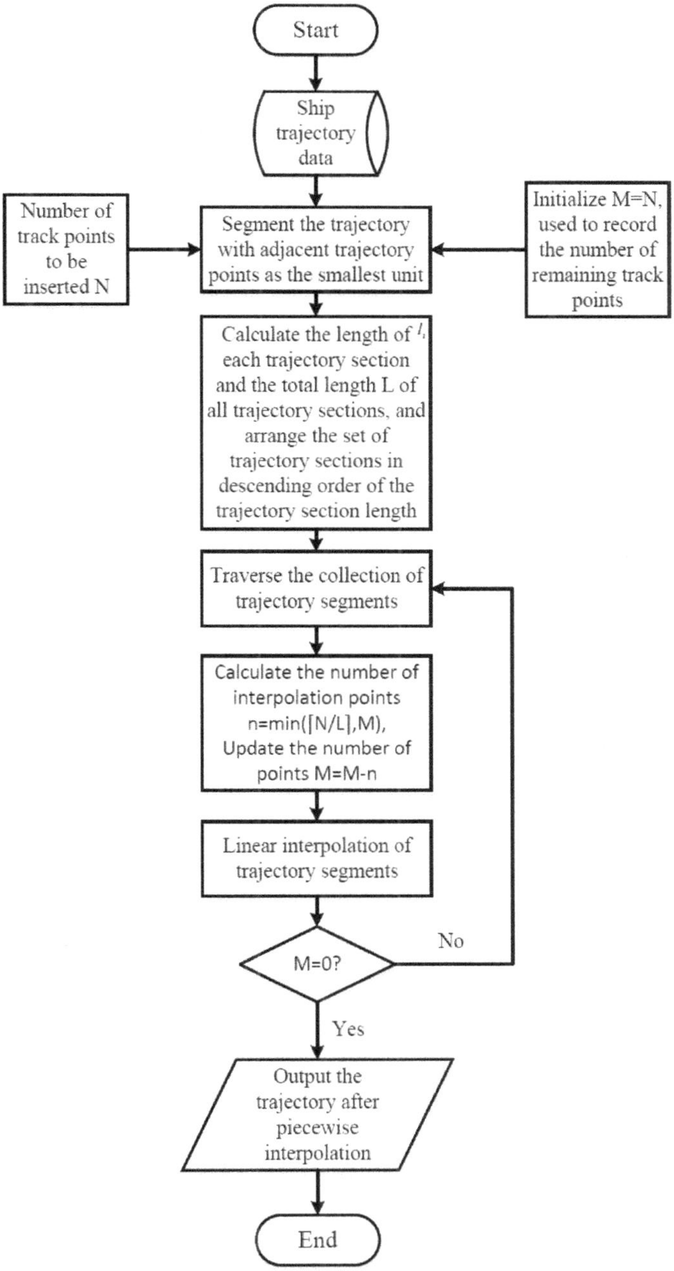

Figure 5. Low chart of segmented interpolation.

2.2. Ship Trajectory Classification Module

In the key monitoring areas of ports, seaports, and other regulatory agencies, as the flow of ships increases, an efficient ship trajectory classification algorithm is needed to classify ships in the jurisdiction, improve the level of intelligent management and supervision efficiency, and reduce busy waters. There is a risk of major and catastrophic traffic accidents. In this section, we use the trajectory clustering results as the training dataset to investigate the classification of ship trajectories and propose a ship trajectory classification method based on deep convolutional neural networks.

2.2.1. Longitude and Latitude Mapping and Coordinate Conversion

The latitude range of the water area where the experimental data in this article are located is 48 degrees 9 min 7.28 s north latitude to 49 degrees 6 min 44.28 s north latitude, and the longitude range is 123 degrees 3 min 43.33 s west longitude to 123 degrees 42 min 2.71 s west longitude, as shown in Figure 6. In this section, we assume that the area is the key monitoring area of the ship supervision department, model the area and convert the latitude and longitude data into image data according to the length and width ratio of the water area where the experimental data are located.

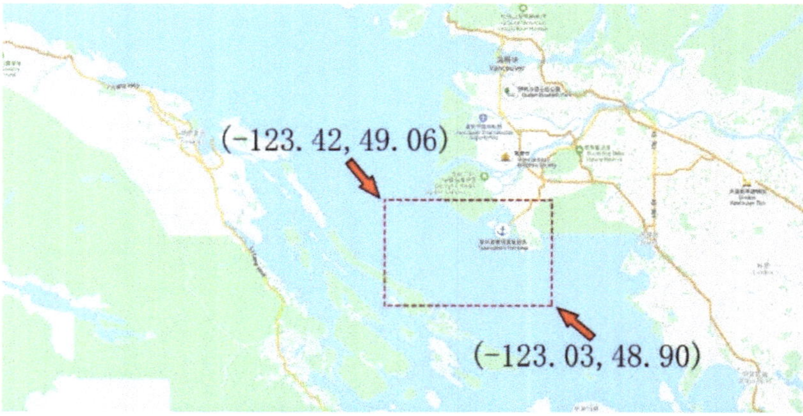

Figure 6. Longitude and latitude information of the water area where the experimental data are located.

2.2.2. Calculation of the Aspect Ratio of the Water Area

The water area where the experimental data are located is a rectangular area, and the aspect ratio is obtained by calculating the distance between the two sides of the rectangular area to determine the image resolution using the Haversine formula [30] to calculate the distance between two longitude and latitude coordinate points. Formula (1) introduces the method for calculating the distance between two longitude and latitude coordinate points when two longitude and latitude coordinate points are known. R is the radius of the earth, and the average value is 6371 km. φ_1 and φ_2 represent the latitude of the two points, and $\Delta\lambda$ represents the difference between the longitudes of the two points. According to this calculation, the length of the experimental area is 28.41 km, the width is 17.82 km, and the approximate ratio is 14:9.

$$haver\sin\left(\frac{d}{R}\right) = haver\sin(\varphi_1 - \varphi_2) + \cos\varphi_1\cos\varphi_2 haver\sin(\Delta\lambda) \tag{1}$$

$$haver\sin(\theta) = \sin^2\left(\frac{\theta}{2}\right) = \frac{(1-\cos(\theta))}{2}$$

The higher the image resolution, the higher the computational cost and the lower the computational efficiency of the deep convolutional neural network. Considering the above problems, in this paper, we set the resolution to 112 ∗ 72, keeping the ratio of the image unchanged at 14:9 so that the latitude value of (49.06, 48.90) is mapped to the range of the pixel point (0, 71) inside, the longitude value of (−123.42, −123.03) is mapped to the pixel point (0, 111) range, as shown in Figure 7.

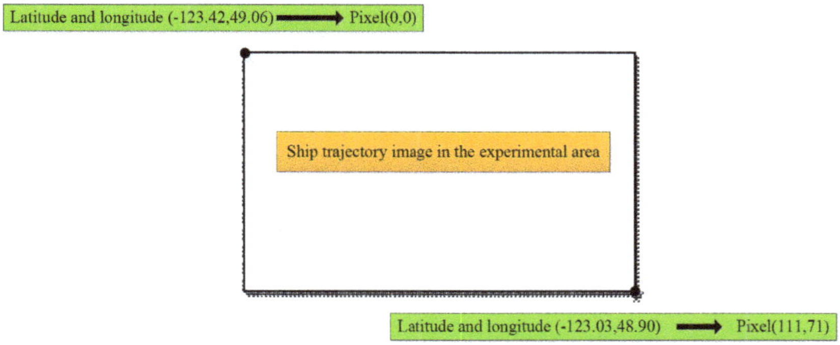

Figure 7. Schematic diagram of image pixels of latitude and longitude mapping.

Figure 8 shows the ship trajectory image data after the latitude and longitude data of the ship trajectory are converted. According to the clustering results in Section 3, there are five types of ship trajectories in the waters where the experimental data are located based on the channel category division, so the label of the dataset is set to 0, 1, 2, 3, 4. The resolution of each ship trajectory image is 112 ⁎ 72, which corresponds to the latitude and longitude range of the water area. The specific dataset details, as well as the division of training set and test set are shown in Table 3.

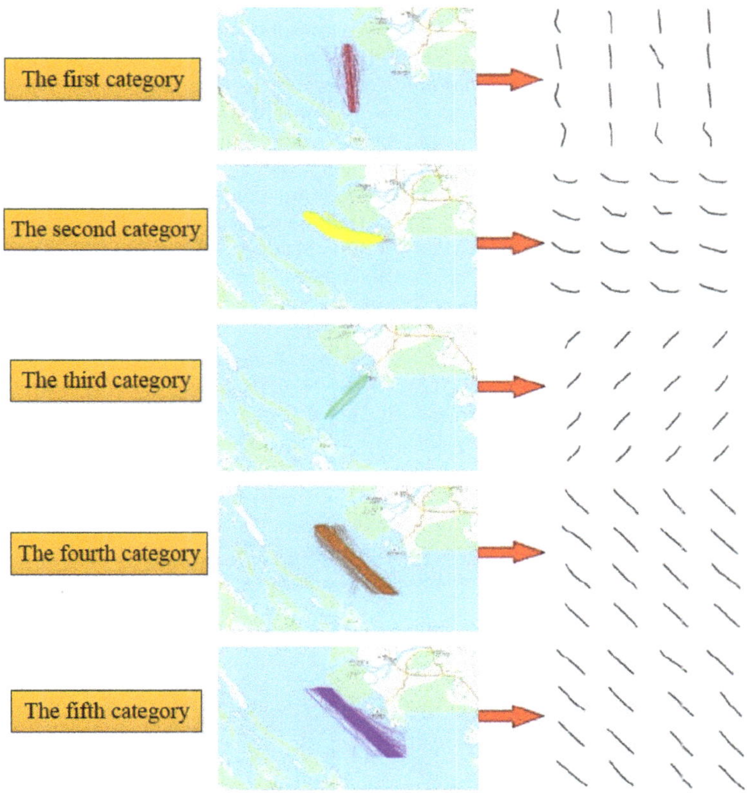

Figure 8. Trajectory image data display diagram.

Table 3. Dataset details.

Type	Quantity	Number of Various Types of Trajectories
Total dataset	2624	The first category: 321 The second category: 380 The third category: 248 The fourth category: 747 The fifth category: 928
Train set	2099	The first category: 256 The second category: 304 The third category: 198 The fourth category: 608 The fifth category: 733
Test set	525	The first category: 65 The second category: 76 The third category: 50 The fourth category: 139 The fifth category: 195

2.2.3. Deep Convolutional Network Model Construction

ResNet (residual network) residual network [31] is widely used in target classification and other fields. It is a part of the classic backbone neural network for computer vision tasks. Typical networks include ResNet50, ResNet101, etc. The ResNet network proves that convolutional neural networks can develop more deeply (including more hidden layers) and verifies that deep convolutional neural networks have better performance.

ResNet50 has a unique residual structure, as shown in Figure 9. One of the core technologies of the residual structure is the use of a shortcut connection. There are two main reasons for the disappearance of the gradient. When the number of network layers is very deep and the layer where the current parameter is located is close to the input of the network, the derivation chain is very long; if some of the intermediate results have a low value, after chain accumulation, the final gradient value will be close to zero, resulting in the parameters not being updated. The input is directly added to the output obtained through the convolution operation, which can avoid the problem of the disappearance of the gradient and can capture small perturbations. In addition, the first and last ends of the residual structure use convolution to reduce and restore data dimensions. The time complexity of the two structures is similar, but it deepens the number of network layers and structures and resolves network degradation and training process performance. As shown in Figures 10 and 11, in the actual processing step, jump connections are divided into two types according to the size of the input and output of the residual block. One is the identity block (ID BLOCK) when the input and output are consistent, and the other is the convolutional block (CONV BLOCK) when the input and output are inconsistent. The jump connection is processed by convolution calculation to achieve unity of input and output dimensions. ResNet50 adopts small-size convolution kernels and uses batch normalization [32] technology. In this paper, we build a ship trajectory and channel classification model based on ResNet50 as a deep convolutional neural network framework.

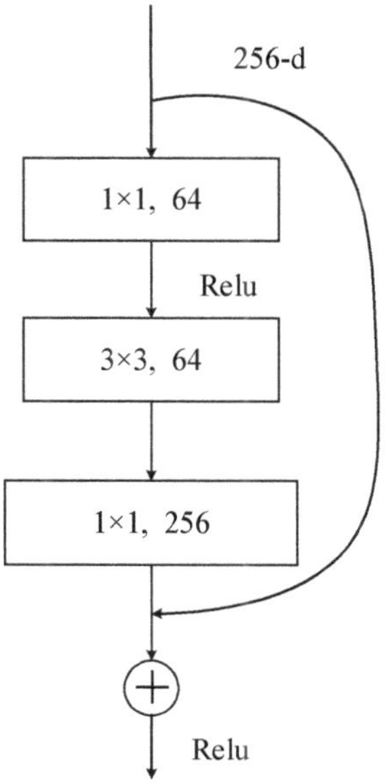

Figure 9. Residual network structure.

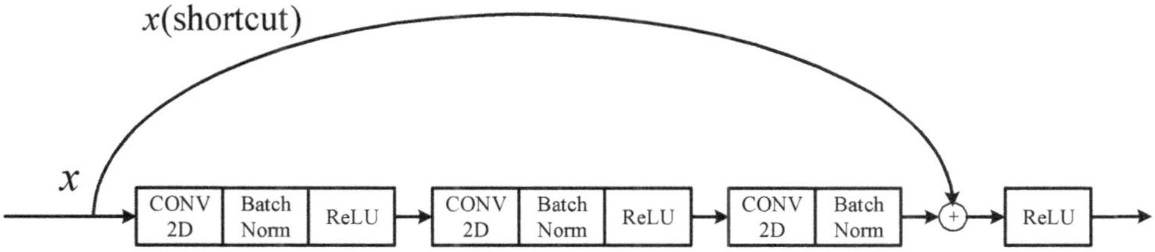

Figure 10. ID BLOCK structure.

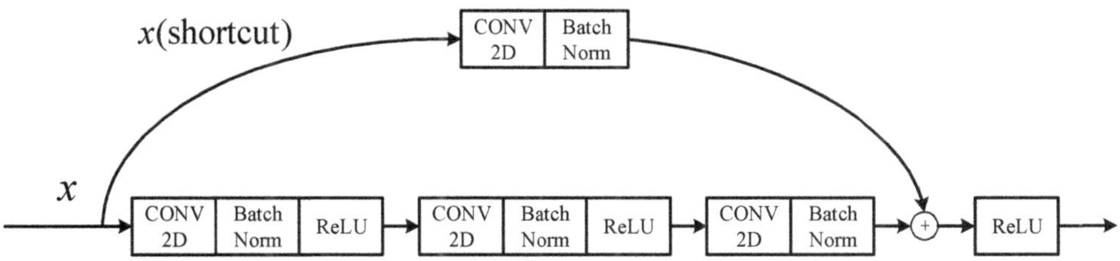

Figure 11. CONV BLOCK structure.

2.2.4. Model Building

The ship trajectory classification network structure proposed in this paper is shown in Figure 12 and Table 4. The structure is composed of five convolution blocks stacked in sequence. Each convolution block contains the residual network substructure shown in Figure 11. The residual network substructure in different convolution blocks has different numbers of convolution kernels. The input layer dimension parameter of the network model is set to $112 * 72 * 3$, the mini batch size is set to 64, and the output layer category is set to 5.

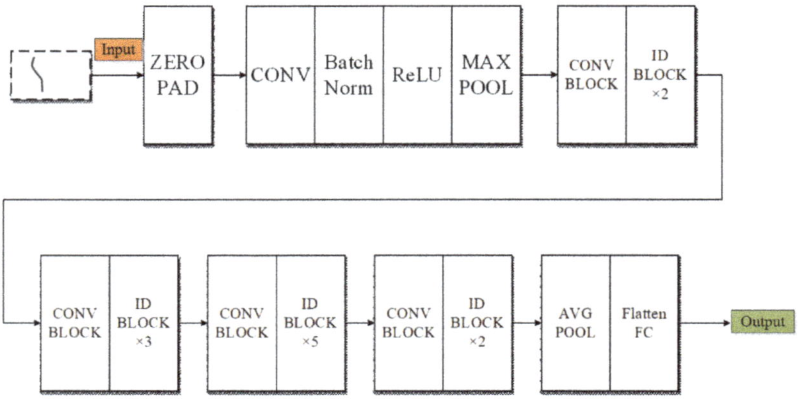

Figure 12. Ship trajectory classification network architecture.

Table 4. Ship trajectory classification network structure.

Layer Name	Output Size	Layer Structure
Conv1	112 × 112	7 × 7, 64, stride = 2 3 × 3 max pool, stride = 2
Conv2_x	56 × 56	$\begin{bmatrix} 1\times1, & 64 \\ 3\times3, & 64 \\ 1\times1, & 256 \end{bmatrix} \times 3$
Conv3_x	28 × 28	$\begin{bmatrix} 1\times1, & 128 \\ 3\times3, & 128 \\ 1\times1, & 512 \end{bmatrix} \times 4$
Conv4_x	14 × 14	$\begin{bmatrix} 1\times1, & 256 \\ 3\times3, & 256 \\ 1\times1, & 1024 \end{bmatrix} \times 6$
Conv5_x	7 × 7	$\begin{bmatrix} 1\times1, & 512 \\ 3\times3, & 512 \\ 1\times1, & 2048 \end{bmatrix} \times 3$
Output layer	1 × 1	Average pool, 1000-d fc, softmax

3. Results

3.1. Model Training

Assuming that the ship trajectory image data point is X, because the pixel values of the image data are in the range of (0, 255), each data point is normalized before model training, as shown in formula (2), and the data format is converted to the float32 data type in the tensorflow framework.

$$X = float\left(\frac{X}{255}\right) \tag{2}$$

3.1.1. Experimental Environment and Hyperparameters

This experiment runs on the tensorflow deep learning framework and uses a GTX1060 graphics card for training. The hyperparameter settings are shown in Table 5.

Table 5. Hyperparameters.

Type	Parameter
Mini batch size	64
Learning rate	0.001
Number of iterations	20
Number of training sets	2099
Number of test sets	520
Label format	One-hot
Shuffle	True

3.1.2. Optimizer and Loss Function

In this paper, we use Radam as the optimizer. Radam is a deep learning optimizer proposed by Chinese doctoral student Liu Yiyuan in 2019. It is designed to solve the problem that SGD has good convergence effect but slow speed, whereas Adam converges quickly, it is not easy to converge to the local optimal solution. Based on variance dispersion, Radam dynamically turns on and off the adaptive learning rate and realizes a method that does not need to warm up the learning rate in the adjustable parameters.

It has the advantages of both Adam and SGD, which can ensure fast convergence speed and does easily fall into the local optimal solution. In the case of a high learning rate, the accuracy of Radam is better than that of SGD. In addition, for the multi-classification problem, the cross-entropy loss function and the Softmax activation function are used.

3.1.3. Training Effect Analysis

In this paper, we use the Tensorboard data analysis visualization tool in Tensorflow to analyze the model training effect. Figure 13 shows the change in loss value during the training process, with the number of iterations set to 20. With the increase in the number of iterations, the loss curve of the ship trajectory classification model shows a convergence trend as a whole, with a small fluctuation between the fourth and eighth iterations but finally converging around 0.04. The training results show that the model has learned the data features of the ship trajectory images and the model training has achieved the expected effect. Figure 14 is a graph of the accuracy rate of the validation set during the model training process. With the increase in training time, the accuracy rate of the model on the validation set classification increases rapidly. After the number of iterations reaches 10, the accuracy rate remains around 98.90%. This also shows that the model already has excellent classification ability for the training set.

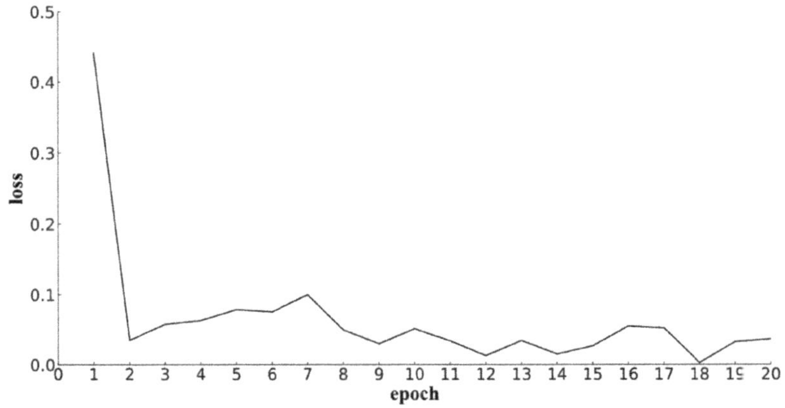

Figure 13. Loss curve.

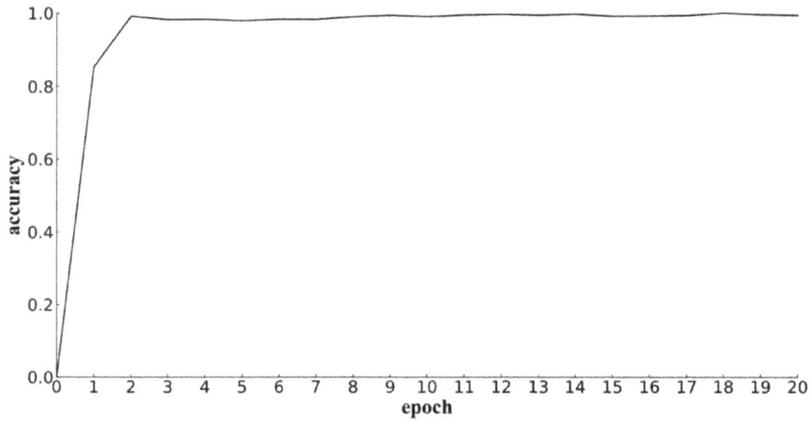

Figure 14. Accuracy graph.

3.2. Results of the Improved QuickBundles Clustering Algorithm

In order to further verify the clustering ability of the improved QuickBundles clustering algorithm for ship trajectory, in this paper, we compare and analyze the clustering effect and execution efficiency of the improved QuickBundles clustering algorithm, the QuickBundles clustering algorithm, and the DBSCAN algorithm. In the experiment, the thresholds of the three clustering algorithms are the best clustering thresholds obtained in Section 3.1.

The comparison results of the contour coefficients of the three clustering algorithms are shown in Figure 15a. The contour coefficient of DBSCAN is 0.5568, the contour coefficient of the QuickBundles clustering algorithm is 0.6173, and the contour coefficient of the improved QuickBundles clustering algorithm is 0.6380. The QuickBundle clustering algorithm surpasses the other two algorithms according to various metrics. Figure 16 shows the statistics of the three clustering algorithms. The distribution of contour coefficients of all trajectory data, the mean, upper quartile, median, and lower quartile of the improved QuickBundles clustering algorithm in the figure are higher than those of the QuickBundles clustering algorithm and the DBSCAN algorithm. The CHI comparison results of the three algorithms are shown in Figure 15b. The CHI of the improved QuickBundles clustering algorithm is 3769.2168, which is significantly higher than the other two comparison algorithms. Figure 15c shows the DBI comparison results of the three algorithms. For the ship trajectory data, the improved QuickBundles clustering algorithm is better than the other two algorithms. In terms of algorithm execution efficiency, as shown in Figure 15d, the improved QuickBundles clustering algorithm is only 305 milliseconds slower than the QuickBundles clustering algorithm, and the execution time of DBSCAN is less than 20.9 s.

The comparison results show that the QuickBundles clustering algorithm has improved the clustering effect of the DBSCAN ship trajectory clustering algorithm on the ship trajectory dataset. In terms of various indicators, the improved QuickBundles clustering algorithm has improved performance compared with the pre-improved QuickBundles algorithm and is more suitable for the clustering of ship trajectory data. In terms of algorithm execution efficiency, the improved sampling method does not significantly affect the algorithm. The improved QuickBundles algorithm is only 305 milliseconds slower than the QuickBundles algorithm, which is slower than DBSCAN's 20.9 s.

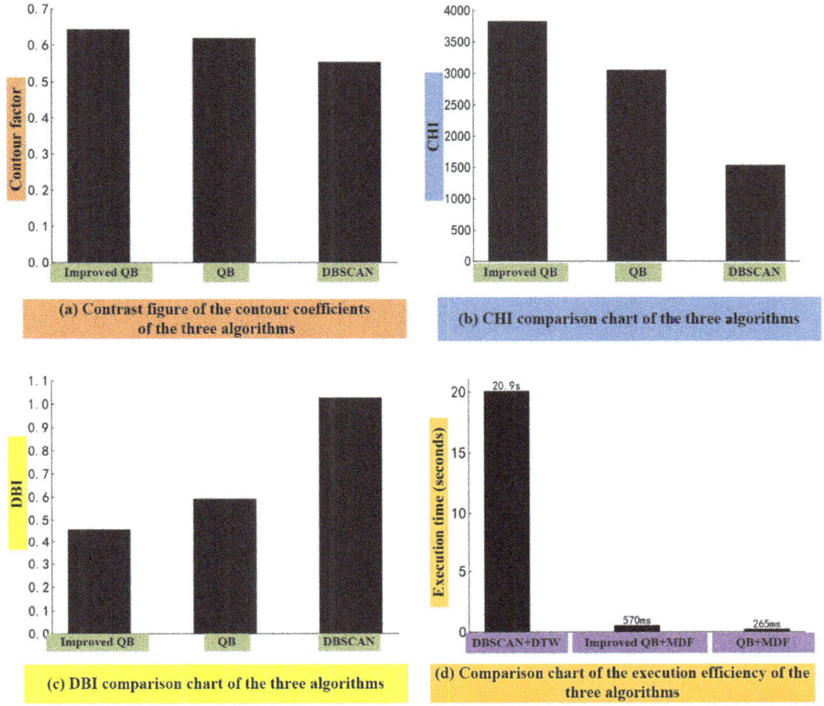

Figure 15. Comparison of the clustering evaluation indicators of the three algorithms.

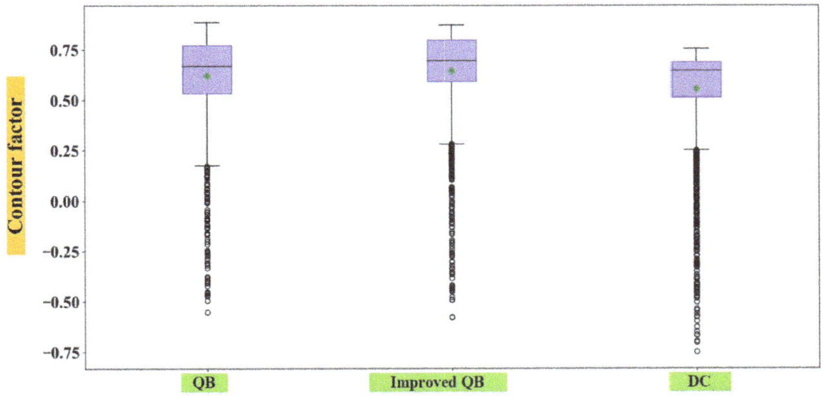

Figure 16. Box diagram of ship trajectory profile coefficients in the three algorithms.

In this section, we take the complete ship trajectories in complex waters as the research object, focusing on the characteristics of ship trajectories and headings in order to achieve fast and accurate clustering based on MDF distance and the QuickBundles clustering algorithm. A ship trajectory clustering method based on the improved QuickBundles algorithm is presented. In this section, first we compare the MDF distance with two classical trajectory metrics and analyze their advantages and disadvantages. We propose a sampling method based on the difference of the heading angle of the ship trajectory, improving the sampling method of the QuickBundles algorithm. Finally, a ship trajectory clustering experiment based on the improved QuickBundles algorithm is carried out, and the rationality and reliability of the research method presented in this section are analyzed by comparison with the DBSCAN algorithm on the public American coastal AIS ship trajectory dataset.

The experimental results show that the method proposed in this section can reasonably cluster the ship trajectories of various waterways and that the improved QuickBundles clustering algorithm has better clustering performance. In addition, using the dataset and without considering the trajectory noise, the improved QuickBundles clustering algorithm is slightly better than the traditional DBSCAN trajectory clustering algorithm in terms of accuracy and algorithm execution efficiency. The ship trajectory clustering results in this section will be used as the basis for subsequent ship trajectory anomaly detection research and ship trajectory classification research.

3.3. Ship Trajectory Classification Network Model Test Set Analysis

After the training of the ship trajectory classification network model, we use the test set to verify the model's channel trajectory classification effect. Table 6 is the confusion matrix of the classification results of the test set. Among them, the classification accuracy of the first, second, and third categories is 100%. Among the 334 samples of the fourth and fifth categories, there are six misidentified samples. Figure 17 is a comparison chart of the trajectories of the fourth and fifth types of ships in the water area where the experimental data are located. It can be seen that in the fourth and fifth types of ship trajectories, a very small part of the ships did not travel in the corresponding channel but sailed in the separation zone between the two channels, which violated the "General Provisions for Ship Routing System" and also led to misclassification of ship trajectories. In spite of this, the ship trajectory classification model presented in this paper still achieves an accuracy of 98.85% on the test set, achieving accurate classification of ship trajectory categories in the analyzed waters.

Table 6. Confusion matrix of ship trajectory image test set.

Actual Value/Predictive Value	First Category	Second Category	Third Category	Fourth Category	Fifth Category
First category	65	0	0	0	0
Second category	0	76	0	0	0
Third category	0	0	50	0	0
Fourth category	0	0	0	136	3
Fifth category	0	0	0	3	192

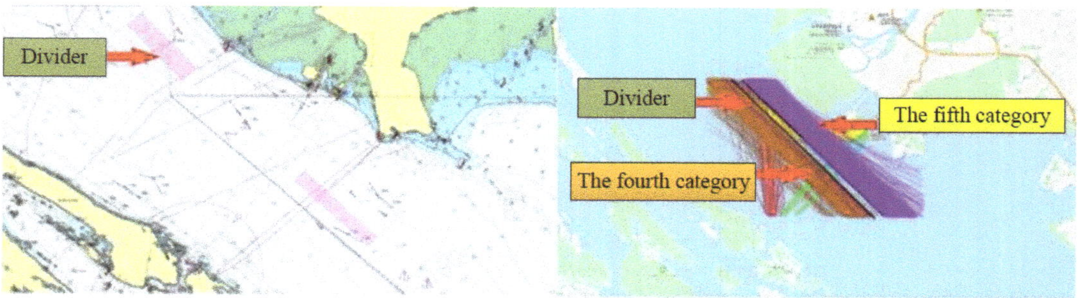

Figure 17. The fourth- and fifth-type channel information map of the water area where the experimental data are located.

3.4. Cross-Validation Comparative Analysis

In order to further verify the classification ability of the ship trajectory classification network model, we select the fully connected neural network model and the multi-category SVM model [33] with the longitude and latitude data format as input as a comparison to carry out comparative analysis. The parameter information of the fully connected neural network comparison model selected in this paper is shown in Table 7. Among them, the fully connected neural network requires a unified data input format, so in this paper, we draw on the method of Chen [34] and others; using cubic spline interpolation, each ship trajectory data are sampled as 50 points.

Table 7. Fully connected neural network model parameters.

Layer Name	Layer Structure	Activation Function
Input layer	$50 * 2$	none
Fully connected 1	64	relu
Fully connected 2	128	relu
Fully connected 2	256	relu
Fully connected 2	512	relu
Output layer	5	softmax

The input data format of the fully connected neural network participating in the comparison is an array of $50 * 2$, where 2 is the longitude and latitude of the ship's trajectory data, and 50 is the trajectory data length after sampling for each trajectory. The input data format of the SVM model participating in the comparison is an array of $100 * 1$; 100 is the result of flattening the above $50 * 2$ data, and the kernel function is RBF. All models have undergone 10-fold cross-validation. Figure 18 shows the results of the ten-fold cross-validation of the three models, and Table 8 shows the average accuracy of the three models' 10-fold cross-validation. It can be seen from the experimental comparison results that for the dataset used in this paper, the proposed ship trajectory classification network model achieves 98.72% accuracy, which is higher than the 93.53% accuracy of the fully connected neural network model and the 91.73% accuracy of the SVM model. The superiority of the ship trajectory and waterway classification model in the performance of ship trajectory classification is verified.

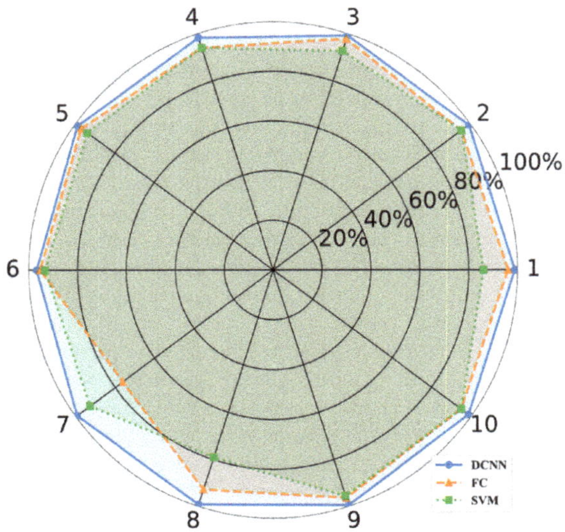

Figure 18. Ten-fold cross-validation results of the three models.

Table 8. The age accuracy of the three models' ten-fold cross-validation.

Model	Average Accuracy
Deep convolutional neural network (DCNN)	98.72%
Fully connected neural network (FC)	93.53%
Support vector machines (SVM)	91.73%

In this section, the ship trajectory classification research fails to use the ship trajectory spatial feature information, and the process of extracting features such as ship trajectory heading and speed is too cumbersome. We propose a ship trajectory classification method based on a deep convolutional neural network. First, based on the clustering results in Section 3, the trajectory longitude and latitude data are converted into high-dimensional trajectory image data, and the spatial characteristics of the ship trajectory are extracted. Then, the ship trajectory classification model is constructed based on the principle of a residual network [35], and ship trajectory classification is realized. Finally, using the latitude and longitude data of the ship trajectory as input, a fully connected neural network model and SVM model are constructed as comparison models, and a comparison experiment with the model proposed in this section is carried out. The experimental results show that for the dataset used in this article, the accuracy of this model on the test set is 98.85%, and the accuracy of the ten-fold cross-validation is 98.72%, which is higher than the 93.53% of the fully connected neural network and the 91.73% of the SVM.

According to the above comparison results, the ship trajectory is converted into image data while retaining its ability to express spatial features, simplifying the data feature extraction work and improving the efficiency of data processing. The deep convolutional neural network used to complete the trajectory classification task can make full use of the spatial characteristics of the trajectory data. Compared with the fully connected neural network model and the SVM model, the method proposed in this section has higher classification accuracy of the ship trajectory and provides support for the identification of the channel category. The research results of this section can provide auxiliary support for the intelligent decision making of the ship supervision department. The results of this study can provide support for the intelligent decision making for the ship navigation safety monitoring [36,37] department.

4. Conclusions

The main research work of this paper can be summarized by the following two points:

- Aiming at the problem of invalid trajectories in the original ship trajectory data, a set of ship trajectory preprocessing methods based on time interval and speed changes are summarized. In order to improve the accuracy and execution efficiency of ship trajectory clustering, research on ship trajectory clustering based on MDF distance and QuickBundles clustering algorithm is carried out. Aiming at the problem that the sampling method of the QuickBundles algorithm does not consider the of local characteristics of ship trajectory, a new method is proposed. Considering the sampling method of heading, the QuickBundles clustering algorithm is improved, and the purpose is to analyze the characteristics of water traffic and ship motion in the water area.
- In order to extract the spatial characteristics of the ship trajectory, complete the research of ship trajectory classification, convert the ship trajectory longitude and latitude data into ship trajectory image data, build a deep convolutional neural network ship trajectory classification model, and propose a deep convolution based on ship trajectory classification model, the neural network-based ship trajectory classification method analyzes the accuracy and reliability of the proposed method through real ship trajectory data and aims to solve the problem of classification of the waterway to which a ship's trajectory belongs.

Future research will be carried out considering the following directions:

- In ship trajectory clustering research, the problem of ship trajectory clusters that contain a large amount of noise data has not been considered. In the future, it is necessary to conduct research on the noise characteristics of ship trajectory data to extract valuable ship trajectory information from it.
- In this paper, we use deep convolutional neural networks to classify ship trajectories. With the rapid development of graph neural networks, the next step can be to consider using graph neural networks to complete ship trajectory classification tasks and compare different algorithms.

Author Contributions: Conceptualization, T.G.; methodology, T.G.; validation, T.G.; resources, L.X.; writing—original draft preparation, T.G.; writing—review and editing, T.G. and L.X.; funding acquisition, L.X. All authors have read and agreed to the published version of the manuscript.

Funding: National Key Research & Development Program of China (no. 2019YFB1600600, 2019YFB1600604); Natural Science Foundation of Hubei Province (2019CFA039, 2020CFB691).

Institutional Review Board Statement: Not applicable.

Informed Consent Statement: Not applicable.

Data Availability Statement: Not applicable.

Acknowledgments: This research is supported by the National Key Research & Development Program of China (no. 2019YFB1600600, 2019YFB1600604), and the Natural Science Foundation of Hubei Province (2019CFA039, 2020CFB691).

Conflicts of Interest: The authors declare no conflict of interest.

References

1. Xin, W.; Liu, Z.; Yao, C. Classification of Vessel Motion Pattern in Inland Waterways Based on Automatic Identification System. *Ocean Eng.* **2017**, *146*, 486–497.
2. Xu, H.; Oliveira, P.; Guedes Soares, C. L1 adaptive backstepping control for path-following of underactuated marine surface ships. *Eur. J. Control* **2020**, *58*, 357–372. [CrossRef]
3. Zhou, Q.; Thai, V.V. Fuzzy and grey theories in failure mode and effect analysis for tanker equipment failure prediction. *Saf. Sci.* **2016**, *83*, 74–79. [CrossRef]
4. Zhao, L.; Shi, G. A method for simplifying ship trajectory based on improved Douglas–Peucker algorithm. *Ocean Eng.* **2018**, *166*, 37–46. [CrossRef]
5. Wei, Z.; Xie, X.; Zhang, X. AIS trajectory simplification algorithm considering ship behaviours. *Ocean Eng.* **2020**, *216*, 108086. [CrossRef]
6. Zhao, L.; Shi, G. A trajectory clustering method based on Douglas-Peucker compression and density for marine traffic pattern recognition. *Ocean Eng.* **2019**, *172*, 456–467. [CrossRef]
7. Zhang, R.; Xie, P.; Wang, C. Classifying transportation mode and speed from trajectory data via deep multi-scale learning. *Comput. Netw.* **2019**, *162*, 106861. [CrossRef]
8. Chen, X.; Liu, Y.; Achuthan, K. A ship movement classification based on Automatic Identification System (AIS) data using Convolutional Neural Network. *Ocean Eng.* **2020**, *218*, 108182. [CrossRef]
9. Guo, S.; Mou, J.; Chen, L.; Chen, P. An Anomaly Detection Method for AIS Trajectory Based on Kinematic Interpolation. *J. Mar. Sci. Eng.* **2021**, *9*, 609. [CrossRef]
10. Liang, M.; Zhan, Y.; Liu, R.W. MVFFNet: Multi-View Feature Fusion Network for Imbalanced Ship Classification. *Pattern Recognit. Lett.* **2021**, *151*, 26–32. [CrossRef]
11. Xiao, Z.; Fu, X.; Zhang, L. Traffic Pattern Mining and Forecasting Technologies in Maritime Traffic Service Networks: A Comprehensive Survey. *IEEE Trans. Intell. Transp. Syst.* **2020**, *21*, 1796–1825. [CrossRef]
12. Huang, L.; Wen, Y.; Guo, W. Mobility pattern analysis of ship trajectories based on semantic transformation and topic model. *Ocean Eng.* **2020**, *201*, 107092. [CrossRef]
13. He, K.; Zhang, X.; Ren, S.; Sun, J. Deep Residual Learning for Image Recognition. In Proceedings of the IEEE Conference on Computer Vision and Pattern Recognition, Las Vegas, NV, USA, 27–30 June 2016; pp. 770–778.
14. Choi, S.H.; Jung, S.H. Stable Acquisition of Fine-Grained Segments Using Batch Normalization and Focal Loss with L1 Regularization in U-Net Structure. *Int. J. Fuzzy Log. Intell. Syst.* **2020**, *20*, 59–68. [CrossRef]
15. Nieto, M.; Garau, B.; Balle, S.; Simarro, G.; Zarruk, G.; Ortiz, A.; Tintoré, J.; Álvarez, E.; Gómez-Pujol, L.; Orfila, A. An opensource, low cost video-based coastal monitoring system. *Earth Surf. Process. Landf.* **2010**, *35*, 1712–1719. [CrossRef]
16. Casella, E.; Rovere, A.; Pedroncini, A.; Stark, C.; Casella, M.; Ferrari, M.; Firpo, M. Drones as tools for monitoring beach topography changes in the Ligurian Sea (NW Mediterranean). *Geo-Mar. Lett.* **2016**, *36*, 151–163. [CrossRef]

17. Luijendijk, A.; Hagenaars, G.; Ranasinghe, R.; Baart, F.; Donchyts, G.; Aarninkhof, S. The State of the World's Beaches. *Sci. Rep.* **2018**, *8*, 6641. [CrossRef] [PubMed]
18. Vos, K.; Splinter, K.; Harley, M.; Simmons, J.; Turner, I. CoastSat: A Google Earth Engine-enabled Python toolkit to extract horelines from publicly available satellite imagery. *Environ. Model. Softw.* **2019**, *122*, 104528. [CrossRef]
19. Rutten, J.; Ruessink, B.G.; Price, T.D. Observations on sandbar behaviour along a man-made curved coast. *Earth Surf. Process. Landf.* **2018**, *43*, 134–149. [CrossRef]
20. De Swart, R.; Ribas, F.; Simarro, G.; Guillen, J.; Calvete, D. The role of bathymetry and directional wave conditions on observed crescentic bar dynamics. In *Earth Surface Processes and Landforms*; Wiley Online Library: Hoboken, NJ, USA, 2021.
21. Blossier, B.; Bryan, K.R.; Daly, C.J.; Winter, C. Nearshore sandbar rotation at single-barred embayed beaches. *J. Geophys. Res. Ocean.* **2016**, *121*, 2286–2313. [CrossRef]
22. Anderson, D.; Bak, A.S.; Brodie, K.L.; Cohn, N.; Holman, R.A.; Stanley, J. Quantifying Optically Derived Two-Dimensional Wave-Averaged Currents in the Surf Zone. *Remote Sens.* **2021**, *13*, 690. [CrossRef]
23. Bouvier, C.; Balouin, Y.; Castelle, B.; Holman, R. Modelling camera viewing angle deviation to improve nearshore video monitoring. *Coast. Eng.* **2019**, *147*, 99–106. [CrossRef]
24. Rodriguez-Padilla, I.; Castelle, B.; Marieu, V.; Morichon, D. A Simple and Efficient Image Stabilization Method for Coastal Monitoring Video Systems. *Remote Sens.* **2019**, *12*, 70. [CrossRef]
25. Simarro, G.; Calvete, D.; Souto, P. UCalib: Cameras Autocalibration on Coastal Video Monitoring Systems. *Remote Sens.* **2021**, *13*, 2795. [CrossRef]
26. Medellín, G.; Torres-Freyermuth, A. Morphodynamics along a micro-tidal sea breeze dominated beach in the vicinity of coastal structures. *Mar. Geol.* **2019**, *417*, 106013. [CrossRef]
27. Franklin, G.; Medellín, G.; Appendini, C.; Gómez, J.; Torres-Freyermuth, A.; López-González, J.; Ruíz-Salcines, P. Impact of port development on the northern Yucatan Peninsula coastline. *Reg. Stud. Mar. Sci.* **2021**, *45*, 101835. [CrossRef]
28. Kurczyn, J.; Appendini, C.; Beier, E.; Sosa-López, A.; López-González, J.; Posada-Venegas, G. Oceanic and atmospheric impact of Central American Cold Surges (Nortes) in the Gulf of Mexico. *Int. J. Climatol.* **2020**, *41*, 1450–1468. [CrossRef]
29. Briggs, T.; Figlus, J.; Torres-Freyermuth, A.; Puleo, J.; Warren, W.; Alrushaid, T. Variability in onshore sediment transport on a natural beach during a central American cold surge event. *J. Coast. Res.* **2020**, *36*, 487–497. [CrossRef]
30. Toscano, D.; Murena, F. Atmospheric ship emissions in ports: A review. Correlation with data of ship traffic. *Atmos. Environ. X* **2019**, *4*, 100050. [CrossRef]
31. Iris, Ç.; Lam, J.S.L. A review of energy efficiency in ports: Operational strategies, technologies and energy management systems. *Renew. Sustain. Energy Rev.* **2019**, *112*, 170–182. [CrossRef]
32. Nunes, R.A.O.; Alvim-Ferraz, M.C.M.; Martins, F.G.; Sousa, S.I.V. The activity-based methodology to assess ship emissions—A review. *Environ. Pollut.* **2017**, *231*, 87–103. [CrossRef]
33. Manisalidis, I.; Stavropoulou, E.; Stavropoulos, A.; Bezirtzoglou, E. Environmental and health impacts of air pollution: A Review. *Front. Public Health* **2020**, *8*, 14. [CrossRef] [PubMed]
34. Venturini, G.; Iris, Ç.; Kontovas, C.A.; Larsen, A. The multi-port berth allocation problem with speed optimization and emission considerations. *Transp. Res. Part D Transp. Environ.* **2017**, *54*, 142–159. [CrossRef]
35. Lee, H.; Park, D.; Choo, S.; Pham, H.T. Estimation of the Non-Greenhouse Gas Emissions Inventory from Ships in the Port of Incheon. *Sustainability* **2020**, *12*, 8231. [CrossRef]
36. Zhang, J.; Wan, C.; He, A.; Zhang, D.; Soares, C.G. A two-stage black-spot identification model for inland waterway transportation. *Reliab. Eng. Syst. Saf.* **2021**, *213*, 107677. [CrossRef]
37. Wan, C.; Yan, X.; Zhang, D.; Qu, Z.; Yang, Z. An advanced fuzzy Bayesian-based FMEA approach for assessing maritime supply chain risks. *Transp. Res. Part E Logist. Transp. Rev.* **2019**, *125*, 222–240. [CrossRef]

Journal of
Marine Science and Engineering

MDPI

Article

Ship Traffic Flow Prediction in Wind Farms Water Area Based on Spatiotemporal Dependence

Tian Xu * and Qingnian Zhang

School of Transportation and Logistics Engineering, Wuhan University of Technology, Wuhan 430063, China; 258553@whut.edu.cn
* Correspondence: yingying1520@whut.edu.cn

Abstract: To analyze the changing characteristics of ship traffic flow in wind farms water area, and to improve the accuracy of ship traffic flow prediction, a Gated Recurrent Unit (GRU) of a Recurrent Neural Network (RNN) was established to analyze multiple traffic flow sections in complex waters based on their traffic flow structure. Herein, we construct a spatiotemporal dependence feature matrix to predict ship traffic flow instead of the traditional ship traffic flow time series as the input of the neural network. The model was used to predict the ship traffic flow in the water area of wind farms in Yancheng city, Jiangsu Province. Autoregressive Integrated Moving Average (ARIMA), Support-Vector Machine (SVM) and Long Short-Term Memory (LSTM) were chosen as the control tests. The GRU method based on the spatiotemporal dependence is more accurate than the current mainstream ship traffic flow prediction methods. The results verify the reliability and validity of the GRU method.

Keywords: complex waters; ship traffic flow; spatiotemporal dependence; gate recurrent unit

Citation: Xu, T.; Zhang, Q. Ship Traffic Flow Prediction in Wind Farms Water Area Based on Spatiotemporal Dependence. *J. Mar. Sci. Eng.* **2022**, *10*, 295. https://doi.org/10.3390/jmse10020295

Academic Editor: Claudio Ferrari

Received: 7 January 2022
Accepted: 15 February 2022
Published: 21 February 2022

1. Introduction

Marine wind energy is a green renewable resource that has the advantages of cleanliness, low development cost, and abundant reserves [1,2]. The development of marine wind energy under the background of encouraging sustainable economic development is conducive to alleviating the energy crisis and preventing climate change. With the development of offshore wind power technology, offshore wind farm construction has gradually become a key development field [3]. However, the offshore wind farm needs to occupy a large area of water in the process of construction, and some ship traffic inevitably flows through the wind farm. With the increasing frequency of offshore wind power construction, the risk assessment of ship navigation in wind farms water area has also attracted the attention of scholars at home and abroad [4].

In the water area of the wind farm, ships need to be in close contact with the wind turbine, and the working frequency is very high. Compared with ships in other water areas, the collision probability of ships in the water area of a wind farm is greater [5,6]. Especially in severe weather—such as strong winds, large waves, and dense fog—the maneuverability of the ship is limited, the visibility in the water area is reduced, and the risk of collision between the ships and the wind turbines is further exacerbated [7]. With the increase in offshore wind power construction, the safety of ship navigation in the waters of offshore wind farms has gradually attracted the extensive attention of relevant scholars. At present, the research on ship navigation safety in the water area of wind farms mainly focuses on exploring the variation law of collision risk between ships and offshore wind turbines [8,9].

However, the impact of wind farm characteristics and ship traffic flow characteristics is not fully considered, and the established collision risk model often cannot reflect the actual situation of ship/wind turbine collisions. Therefore, it is necessary to analyze the ship traffic flow in the wind farms water area as the basis of ship navigation risk assessment [10,11]. By analyzing the variation in characteristics of ship traffic flow, mining the potential regularity of traffic flow data, and predicting the traffic flow state in the future, we can provide a theoretical basis for the formulation of traffic control measures in the wind farms water area.

Therefore, we propose a method to predict the traffic volume in wind farm areas based on spatiotemporal dependence. A feature matrix is constructed to represent the spatial relationship of traffic flow based on Pearson's Correlation Coefficient (PCC), and then GRU is established to predict the ship traffic flow. The construction in this paper has two folders: Firstly, the spatial effects of traffic flow on different routes are considered. Secondly, the matrix is used as the input of the neural network instead of a single value. The method proposed in this paper can support the construction of safety supervision in wind farms water area.

The remainder of this paper is organized as follows: Section 2 reviews research related to the research and safety of wind farm water and traffic flow prediction. Section 3 elaborates on the framework construction and methodological development, followed by case studies in Section 4. Finally, Section 5 discusses the major findings and potential research improvements, and conclusions are summarized in Section 5

2. Literature Review

Traffic flow prediction is an important topic in maritime traffic research. Recently, the study of ship traffic flow prediction can be divided into statistical methods and data-driven methods.

Several statistical models have been applied extensively in traffic flow prediction, including linear regression, ARIMA, Kalman filtering, Bayesian networks, Markov models, etc. Sun et al. (2003) introduced local linear regression into traffic flow prediction research [12]. Williams and Hoel (2003) adopted the seasonal ARIMA process to forecast vehicular traffic flow [13]. Getahun (2021) modeled a time series of road traffic accidents based on ARIMA [14]. Guo et al. (2014) developed an adaptive Kalman filter method for stochastic traffic flow rate prediction [15], while Xie et al. (2007) researched traffic flow prediction using a Kalman filter with discrete wavelet decomposition [16]. Saeedmanesh et al. (2021) developed an extended Kalman filter approach for real-time state estimation in multiregion MFD urban networks [17]. Smith et al. (1997, 2002) proposed a nonparametric method for traffic flow forecasting, and compared parametric and nonparametric models [18,19]. Zheng and Su (2014) researched traffic flow forecasting using a constrained linearly sewing principal component algorithm [20]. Wang et al. (2021) proposed a non-parametric model with an optimized training strategy for vehicle traffic flow prediction [21]. Castillo et al. (2008) introduced a Bayesian network for traffic flow prediction [22]. Wang et al. (2014) designed an architecture for traffic flow prediction using a new Bayesian combination method [23]. Afrin et al. (2021) estimated traffic congestion based on a Bayesian network [24]. Qi and Ishak (2014) developed a hidden Markov model for the prediction of traffic flow on freeways [25]. Rajawat et al. (2021) developed a comprehensive framework for the prediction of human mobility patterns based on a hidden Markov model [26].

In recent years, data-driven methods have been widely used in traffic flow prediction. The advantage of these methods is their ability to predict future traffic flow directly from the given big data without modeling the traffic flow phenomenon. Many researchers have applied SVM regression models to traffic flow prediction and achieved good results [27–29]. Some improved SVM methods are also widely used [30,31]. Yao et al. (2014) developed a framework in multistep-ahead prediction for rock displacement surrounding a tunnel, using an improved SVM [32]. Toan et al. (2021) applied an SVM for short-term traffic flow prediction [33].

Another widely used method is the K-nearest neighbor model, which is easy to implement because the processes of training data and estimating parameters are simple. Hong et al. (2015) developed a hybrid multimetric K-nearest neighbor regression model for traffic flow prediction [34]. Akbari et al. (2011) applied the K-nearest neighbor algorithm for daily inflow forecasting [35]. Yu et al. (2016) designed a prediction model for multiple-timestep prediction of traffic conditions [36]. Another very important method is machine learning. Many kinds of research for traffic prediction based on machine learning have been published [37–42]. Li (2016) applied dynamic fuzzy neural networks for traffic flow prediction [43]. Huang et al. (2014) proposed a Deep Belief Network (DBN) for traffic flow prediction, which is a deep architecture [44]. Yang et al. (2016) designed a type of unsupervised learning architecture of the neural network approach for traffic flow prediction using the Taguchi method [45]. Lu et al. (2021) proposed a combined method for short-term traffic flow prediction based on a recurrent neural network [46].

In summary, many researchers have made great progress in the research of traffic flow prediction, but the assumption that the model parameterization performs relatively badly in variable traffic conditions affects the majority of statistical models. On the other hand, data-driven models such as deep learning techniques are often used to make predictions. Although the accuracy of the deep learning method is higher than that of other algorithms, the training time is much longer than that of other algorithms. The GRU has fewer parameters than other models, reducing the risk of overfitting, and has a shorter training time. Additionally, the GRU can simultaneously consider the influence of features and historical time series. At present, GRU networks are mainly used in classification, regression, and time-series prediction problems. Therefore, a GRU was established to predict the ship traffic flow in this research.

Several studies have emerged over recent years covering many aspects of wind farms, such as site selection [47,48], operation and maintenance [49–51], and wildlife impact [52,53], among others. However, research on traffic flow prediction in the water area of wind farms is rare at present, and most such studies take a single port or channel section as their research object, not considering the spatiotemporal dependence of ship traffic flow. Therefore, taking the supervision of maritime traffic safety in complex waters as the starting point, the water area of Yancheng wind farms was selected as the research object to predict ship traffic flow in different routes in the water area of the wind farms.

3. Methodology

The logical framework for the prediction of ship traffic flow in wind farm waters is depicted in Figure 1; it consists of three components: data preparation, spatial relationship analysis, and time-series prediction model.

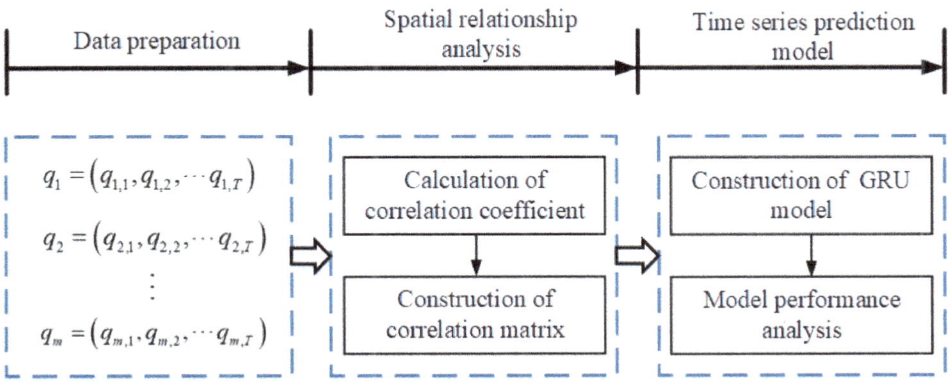

Figure 1. Logical framework in this study.

3.1. Data Preparation

As of December 2004, the International Maritime Organization (IMO) required all vessels over 299 GT to install an Automatic Identification System (AIS) transponder on board [54]. The increasing number of ships equipped with AISs provides a lot of basic data for traffic flow prediction research. AIS data are received as a series of messages following a nonstandard pattern of irregular time intervals. Since there is a lot of noise in the raw AIS data, it is necessary to preprocess the data. Firstly, the data of ship position abnormality, speed abnormality, and course abnormality in the experimental dataset are removed based on the algorithm proposed by [55]. Then, the linear interpolation method should be used to interpolate the ship's AIS data per 10 s.

3.2. Analysis of Spatial Dependence of Ship Traffic Flow

For complex waters, such as wind farms water areas, the ship traffic flow may be affected by the traffic conditions of adjacent routes. Therefore, considering the spatial ship traffic flow dependence, Pearson's Correlation Coefficient (PCC) method was used to calculate the correlation coefficient between the traffic flows of adjacent routes [56,57]. PCC is the covariance of the two variables divided by the product of their standard deviations.

Adding the route information with high correlation to the prediction model improves the accuracy of the marine traffic flow prediction model. If the traffic flow sequence of section x and section y is as follows:

$$q_x = \left(q_{x,1}, q_{x,2}, \cdots q_{x,T}\right), \tag{1}$$

$$q_y = \left(q_{y,1}, q_{y,2}, \cdots q_{y,T}\right), \tag{2}$$

then the correlation coefficient of the two sections is given by:

$$\rho_{x,y} = \frac{\sum_{i=1}^{T}\left(q_{x,i} - \overline{q_x}\right)\left(q_{y,i} - \overline{q_y}\right)}{\sqrt{\sum_{i=1}^{T}\left(q_{y,i} - \overline{q_y}\right)^2}\sqrt{\sum_{i=1}^{T}\left(q_{y,i} - \overline{q_y}\right)^2}}, \tag{3}$$

where $\rho_{x,y}$ is the degree of correction between x and y, referred to as the correlation coefficient, while T represents the length of the time series. The closer $\rho_{x,y}$ is to 1, the greater the correlation between the target section and the adjacent section. When $\rho_{x,y} = 0$, there is no correlation between the target section and adjacent sections. When $0 < \rho_{x,y} < 0.5$, this indicates that the correlation between the target section and adjacent sections is low. To ensure the prediction accuracy of the model, $0.5 \leq \rho_{x,y} < 1$ was set as the spatial threshold range.

3.3. Time-Series Prediction Model Based on an Improved Recurrent Neural Network

A GRU can be used to mine the time characteristics of traffic flow and capture the time dependence of ship traffic flow [58–60]. Cho et al. introduced a GRU as a gating mechanism in recurrent neural networks [61]. The GRU's functions are similar to those of a Long Short-Term Memory (LSTM) network with a forget gate, but with fewer parameters. The GRU outperformed LSTM on certain tasks, such as polyphonic music modeling, speech signal modeling, and natural language processing. The GRU also uses only two parameters, which can help to reduce the risk of overfitting. The basic structure of the GRU is shown in Figure 2.

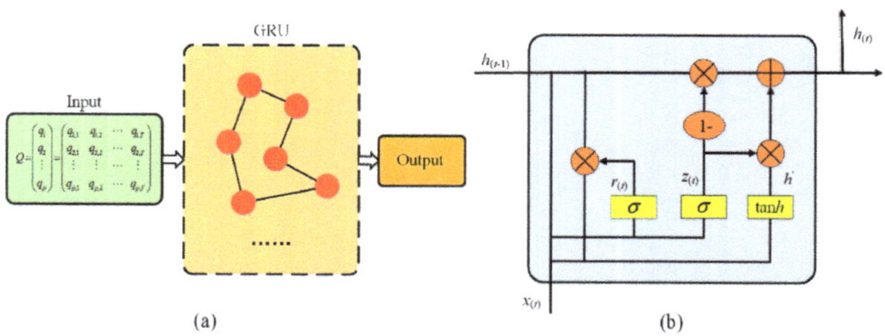

Figure 2. Diagram of the model structure used in this research: (**a**) Whole structure of the model. (**b**) Basic structure of the GRU.

$x_{(t)}$ represents the input signal vector of the current node; $h_{(t-1)}$ represents the hidden state vector passed down from the previous node, $h_{(t)}$ is the output vector, and $h'_{(t)}$ is candidate activation vector; $r_{(t)}$ is the reset gate vector, $z_{(t)}$ is the update gate vector, while σ represents the original function, and is a sigmoid function. The GRU uses $x_{(t)}$ and $h_{(t-1)}$ to obtain two gating states; the functions of the reset gate and update gate are as follows:

$$r_{(t)} = \text{Sigmoid\_1}(W_r(h_{(t-1)}, x_{(t)})), \tag{4}$$

$$z_{(t)} = \text{Sigmoid\_2}(W_z(h_{(t-1)}, x_{(t)})), \tag{5}$$

where W_r is the weight matrix of reset gate, while W_z is the weight matrix of the update gate. After obtaining the gating signal, use $h'_{(t-1)} = r_{(t)} \odot h_{(t-1)}$ to reset the data, then splice $h'_{(t-1)}$ and $x_{(t)}$. The activation function tanh is used to standardize the data to obtain $h'_{(t)}$; the calculation formula is as follows:

$$h'_{(t)} = \tanh\left(W_{h'}\left(h_{(t-1)} \odot r_{(t)}, x_{(t)}\right)\right), \tag{6}$$

where the operator $\odot$ denotes the Hadamard product, $W_{h'}$ is the weight matrix of candidate activation in the GRU, and $h'_{(t)}$ mainly contains the current input $x_{(t)}$, which adds $h'_{(t)}$ to the current hidden state and effectively remembers the current state.

In the last memory update stage, forgetting and memorizing steps are both used, as is the previously obtained update gate z; the update expression is as follows:

$$h_{(t)} = z_{(t)} \odot h_{(t-1)} + (1 - z_{(t)}) \odot h'_{(t)}, \tag{7}$$

where $h_{(t)}$ represents the output of the network at the moment t and $z_{(t)} \in (0,1)$. The input of each layer of the GRU considers the output of the previous layer of the GRU, so as to capture the timing relationship of ship traffic flow. After constructing a GRU, it is used as the basic unit from sequence to sequence as a model to generate the final prediction result.

The ship traffic flow in an area is related not only to the actual traffic flow in the given moments, but also to the spatially related route. Therefore, compared with the traditional ship traffic flow prediction method, a PCC algorithm can be introduced to calculate the correlation of traffic flow between different routes. This method can screen the areas with high correlation and reconstruct the spatiotemporal dependence matrix, in order to improve the input of the GRU model and to predict the ship traffic flow in complex waters more accurately. The algorithm flow is as follows:

(1) According to the characteristics of traffic flow structure in complex waters, the temporal and spatial characteristic matrix of multiple observation sections is defined as Q.

$$Q = \begin{pmatrix} q_1 \\ q_2 \\ \vdots \\ q_m \end{pmatrix} = \begin{pmatrix} q_{1,1} & q_{1,2} & \cdots & q_{1,T} \\ q_{2,1} & q_{2,2} & \cdots & q_{2,T} \\ \vdots & \vdots & \cdots & \vdots \\ q_{m,1} & q_{m,2} & \cdots & q_{m,T} \end{pmatrix}, \tag{8}$$

where q_m is the time series of the observation section, and $q_{m,T}$ is the traffic volume of observation section m at the moment T;

(2) Calculate the correlation coefficient between the traffic flow at the observation section of two routes, constructing a spatial correlation matrix as follows:

$$\rho_{xy} = \begin{pmatrix} \rho_{1,1} & \rho_{1,2} & \cdots & \rho_{1,m} \\ \rho_{2,1} & \rho_{2,2} & \cdots & \rho_{2,m} \\ \vdots & \vdots & \cdots & \vdots \\ \rho_{m,1} & \rho_{m,2} & \cdots & \rho_{m,m} \end{pmatrix}; \tag{9}$$

(3) Rank traffic flow correlation between sections. Select the sections to be predicted and sort the other sections by correlation;

(4) Set the correlation threshold δ. The sections with a correlation greater than δ are reconstructed into a new spatiotemporally dependent characteristic matrix;

(5) Build the GRU model. The input of the GRU model is improved by using the reconstructed spatiotemporal dependence characteristic matrix, and the improved GRU is used to predict ship traffic flow;

(6) Calculate the Mean Absolute Percentage Error (MAPE); taking the minimum MAPE as the final optimization goal of the model, the optimal spatial threshold is determined;

(7) Results and analysis: To quantitatively analyze the prediction results and the performance of the model, the Mean Absolute Error (MAE), Root-Mean-Square Error (RMSE), Mean Absolute Percentage Error (MAPE), and R^2 are used as prediction and evaluation indicators. The calculation method is as follows:

$$\mathrm{MAE} = \frac{1}{n}\sum\nolimits_{1}^{n}|p_i - x_i|, \tag{10}$$

$$\mathrm{RMSE} = \sqrt{\frac{1}{n}\sum\nolimits_{1}^{n}(p_i - x_i)^2}, \tag{11}$$

$$\mathrm{MAPE} = \frac{1}{n}\sum\nolimits_{1}^{n}\frac{|p_i - x_i|}{x_i} \times 100, \tag{12}$$

$$\mathrm{R}^2 = 1 - \frac{\sum_1^n (p_i - x_i)^2}{\sum_1^n (\overline{x_i} - x_i)^2}, \tag{13}$$

where p_i is the predicted value obtained by PCC–GRU and by SVM, LSTM, and ARIMA in the control experiment; x_i is the actual value; $\overline{x_i}$ is the average of the actual flow section values.

4. Case Study

4.1. Research Area and Data

Due to the influence of meteorological and hydrological conditions, construction conditions, and water area location, the water area traffic environment of wind farms presents complex temporal and spatial characteristics. Based on AIS data, real-time prediction of ship traffic flow changes in different stages of wind farm groups can provide an effective means for daily maintenance of wind farms and safety monitoring in the process of construction.

To verify the effectiveness and feasibility of the model, the wind farm water area in Jiangsu Province was selected as the research object, as shown in Figure 3. The cross-section flow statistics of complex routes in the region were evaluated. The routes observed in this research were routes recommended in nautical charts, and the observation section was set in these routes. If a ship passes through the observation section, the traffic volume increases by 1.

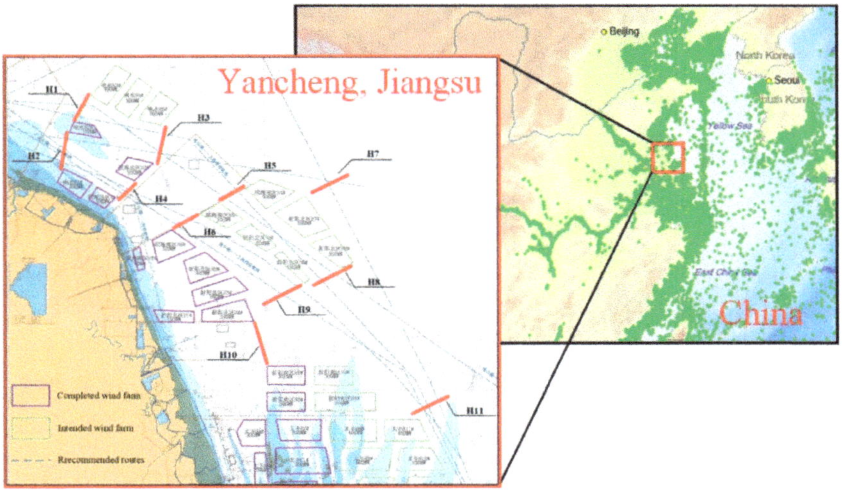

Figure 3. Schematic diagram of the observation section in the Yancheng wind farm area.

The cross-sectional diagram is shown in Figure 3. The location information of the section is shown in Table 1. The traffic flow time series from 1 to 29 March 2021 was selected as the training data, and the traffic flow data on 30 March were used as the verification set. The ship traffic flow data of sections H1–H11 were collected in hours, and the ship traffic flow data for one day were obtained, as shown in Table 2. q_1–q_{11} represent the traffic volume through sections H1–H11, respectively. Traffic volume is measured as the number of ships in the section during the given time.

Taking section H5 as an example, the model proposed in this paper was used to predict ship traffic flow. Spatial state variables and temporal state variables were the two input parameters of the model. The spatial state variable is a matrix formed by the correlation coefficients between sections, while the time state variable refers to the time interval adopted by the spatial matrix. When the neural network model was built, the recurrent neural network of each layer of encoder and decoder had several GRUs. The number of GRUs is always a multiple of 2—generally between 16 and 128; if it is too large, it will increase the computational complexity and make the training time too long; if it is too small, it will also affect the performance of the model. Figure 4 provides the MEA and RMSE of the experiments using different numbers of GRUs. The GRU number of 64 obtained the lowest error. Therefore, we set the number of GRUs to 64 in our experiments. The validation set loss of the experiment using different batch sizes is shown in Figure 5. The batch size of 8 yielded the lowest error. As a result, the batch size of our model was set to 8. When a complete dataset passes through the neural network and returns, the process is called an epoch; we set the number of epochs to 100. Table 3 shows the hyperparameters used by PCC–GRU in the experiment. All experiments were conducted in Keras on a laptop with an NVIDIA 2080ti GPU, an Intel Core i9-9900KF CPU (3.6 GHz), 16 GB RAM, and the Windows 10 operating system.

Table 1. Coordinates of the observation section.

Sections	P_1	P_2
H1	120.06495° E 34.75165° N	120.13086° E 34.96235° N
H2	120.02375° E 34.69723° N	119.96195° E 34.42275° N
H3	120.35502° E 34.76512° N	120.27413° E 34.64402° N
H4	120.26468° E 34.49281° N	120.20728° E 34.42028° N
H5	120.57021° E 34.48273° N	120.47371° E 34.43313° N
H6	120.47302° E 34.35673° N	120.57023° E 34.30368° N
H7	121.30618° E 34.65731° N	121.08612° E 34.45872° N
H8	121.04723° E 34.20372° N	120.93305° E 34.15846° N
H9	120.88817° E 34.13275° N	120.77619° E 34.08367° N
H10	120.72357° E 34.06888° N	120.78652° E 33.81706° N
H11	121.32603° E 33.78164° N	121.21804° E 33.26723° N

Table 2. Ship traffic flow per hour on 30 March 2021.

Time	q_1	q_2	q_3	q_4	q_5	q_6	q_7	q_8	q_9	q_{10}	q_{11}
0:00	2	2	3	3	2	2	5	5	1	3	5
1:00	2	1	2	4	6	3	4	3	3	3	10
2:00	5	3	4	3	4	4	3	2	2	3	6
3:00	2	3	6	4	5	3	4	5	6	3	6
4:00	4	4	8	4	7	6	6	2	7	6	9
5:00	4	5	5	2	6	3	2	4	4	3	6
6:00	5	2	6	5	9	4	3	6	2	5	11
7:00	5	3	3	2	3	5	4	5	3	2	4
8:00	7	4	4	4	4	5	2	5	3	5	5
9:00	5	2	3	3	4	4	5	7	8	6	2
10:00	4	5	3	4	6	3	3	5	2	5	9
11:00	6	4	2	3	7	4	5	4	4	3	13
12:00	7	2	7	5	2	2	1	7	6	4	6
13:00	6	3	4	4	2	5	5	2	7	4	8
14:00	5	4	6	2	6	4	4	6	3	5	4
15:00	6	2	5	3	6	6	6	7	4	3	6
16:00	8	1	5	2	9	5	7	4	2	3	11
17:00	6	2	6	3	7	4	4	7	2	2	8
18:00	3	3	4	3	6	6	4	5	3	6	7
19:00	4	6	3	5	3	6	6	4	5	7	4
20:00	6	3	3	3	2	3	2	6	4	1	3
21:00	4	5	2	6	5	3	6	5	6	2	9
22:00	2	2	3	3	4	5	5	3	3	4	6
23:00	3	3	2	1	3	4	3	5	2	3	8

Table 3. Hyperparameters used in PCC–GRU.

Hyperparameter	Value
Epochs	100
Dropout Rate	Rate
Learning Rate	0.0005
Batch Size	8
Hidden Unit	64

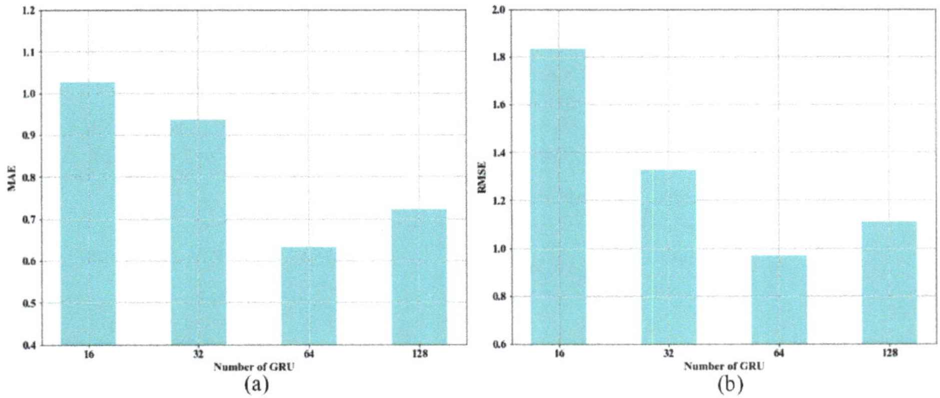

Figure 4. MAE and RMSE with the same number of GRUs: (**a**) MAE of the experiment. (**b**) RMSE of the experiment.

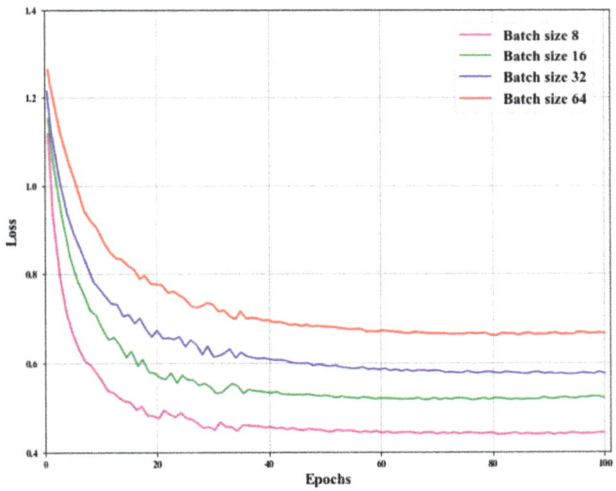

Figure 5. The loss of different batch sizes.

4.2. Results and Analysis

The data of four sections (H3, H5, H9, and H11) were selected to verify the method proposed in this research. In order to set the optimal threshold of PCC and time steps in the GRUs, many experiments were carried out in this research. Indicators for evaluating the experimental results can be calculated using Equations (10)–(13). In these equations, x_i is the temporary ship traffic volume value measured on 30 March 2021, while p_i is the predicted value by PCC–GRU, calculated with different parameters.

To study the influence of spatiotemporal state variables on prediction accuracy and determine the optimal parameters, the thresholds of PCC were set as 0.8, 0.7, 0.6, and 0.5, because when the threshold is greater than or equal to 0.8, there is no section related to H3, H5, H9, or H11 in space. To improve the prediction accuracy, the PCC algorithm was used to select spatial characteristic variables. From the position of each observation section, it can be seen that the sections with a large correlation and with the target section are located in the water area around the target, indicating that the traffic flow of the surrounding water area has a great impact on the target section. According to the actual situation, the effectiveness of the PCC algorithm in spatial feature variable selection was verified.

Timesteps are an import parameter in neural networks. If the number of timesteps is *n*, that means that we think that each value is related to *n* values in front of it. If the timestep used in prediction is too large, it will lead to time information redundancy and affect the accuracy of the prediction results. If the timestep is too small, it will lead to overfitting of the model. The performance of the model can be seen in Tables 4–7, where the timestep of 7 obtains the lowest error. Therefore, we set the timestep to 7 in our experiments.

Table 4. Prediction performance of different thresholds in H3.

Threshold	Section Related to H3	Time Steps	MAE	RMSE	MAPE	R^2
0.8	/	3	1.61207	1.75961	42.58056	0.16006
		4	1.16523	1.38184	34.99692	0.48199
		5	0.69115	0.82319	18.32865	0.71491
		6	0.67851	0.80884	19.62209	0.81716
		7	0.65454	0.78254	18.45094	0.82826
0.7	H1 H5	3	1.47985	1.69798	41.62103	0.15091
		4	1.16954	1.38582	33.64929	0.47622
		5	0.70286	0.88204	20.64433	0.82194
		6	0.66256	0.83853	19.73351	0.80314
		7	0.65273	0.74592	18.75563	0.81522
0.6	H1 H5 H8	3	1.31057	1.68343	39.71651	0.13742
		4	1.20977	1.30272	36.35331	0.45431
		5	0.66667	0.81497	20.85102	0.81914
		6	0.66667	0.76497	19.59404	0.88914
		7	0.53453	0.68563	10.65256	0.90365
0.5	H1 H5 H8	3	1.31057	1.68343	39.71651	0.13742
		4	1.20977	1.30272	36.35331	0.45431
		5	0.66667	0.81497	20.85102	0.81914
		6	0.66667	0.76497	19.59404	0.88914
		7	0.53453	0.68563	10.65256	0.90365

Table 5. Prediction performance of different thresholds in H5.

Threshold	Section Related to H5	Time Steps	MAE	RMSE	MAPE	R^2
0.8	/	3	1.58275	1.76979	38.51897	0.22023
		4	1.21120	1.41877	30.82569	0.49887
		5	0.67241	0.82000	16.96503	0.83260
		6	0.66235	0.81385	16.22976	0.83510
		7	0.64080	0.80765	16.31632	0.83760
0.7	H3 H8	3	1.41724	1.75878	38.06598	0.39443
		4	1.16666	1.39777	29.46423	0.51568
		5	0.66816	0.81505	16.21954	0.83117
		6	0.63662	0.79003	15.56588	0.84225
		7	0.63345	0.72873	15.60616	0.85674
0.6	H1 H3 H8	3	1.39669	1.69558	36.73457	0.32426
		4	1.16373	1.33858	28.51531	0.59998
		5	0.66104	0.80524	15.78878	0.82455
		6	0.64689	0.78769	13.30716	0.89652
		7	0.62234	0.71074	9.47337	0.91359
0.5	H1 H3 H8	3	1.39669	1.69558	36.73457	0.32426
		4	1.16373	1.33858	28.51531	0.59998
		5	0.66104	0.80524	15.78878	0.82455
		6	0.64689	0.78769	13.30716	0.89652
		7	0.62234	0.71074	9.47337	0.91359

Table 6. Prediction performance of different thresholds in H9.

Threshold	Section Related to H9	Time Steps	MAE	RMSE	MAPE	R^2
0.8	/	3	1.48127	1.86125	30.61857	0.23653
		4	1.22701	1.42535	25.32567	0.30146
		5	0.68241	0.83008	14.76454	0.76802
		6	0.67965	0.82055	13.19848	0.77274
		7	0.66523	0.81516	12.53072	0.78172
0.7	H6	3	1.46275	1.85476	29.59729	0.25606
		4	1.12356	1.37632	23.55022	0.34897
		5	0.66839	0.82769	13.284	0.77004
		6	0.61092	0.82319	13.10851	0.78238
		7	0.60939	0.80334	12.46232	0.79934
0.6	H4 H6	3	1.45873	1.80834	31.70823	0.272437
		4	1.10492	1.37089	23.86203	0.367155
		5	0.63366	0.82304	14.44496	0.81392
		6	0.61264	0.82218	14.5587	0.83497
		7	0.60034	0.80091	11.53072	0.85034
0.5	H2 H4 H6	3	1.42701	1.76694	30.44643	0.27749
		4	1.10471	1.36167	24.53903	0.36465
		5	0.62425	0.78569	12.67354	0.84606
		6	0.58977	0.77932	11.04543	0.89926
		7	0.56942	0.75345	9.59432	0.92342

Table 7. Prediction performance of different thresholds in H11.

Threshold	Section Related to H11	Time Steps	MAE	RMSE	MAPE	R^2
0.8	/	3	1.51585	1.78957	21.30284	0.39355
		4	1.21831	1.42137	16.93147	0.58054
		5	0.67381	0.82084	10.42556	0.69343
		6	0.64799	0.80493	9.12566	0.79752
		7	0.62345	0.79238	8.70294	0.80071
0.7	H8	3	1.50861	1.77856	21.14639	0.45469
		4	1.17674	1.39735	16.42899	0.61225
		5	0.68965	0.81455	10.25764	0.79094
		6	0.66667	0.80497	9.05628	0.74576
		7	0.61225	0.75934	8.57453	0.83634
0.6	H8 H9	3	1.48741	1.74551	20.91039	0.48333
		4	1.12575	1.33582	15.97794	0.67957
		5	0.65062	0.80676	10.17630	0.79775
		6	0.62077	0.79132	8.80699	0.83894
		7	0.60225	0.76385	8.53494	0.85846
0.5	H5 H8 H9	3	1.42989	1.76357	18.58104	0.56642
		4	1.06954	1.31858	13.15588	0.69508
		5	0.61793	0.76473	9.267573	0.85563
		6	0.58345	0.72785	8.238327	0.89162
		7	0.52425	0.68346	7.33485	0.92125

Further analysis of the impact of timestep on prediction accuracy shows that the selection of different parameter combinations will change the prediction accuracy of the model. Within a certain range, the prediction accuracy is positively correlated with the timestep. In a certain range, with the decrease in the spatial threshold, the spatial correlation between observation sections is lower. In the case of a high threshold, too little information on other sections leads to a poor prediction effect. In the case of a low threshold, too much irrelevant information is added to interfere with the prediction effect. Based on the prediction performance of different thresholds in Tables 4–7, the threshold was set to 0.5, 0.5, 0.6, and 0.6, respectively. The predicted and true values of observation sections H3, H5, H9, and H11 in the training dataset are shown in Figure 6; the predicted results are in good agreement with the actual values, indicating that the PCC–GRU model is feasible in the actual traffic scenario.

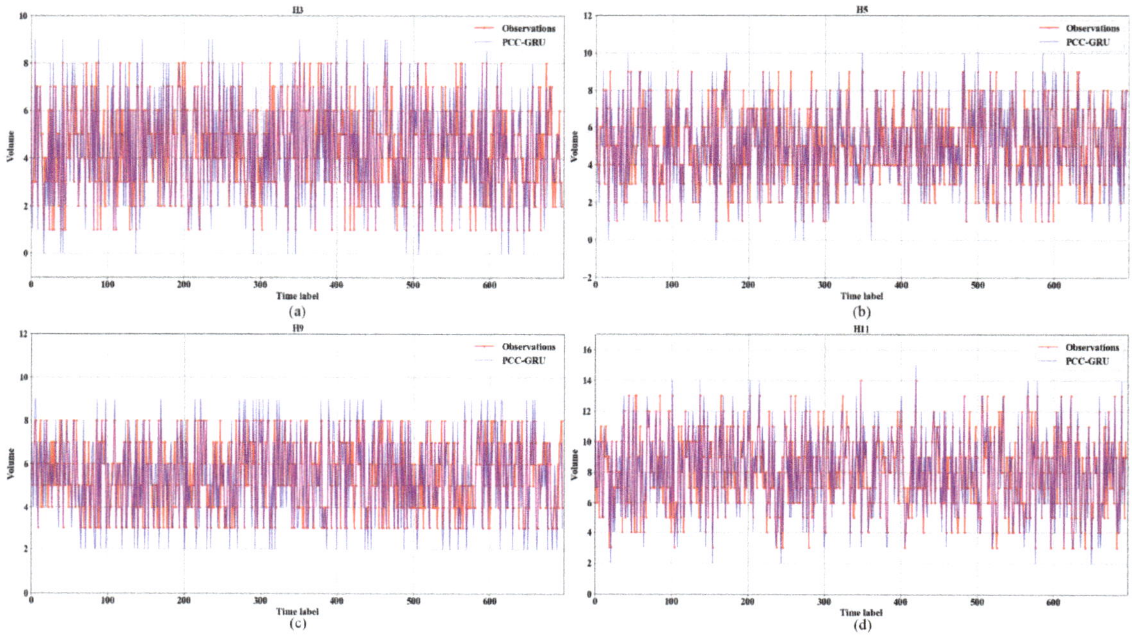

Figure 6. Prediction results of the PCC–GRU model: (**a**) Training data in H3. (**b**) Training data in H5. (**c**) Training data in H9. (**d**) Training data in H11.

To verify the prediction performance of the PCC–GRU model, comparative experiments were set up in this study. LSTM, SVM, and ARIMA were used to predict the traffic flow of H3, H5, H9, and H11. These methods are commonly used in maritime traffic flow prediction [62–64]; therefore, these methods were used as benchmarking methods. After many experiments, the parameters of each model in the experiment were determined, as shown in Table 8. The prediction results of the ship traffic flow of each model are shown in Figure 7.

Table 8. Parameter setting of models.

Model	Parameter	
LSTM	Neuron	12
	Timesteps	5
	Number of Iterations	300
SVM	Kernel Function	Radial Basis Function
	Penalty Factor	0.8
	Number of Iterations	500
ARIMA	Autoregressive Terms	2
	Moving average Terms	6
	Difference Items	1

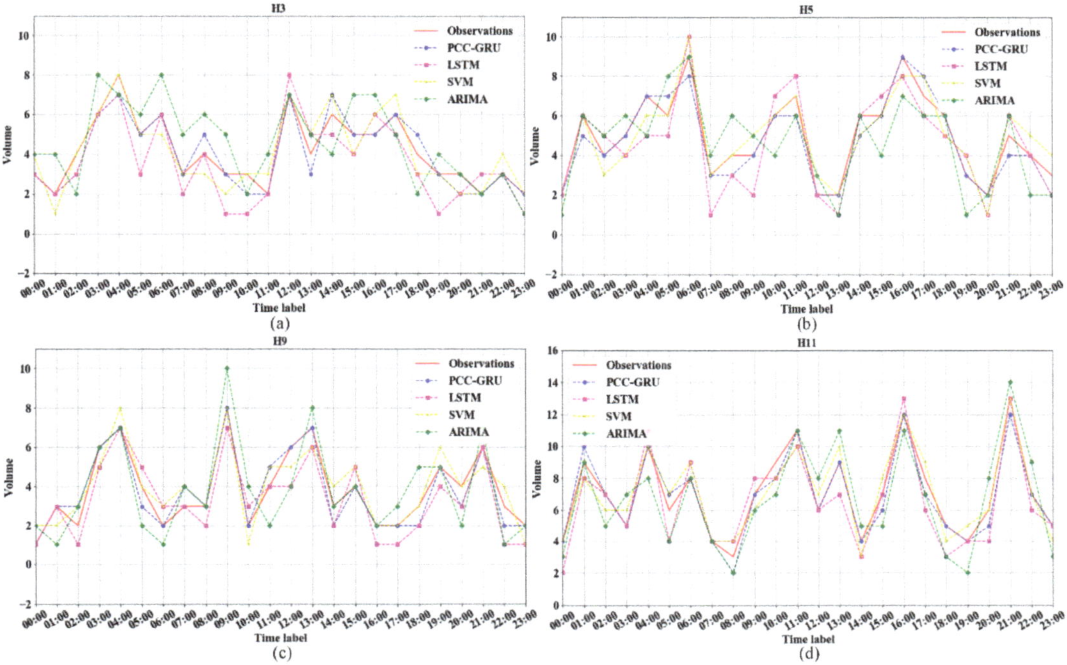

Figure 7. Comparison of prediction results of different models: (**a**) Prediction results in H3. (**b**) Prediction results in H5. (**c**) Prediction results in H9. (**d**) Prediction results in H11.

It can be seen from Table 9 that different models have different results in the traffic flow prediction experiment in the wind farms water area. In our case study, the LSTM model did not take into account the spatial information in the data, so its prediction accuracy was lower than that of the PCC–GRU model, and the time taken to train the LSTM model was too long for real-time prediction of maritime traffic flow. The machine learning algorithm SVM is suitable for short-term traffic forecasting, but the output of the SVM model will oscillate as the training data and the forecasting time increase. Meanwhile, the linear model ARIMA is unable to identify the randomness and nonlinearity in the data, making it difficult to make accurate predictions of random changes in traffic. The MAE of the PCC–GRU model in the prediction experiments of the four sections (H3, H5, H9, and H11) was 0.3333, 0.3750, 0.3333, and 0.3333, respectively; the RMSE of the PCC–GRU model of the four sections was 0.5774, 0.6124, 0.5774, and 5774, respectively, while the MAPE was 8.1597, 7.0006, 10.7639, and 5.1403, respectively, which is the smallest in the comparative analysis of the models. R^2 is usually used to describe the fitting degree of the data; the closer it is to 1, the better the fitting degree, and the smaller the deviation between the fitted curve and the original data points. The R^2 of the PCC–GRU model was 0.8799, 0.9116, 0.9063, and 0.9521, respectively, which was greater than that of other models. The bar chart of the error analysis indicators can be seen in Figure 8. In conclusion, the prediction results of the PCC–GRU model are closer to the observed values, and the performance of its algorithm is better than that of other traditional traffic flow prediction methods. The analysis of vessel traffic flow is the basis for evaluating the safety of vessel navigation in a given area of water. By predicting the ship traffic flow in the wind farms water area and maintaining real-time monitoring, it is possible to predict the business of vessel navigation and to control the traffic flow to avoid close-quarters situations.

Table 9. Performance comparison of different models.

	Models	Evaluation Indexes			
		MAE	RMSE	MAPE	R^2
H3	PCC–GRU	0.3333	0.5774	8.1597	0.8799
	ARIMA	0.8750	1.0992	24.2411	0.5647
	SVM	0.5833	0.7638	16.5972	0.7899
	LSTM	1.3333	1.5275	36.9792	0.1595
H5	PCC–GRU	0.3750	0.6124	7.0006	0.9116
	ARIMA	0.9167	1.0801	21.6402	0.7250
	SVM	0.6667	0.8165	17.3247	0.8429
	LSTM	1.1250	1.3385	27.1957	0.5774
H9	PCC–GRU	0.3333	0.5774	10.7639	0.9063
	ARIMA	0.8333	1.0000	26.6022	0.7188
	SVM	0.7083	0.8416	24.2640	0.8008
	LSTM	1.0417	1.3385	34.2758	0.4961
H11	PCC–GRU	0.3333	0.5774	5.1403	0.9521
	ARIMA	0.9167	1.1902	14.7058	0.7964
	SVM	0.7083	0.8416	11.9562	0.8982
	LSTM	1.3750	1.5679	22.7626	0.6466

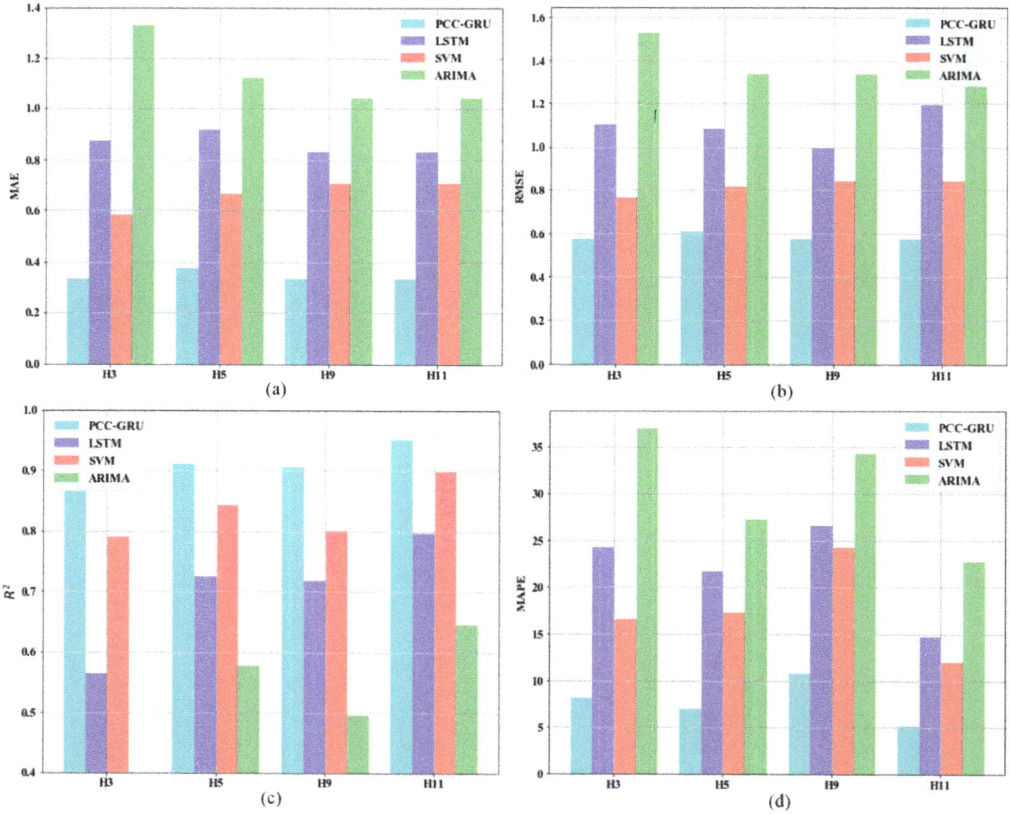

Figure 8. Bar chart of the error analysis indicators: (**a**) MAE of five prediction results. (**b**) RMSE of five prediction results. (**c**) R^2 of five prediction results. (**d**) MASE of five prediction results.

5. Conclusions

According to the needs of traffic characteristic analysis and traffic supervision in complex waters, this study introduced the spatial correlation of ship traffic flow structure into machine learning based on the variant GRU of the recurrent neural network, constructing a spatiotemporal dependence characteristic matrix of ship traffic flow, and improving the prediction accuracy of the ship traffic flow by the neural network. The PCC algorithm is simple to implement and fast to compute, taking into account the degree of correlation of time series at different times, and it is suitable for determining correlation between random variables. The GRU uses a unique memory module instead of implicit nodes, which increases the robustness of the model and can effectively compensate for the shortcomings of traditional neural networks that cannot effectively process long time series. Taking the water area of the wind farm in the Yancheng sea area of Jiangsu Province as an example, the traffic flow in this area was predicted. Compared to results of the commonly used ship traffic flow prediction models ARIMA, SVM, and LSTM, the GRU method's prediction was proven to be effective. The machine learning algorithm SVM is suitable for short-term prediction of traffic flows, but as the prediction time increases, the output of the SVM model will oscillate. The linear model ARIMA cannot identify the randomness and nonlinearity in the data, making it difficult to accurately predict the random changes in traffic. The LSTM model does not take into account the spatial information in the data, and therefore has a lower prediction accuracy than the PCC–GRU model.

By predicting the ship traffic flow and analyzing the temporal and spatial characteristics of traffic flow in complex waters, the marine traffic situation can be evaluated in real time, providing a theoretical basis for the risk evaluation and navigation safety guarantee of complex waters—especially the waters in wind farms with multi-route intersection—and reduces the risks of navigation and operation in the waters. Modelling the spatiotemporal dependence of ship traffic flow is a key area of future research. There are many factors affecting traffic flow prediction that have not been fully considered in this paper. The model proposed in this paper learns static spatial dependencies; however, the dependencies between locations may change over time. In subsequent research, dynamic structures should be considered in order to further improve the prediction performance.

Author Contributions: Conceptualization, T.X. and Q.Z.; methodology, T.X. and Q.Z.; software, T.X.; validation, T.X. and Q.Z.; formal analysis, T.X.; investigation, T.X.; resources, Q.Z.; data curation, T.X.; writing—original draft preparation, T.X.; writing—review and editing, T.X. and Q.Z.; visualization, T.X.; supervision, Q.Z.; project administration, Q.Z.; funding acquisition, Q.Z. All authors have read and agreed to the published version of the manuscript.

Funding: This research was funded by National Natural Science Foundation of China grant number 51879211. And The APC was funded by 51879211.

Institutional Review Board Statement: Not applicable.

Informed Consent Statement: Not applicable.

Data Availability Statement: Not applicable.

Acknowledgments: The authors would like to thank the anonymous reviewers and editors for their constructive comments, which were very helpful in improving this paper. This work was supported by the National Natural Science Foundation of China (NSFC) through Grant No.51879211.

Conflicts of Interest: The authors declare no conflict of interest.

References

1. Premalatha, M.; Abbasi, T.; Abbasi, S.A. Wind energy: Increasing deployment, rising environmental concerns. *Renew. Sustain. Energy Rev.* **2014**, *31*, 270–288.
2. Kusiak, A.; Song, Z. Design of wind farm layout for maximum wind energy capture. *Renew. Energy* **2010**, *35*, 685–694. [CrossRef]
3. Wang, X.Y.; Vilathgamuwa, D.M.; Choi, S.S. Determination of battery storage capacity in energy buffer for wind farm. *IEEE Trans. Energy Convers.* **2008**, *23*, 868–878. [CrossRef]

4. Jongbloed, R.H.; van der Wal, J.T.; Lindeboom, H.J. Identifying space for offshore wind energy in the North Sea. Consequences of scenario calculations for interactions with other marine uses. *Energy Policy* **2014**, *68*, 320–333. [CrossRef]
5. Frandsen, S.; Barthelmie, R.; Pryor, S.; Rathmann, O.; Larsen, S.; Højstrup, J.; Thøgersen, M. Analytical modelling of wind speed deficit in large offshore wind farms. *Wind. Energy Int. J. Prog. Appl. Wind. Power Convers. Technol.* **2006**, *9*, 39–53. [CrossRef]
6. Díaz, H.; Soares, C.G. Review of the current status, technology and future trends of offshore wind farms. *Ocean Eng.* **2020**, *209*, 107381. [CrossRef]
7. Hüppop, O.; Dierschke, J.; Exo, K.M.; Fredrich, E.; Hill, R. Bird migration studies and potential collision risk with offshore wind turbines. *Ibis* **2006**, *148*, 90–109. [CrossRef]
8. Dai, L.; Ehlers, S.; Rausand, M.; Utne, I.B. Risk of collision between service vessels and offshore wind turbines. *Reliab. Eng. Syst. Saf.* **2013**, *109*, 18–31. [CrossRef]
9. Ren, N.X.; Ou, J.P. A crashworthy device against ship-OWT collision and its protection effects on the tower of offshore wind farms. *China Ocean Eng.* **2009**, *23*, 594–602.
10. Furness, R.W.; Wade, H.M.; Masden, E.A. Assessing vulnerability of marine bird populations to offshore wind farms. *J. Environ. Manag.* **2013**, *119*, 56–66. [CrossRef]
11. Shafiee, M. A fuzzy analytic network process model to mitigate the risks associated with offshore wind farms. *Expert Syst. Appl.* **2015**, *42*, 2143–2152. [CrossRef]
12. Sun, H.; Liu, H.; Xiao, H.; He, R.; Ran, B. Use of local linear regression model for short-term traffic forecasting. *Transp. Res. Rec.* **2003**, *1836*, 143–150. [CrossRef]
13. Williams, B.M.; Hoel, L.A. Modeling and forecasting vehicular traffic flow as a seasonal ARIMA process: Theoretical basis and empirical results. *J. Transp. Eng.* **2003**, *129*, 664–672. [CrossRef]
14. Getahun, K.A. Time series modeling of road traffic accidents in Amhara Region. *J. Big Data* **2021**, *8*, 1–15. [CrossRef]
15. Guo, J.; Huang, W.; Williams, B.M. Adaptive Kalman filter approach for stochastic short-term traffic flow rate prediction and uncertainty quantification. *Transp. Res. Part C Emerg. Technol.* **2014**, *43*, 50–64. [CrossRef]
16. Xie, Y.; Zhang, Y.; Ye, Z. Short-term traffic volume forecasting using kalman filter with discrete wavelet decomposition. *Comput.-Aided Civ. Infrastruct Eng.* **2007**, *22*, 326–334. [CrossRef]
17. Saeedmanesh, M.; Kouvelas, A.; Geroliminis, N. An extended Kalman filter approach for real-time state estimation in multi-region MFD urban networks. *Transp. Res. Part C Emerg. Technol.* **2021**, *132*, 103384. [CrossRef]
18. Smith, B.L.; Demetsky, M.J. Forecasting freeway traffic flow for intelligent transportation systems application. *Transp. Res. Part A Policy Pract.* **1997**, *31*, 61.
19. Smith, B.L.; Williams, B.M.; Oswald, R.K. Comparison of parametric and nonparametric models for traffic flow forecasting. *Transp. Res. Part C Emerg. Technol.* **2002**, *10*, 303–321. [CrossRef]
20. Zheng, Z.; Su, D. Short-term traffic volume forecasting: A k-nearest neighbor approach enhanced by constrained linearly sewing principle component algorithm. *Transp. Res. Part C Emerg. Technol.* **2014**, *43*, 143–157. [CrossRef]
21. Wang, J.; Boukerche, A. Non-parametric models with optimized training strategy for vehicles traffic flow prediction. *Comput. Netw.* **2021**, *187*, 107791. [CrossRef]
22. Castillo, E.; Menéndez, J.M.; Sánchez-Cambronero, S. Predicting traffic flow using bayesian networks. *Transp. Res. Part B Meth.* **2008**, *42*, 482–509. [CrossRef]
23. Wang, J.; Deng, W.; Guo, Y. New Bayesian combination method for short-term traffic flow forecasting. *Transp. Res. Part C Emerg. Technol.* **2014**, *43*, 79–94. [CrossRef]
24. Afrin, T.; Yodo, N. A probabilistic estimation of traffic congestion using Bayesian network. *Measurement* **2021**, *174*, 109051. [CrossRef]
25. Qi, Y.; Ishak, S. A hidden Markov model for short term prediction of traffic conditions on freeways. *Transp. Res. Part C Emerg. Technol.* **2014**, *43*, 95–111. [CrossRef]
26. Rajawat, N.; Gupta, N.; Lalwani, S. A comprehensive review of hidden Markov model applications in prediction of human mobility patterns. *Int. J. Swarm Intell.* **2021**, *6*, 24–47. [CrossRef]
27. Zhang, C.; Chen, L. Traffic flow combining forecast model based on least squares support vector machine. *J. Hunan Inst. Eng.* **2010**, *20*, 56–58.
28. Wang, J.; Shi, Q. Short-term traffic speed forecasting hybrid model based on Chaos-Wavelet analysis-support vector machine theory. *Transp. Res. Part C Emerg. Technol.* **2013**, *27*, 219–232. [CrossRef]
29. Hu, Y.; Wu, C.; Liu, H. Prediction of passenger flow on the highway based on the least square support vector machine. *Transp. Res. J. Vilnius Gedim. Tech. Univ. Lith. Acad Sci.* **2011**, *26*, 197–203.
30. Castro-Neto, M.; Jeong, Y.S.; Jeong, M.K.; Han, L.D. Online-SVR for short-term traffic flow prediction under typical and atypical traffic conditions. *Expert Syst. Appl.* **2009**, *36*, 6164–6173. [CrossRef]
31. Asif, M.T.; Dauwels, J.; Chong, Y.G.; Oran, A.; Fathi, E.; Xu, M. Spatiotemporal patterns in large-scale traffic speed prediction. *IEEE Trans. Intell. Transp. Syst.* **2014**, *15*, 794–804. [CrossRef]
32. Yao, B.; Yao, J.; Zhang, M.; Yu, L. Improved support vector machine regression in multi-step-ahead prediction for rock displacement surrounding a tunnel. *Sci. Iran.* **2014**, *21*, 1309–1316.
33. Toan, T.D.; Truong, V.H. Support vector machine for short-term traffic flow prediction and improvement of its model training using nearest neighbor approach. *Transp. Res. Rec.* **2021**, *2675*, 362–373. [CrossRef]

34. Hong, H.; Huang, W.; Xing, X.; Zhou, X. Hybrid multi-metric K-nearest neighbor regression for traffic flow prediction. In Proceedings of the 2015 IEEE 18th International Conference on Intelligent Transportation Systems, Gran Canaria, Spain, 15–18 September 2015; pp. 2262–2267.
35. Akbari, M.; Overloop, P.J.V.; Afshar, A. Clustered K nearest neighbor algorithm for daily inflow forecasting. *Water Resour. Manag.* **2011**, *25*, 1341–1357. [CrossRef]
36. Yu, B.; Song, X.; Guan, F.; Yang, Z.; Yao, B. K-nearest neighbor model for multiple-time-step prediction of short-term traffic condition. *J. Transp. Eng.* **2016**, *142*, 04016018. [CrossRef]
37. Du, W.; Zhang, Q.; Chen, Y.; Ye, Z. San urban short-term traffic flow prediction model based on wavelet neural network with improved whale optimization algorithm. *Sustain. Cities Soc.* **2021**, *69*, 102858. [CrossRef]
38. Li, L.; Yang, Y.; Yuan, Z.; Chen, Z. A spatial-temporal approach for traffic status analysis and prediction based on bi-lstm structure. *Mod. Phys. Lett. B* **2021**, *35*, 2150481. [CrossRef]
39. Wang, K.; Ma, C.; Qiao, Y.; Lu, X.; Hao, W.; Dong, S. A hybrid deep learning model with 1DCNN-LSTM-Attention networks for short-term traffic flow prediction. *Phys. A* **2021**, *583*, 126293. [CrossRef]
40. Chen, X.; Chen, H.; Yang, Y.; Wu, H.; Zhang, W.; Zhao, J.; Xiong, Y. Traffic flow prediction by an ensemble framework with data denoising and deep learning model. *Phys. A* **2021**, *565*, 125574. [CrossRef]
41. Cheng, Z.; Lu, J.; Zhou, H.; Zhang, Y.; Zhang, L. Short-Term Traffic Flow Prediction: An Integrated Method of Econometrics and Hybrid Deep Learning. *IEEE Trans. Intell. Transp. Syst.* **2021**, 1–14. [CrossRef]
42. Zheng, H.; Lin, F.; Feng, X.; Chen, Y. A hybrid deep learning model with attention-based conv-LSTM networks for short-term traffic flow prediction. *IEEE Trans. Intell. Transp. Syst.* **2020**, *22*, 6910–6920. [CrossRef]
43. Li, H. Research on prediction of traffic flow based on dynamic fuzzy neural networks. *Neural Comput. Appl.* **2016**, *27*, 1969–1980. [CrossRef]
44. Huang, W.; Song, G.; Hong, H.; Xie, K. Deep architecture for traffic flow prediction: Deep belief networks with multitask learning. *IEEE Trans. Intell. Transp. Syst.* **2014**, *15*, 2191–2201. [CrossRef]
45. Yang, H.F.; Dillon, T.S.; Chen, Y.P. Optimized structure of the traffic flow forecasting model with a deep learning approach. *IEEE Trans. Neural Netw. Learn. Syst.* **2016**, *28*, 2371–2381. [CrossRef]
46. Lu, S.; Zhang, Q.; Chen, G.; Seng, D. A combined method for short-term traffic flow prediction based on recurrent neural network. *Alex Eng. J.* **2021**, *60*, 87–94. [CrossRef]
47. Ho, L.W.; Lie, T.T.; Leong, P.T.; Clear, T. Developing offshore wind farm siting criteria by using an international Delphi method. *Energy Policy* **2018**, *113*, 53–67. [CrossRef]
48. Noorollahi, Y.; Yousefi, H.; Mohammadi, M. Multi-criteria decision support system for wind farm site selection using GIS. *Sustain. Energy Technol. Assess.* **2016**, *13*, 8–50. [CrossRef]
49. Florian, M.; Sørensen, J.D. Risk-based planning of operation and maintenance for offshore wind farms. *Energy Procedia* **2017**, *137*, 261–272. [CrossRef]
50. Martin, R.; Lazakis, I.; Barbouchi, S.; Johanning, L. Sensitivity analysis of offshore wind farm operation and maintenance cost and availability. *Renew. Energy* **2016**, *85*, 1226–1236. [CrossRef]
51. Santos, F.P.; Teixeira, A.P.; Guedes Soares, C. Maintenance planning of an offshore wind turbine using stochastic Petri nets with predicates. *J. Offshore Mech. Arct. Eng.* **2018**, *140*, 021904. [CrossRef]
52. Höfer, T.; Sunak, Y.; Siddique, H.; Madlener, R. Wind farm siting using a spatial Analytic Hierarchy Process approach: A case study of the Städteregion Aachen. *Appl. Energy* **2016**, *163*, 222–243. [CrossRef]
53. Hooper, T.; Hattam, C.; Austen, M. Recreational use of offshore wind farms: Experiences and opinions of sea anglers in the UK. *Mar. Policy* **2017**, *78*, 55–60. [CrossRef]
54. Cervera, M.A.; Ginesi, A.; Eckstein, K. Satellite-based vessel Automatic Identification System: A feasibility and performance analysis. *Int. J. Satell. Commun. Netw.* **2011**, *29*, 117–142. [CrossRef]
55. Sang, L.Z.; Wall, A.; Mao, Z.; Yan, X.P.; Wang, J. A novel method for restoring the trajectory of the inland waterway ship by using AIS data. *Ocean Eng.* **2015**, *110*, 183–194. [CrossRef]
56. Benesty, J.; Chen, J.; Huang, Y.; Cohen, I. Pearson correlation coefficient. In *Noise Reduction in Speech Processing*; Springer: Berlin/Heidelberg, Germany, 2009; pp. 1–4.
57. Zuo, F.; Li, Y.; Johnson, S.; Johnson, J.; Varughese, S.; Copes, R.; Chen, H. Temporal and spatial variability of traffic-related noise in the City of Toronto, Canada. *Sci. Total Environ.* **2014**, *472*, 1100–1107. [CrossRef] [PubMed]
58. Fu, R.; Zhang, Z.; Li, L. Using LSTM and GRU neural network methods for traffic flow prediction. In Proceedings of the 2016 31st Youth Academic Annual Conference of Chinese Association of Automation (YAC), Wuhan, China, 11–13 November 2016; IEEE: Piscataway, NJ, USA; pp. 324–328.
59. Zhang, D.; Kabuka, M.R. Combining weather condition data to predict traffic flow: A GRU-based deep learning approach. *IET Intell. Transp. Syst.* **2018**, *12*, 578–585. [CrossRef]
60. Wang, S.; Zhao, J.; Shao, C.; Dong, C.D.; Yin, C. Truck traffic flow prediction based on LSTM and GRU methods with sampled GPS data. *IEEE Access* **2020**, *8*, 208158–208169. [CrossRef]
61. Cho, K.; Van Merriënboer, B.; Bahdanau, D.; Bengio, Y. On the properties of neural machine translation: Encoder-decoder approaches. *arXiv* **2014**, arXiv:1409.1259.

62. Liang, M.; Zhan, Y.; Liu, R.W. MVFFNet: Multi-view feature fusion network for imbalanced ship classification. *Pattern Recognit. Lett.* **2021**, *151*, 26–32. [CrossRef]
63. Wang, Y.; Liao, W.; Chang, Y. Gated recurrent unit network-based short-term photovoltaic forecasting. *Energies* **2018**, *11*, 2163. [CrossRef]
64. Zheng, G.; Chai, W.K.; Katos, V.; Walton, M. A joint temporal-spatial ensemble model for short-term traffic prediction. *Neurocomputing* **2021**, *457*, 26–39. [CrossRef]

Journal of
Marine Science and Engineering

MDPI

Article

Ship Intention Prediction at Intersections Based on Vision and Bayesian Framework

Qianqian Chen [1], Changshi Xiao [1,2,3,*], Yuanqiao Wen [2,4], Mengwei Tao [1] and Wenqiang Zhan [3]

[1] School of Navigation, Wuhan University of Technology, Wuhan 430063, China; chenqq@whut.edu.cn (Q.C.); tao_mw@163.com (M.T.)
[2] National Engineering Research Center for Water Transport Safety, Wuhan 430063, China; yqwen@whut.edu.cn
[3] Institute of Ocean Information Technology, Shandong Jiaotong University, Weihai 264200, China; 210127@sdjtu.edu.cn
[4] Intelligent Transportation Systems Research Center, Wuhan University of Technology, Wuhan 430063, China
* Correspondence: cs_xiao@hotmail.com

Abstract: Due to the high error frequency of the existing methods in identifying a ship's navigational intention, accidents frequently occur at intersections. Therefore, it is urgent to improve the ability to perceive ship intention at intersections. In this paper, we propose an algorithm based on the fusion of image sequence and radar information to identify the navigation intention of ships at intersections. Some existing algorithms generally use the Automatic Identification System (AIS) to identify ship intentions but ignore the problems of AIS delay and data loss, resulting in unsatisfactory effectiveness and accuracy of intention recognition. Firstly, to obtain the relationship between radar and image, a cooperative target composed of a group of concentric circles and a central positioning radar angle reflector is designed. Secondly, the corresponding relationship of radar and image characteristic matrix is obtained after employing the RANSAC method to fit radar and image detection information; then, the homographic matrix is solved to realize radar and image data matching. Thirdly, the YOLOv5 detector is used to track the ship motion in the image sequence. The visual measurement model based on continuous object tracking is established to extract the ship motion parameters. Finally, the motion intention of the ship is predicted by integrating the extracted ship motion features with the position information of the shallow layer using a Bayesian framework. Many experiments on real data sets show that our proposed method is superior to the most advanced method for ship intention identification at intersections.

Keywords: ship intention identification; AIS; RANSAC; Bayesian framework; YOLO; intersection

Citation: Chen, Q.; Xiao, C.; Wen, Y.; Tao, M.; Zhan, W. Ship Intention Prediction at Intersections Based on Vision and Bayesian Framework. *J. Mar. Sci. Eng.* **2022**, *10*, 639. https://doi.org/10.3390/jmse10050639

Academic Editor: Md Jahir Rizvi

Received: 15 March 2022
Accepted: 27 April 2022
Published: 7 May 2022

1. Introduction

Due to the combination of information technology and ship technology, ship safety has experienced rapid developments. Although a variety of water situation awareness and intelligent collision avoidance technologies have been developed, ship collision accidents still occur frequently in ports and other complex waters, resulting in severe economic losses and environmental pollution. According to the Shanghai Maritime Safety Administration report, 47% of ship collisions occurred at waterway intersections (Shanghai Maritime Court, Shanghai, China, 2018). These accidents occur mainly because the seafarers are unable to correctly identify or predict the motion intentions of other ships in the intersections. Therefore, it is of significance to study the vessel intention perception methods at intersections and channels.

Although intention prediction has received increasing research in the field of road traffic in recent years, there has been little research on the intention prediction of maritime safety of ships. Moreover, the existing research on the prediction of ground vehicle driving intention cannot be directly applied to ships. For example, ships can travel over relatively wide areas without being restricted by roads. In addition, there is no fixed channel to separate ships with different maneuvers, which increases the uncertainty of the ship's trajectory, thus increasing the difficulty of ship navigation prediction. Beyond that, the movement of ships is susceptible to environmental conditions (such as wind, waves, and ocean currents) and surrounding ships in waterways. As a result, ships may exhibit different motion patterns even if they travel along the same route, further complicating the prediction of ship intentions. Most importantly, ships cannot perform maneuvering operations such as sudden stops, turns, or reversals unlike ground vehicle movements. Furthermore, it takes more time and space for a ship to transition from one state of motion to another. Therefore, the intention prediction method of ground vehicles cannot correctly describe the long-distance movement of ships.

In order to solve the problem of rapid identification of ship intentions, we propose a scheme to update and predict ship intentions based on channel video surveillance and radar data under the Bayesian framework. Firstly, a cooperative target composed of a group of concentric circles and a centrally positioned radar angle reflector was designed to obtain the relationship between radar and image data in this paper. Secondly, the RANSAC method was used to fit the radar and the image detection information, and then the homographic matrix from the radar coordinate system to the image coordinate system was obtained. Thirdly, a visual measurement model based on continuous object tracking was established to extract the motion parameters of the ship. Finally, the motion intention of the ship was predicted by integrating the extracted ship motion features with the position information of the shallow layer using the Bayesian framework. The main contributions of this work are threefold:

1. A ship motion model based on monocular vision is established, which can extract ship motion parameters real-time and is not affected by AIS delay;
2. The accuracy of ship intention prediction can be further improved by using the environmental information of the channel effectively;
3. A dynamic Bayesian model is established that can accurately identify the ship's intention.

The remainder of this paper is organized as follows. Some related works are introduced in Section 2. In Section 3, we will discuss the methodology of our algorithm. Experimental and model prediction performances are reported in Section 4. Finally, the work is concluded in Section 5.

2. Related Work

There are several relevant studies in the literature on the navigational intention prediction of ships. For example, Tang Huang et al. [1] from Dalian Maritime University found that there were field errors and obvious noise in the original AIS track data set of ships and that these data were often irregular timing data. Therefore, they proposed an improved trajectory similarity measure with a directional resolution to improve the accuracy of trajectory clustering. On this basis, they used the Long Short-Term Memory (LSTM) neural network to accurately predict ship behavior. Steidel et al. [2] believe that traditional maritime abnormal behavior detection and the prediction of a ship's navigational intention mainly focus on the development of methods to extract typical ship motion patterns from historical traffic data without considering contextual information. Therefore, he proposed a method to predict a ship's intention by combining historical ship traffic data with information about shipping routes. Later, Pietrzykowski used AIS online data to analyze if the ship's operation and identification behavior were potentially risky [3]. Based on this, Zhang Hong used a data mining method to analyze the AIS data of tuna purse-seine fishing ships to identify the operation state of tuna purse-seine fishing ships [4]. In addition, Gao et al., for example, constructed a Bi-LSTM-RNN model, which can be used for AIS date and time series feature extraction and online parameter adjustment to realize online real-time prediction of ship behavior. This algorithm enhanced the correlation between historical data and future data, thus improving the prediction accuracy [5]. Later, Kawamura used a GPU SPH simulator to predict the motion of a 6-dOF ship in harsh water transport conditions [6]. The study of Murray et al. deconstructed ship behaviors into clusters according to specific regions, and each cluster contained similar trajectory behavior characteristics. A deep learning framework was used to predict ship behavior for each specific cluster. This algorithm believes that the future navigation trajectory of a ship can be predicted by inputting the past trajectory of a ship in a specific area. However, the influence of environmental factors was not considered in this method, which is somewhat different from the actual situation of ship navigation [7].

Ma et al. conducted two valuable studies [8,9]. First, they proposed to extract the mutual behavior between ships from AIS tracking data to capture the spatial dependence between ships that meet and then predicted the ship's intention and collision risk based on a LSTM network. Later, they devised a data-driven approach that linked movement behavior to future and early risks, and predicted a ship's collision risk by classifying behavior into appropriate risk levels. In reference [10], it was assumed that all ships in the scene could not share the same motive and decision of motion, and the observation-inference-prediction-decision (OIPD) model was proposed to avoid collisions by repeatedly iterating the difference between observation and prediction information. Tang [11] used a grid-based method to discretize the historical AIS data into track segments and established a probabilistic directed graph model. Through this model, the state characteristics of each node ship can be counted, and the navigation state of the ship can be detected by the probability graph obtained.

Some recent studies include, for example, [12] a proposed a new spatial-temporal geographic method to solve the risk behavior of multi-ship collision based on ship movement. The direction-constrained space-time prism was used to characterize the possibility of the ship's interaction, which enabled the assessment of the ship's potential collision risk. Suo proposed a modeling method based on a cellular automata simulation to analyze and evaluate maritime traffic risks in a port environment in real-time [13]. Then, Alvarellos et al. established a deep neural network to predict the damage to ships anchored in ports caused by environmental effects within 72 h by considering the influence of ship size, sea state, and weather conditions on ship motion [14]. Based on the above work, Xue proposed a knowledge learning model under multienvironment constraints to analyze ship risks in port waters and improve the decision-making basis for autonomous navigation of intelligent ships [15].

Some other works are also of great research value in predicting ship intentions. Alizadeh proposed a point-based and track-based model, considering the constant distance between the target and the sample trajectory. The LSTM method was then used to measure the dynamic distance between the target and the sample trajectory, and predict the short-term and long-term trajectory of the ship [16]. Subsequently, Praczyk et al. extracted spatial direction (Euler Angle) from the inertial navigation system and used an improved neural network to predict ship behavior [17]. Zissis used an artificial neural network to predict the future position, speed, and heading behaviors of ships on a large scale based on historical AIS data [18].

Mining-related behavior characteristics from AIS data is a general method to study ship behavior. However, due to the large ship flow in inland river confluence waters, AIS data bandwidth is insufficient, and uploading is not timely, which leads to the low real-time prediction of ship navigation intention. It is easy to miss the best decision time and increase the risk of ship collision. In this paper, the motion intention of ships is predicted by combining visual and radar data based on the Bayesian framework, and the problem of accuracy and real-time intention prediction is successfully solved. Firstly, to obtain the relationship between radar and image, a cooperative target composed of a group of concentric circles and a central positioning radar angle reflector is designed. Secondly, the corresponding relationship of radar and image characteristic matrix was obtained after employing the RANSAC method to fit the radar and the image detection information. Then, the homography matrix was solved to realize the radar and the image data matching. Thirdly, the extended YOLO detector was used to track the ship motion in the image sequence, and the visual measurement model based on continuous object tracking was established to extract the ship motion parameters. Finally, the motion intention of the ship was predicted by integrating the ship motion features extracted with the position information of the shallow layer using the Bayesian framework. At the same time, in order to verify the feasibility of the proposed method, experimental scenarios were designed according to the scene characteristics of the intersection area of Wuhan Yangtze River and Han River, and the ship intention recognition algorithm was verified successfully.

3. Methodology

The proposed method takes a dynamic Bayesian algorithm as the main framework and introduces a new image measurement method to extract the motion characteristics of the ship. To better introduce the idea of the proposed algorithm, this section is divided into three sub-sections. In Section 3.1, we will describe how to process sequential images to measure ship speed and angular velocity, in which the measurement model will also be introduced in detail. In Section 3.2, we will discuss how to map an object in the image sequence to an electronic chart. In Section 3.3, how to use a dynamic Bayesian algorithm to estimate the navigation intention will be introduced in detail. The overall description of our algorithm is shown in Figure 1.

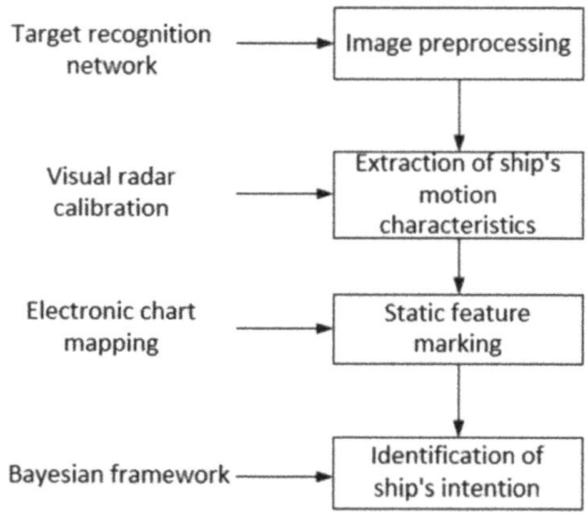

Figure 1. Framework based on vision and Bayesian framework for intent prediction of vessels.

3.1. Radar and Vision Fusion Calibration

Monocular vision and millimeter-wave radar signal have their own characteristics; for example, monocular vision has the advantages of simple structure and strong robustness, while millimeter-wave radar has the advantages of accurate positioning, etc. [19]. Compared with the measurement results of a single sensor, radar and vision fusion measurements can obtain more ship attitude and motion information [20]. However, this method usually requires calibrating radar and visual measurement results first. Therefore, it is necessary to design a cooperative objective for calibration that meets the following requirements:

1. Visual signals are sensitive and easy to accurately detect the position of fixed points;
2. The radar echo signals are strong and easy to accurately detect the position of the fixed point;

Therefore, we creatively designed a radar and vision cooperative target, as shown in Figure 2. The cooperative target consists of a set of concentric circles and a centrally located radar reflector. In order to obtain the sub-pixel coordinates of the center projection point, a recursive algorithm based on harmonic relation was proposed. Then, the radar points and image points were fitted to obtain the linear correspondence to accurately obtain the homographic matrix of radar coordinates to image coordinates.

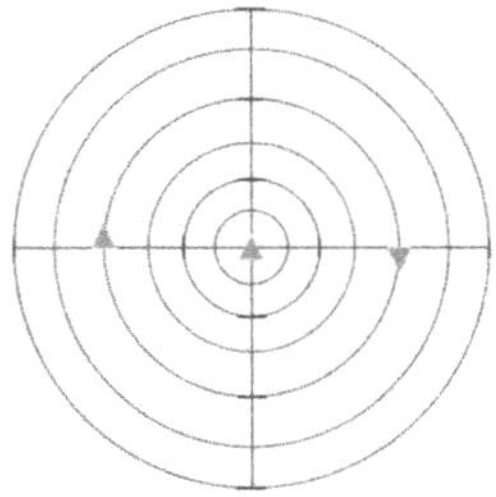

Figure 2. Radar-Vision cooperative target.

Millimeter-wave radar is the main component of the surface target information acquisition system. The millimeter-wave radar used in this paper is the ARS300 series radar provided by Continental, Germany, which operated at 77 GHz and can detect up to 40 targets simultaneously, and is equipped with a special controller. This millimeter-wave radar has the characteristics of small size, strong anti-interference ability, and stable detection. Its performance indicators are shown in Table 1.

Table 1. Radar performance indicators.

Indicators	Detection Distance	Working Frequency	Range Accuracy	Speed Range	Detection Range
Performance	2 km	77 GHz	0.5 m	265 km/h	near 60° far 20°

The millimeter wave radar detection is divided into short-range wave and long-range wave ranges. The long-range wave mainly captures distant targets and improves detection distance; The short-range wave mainly expands the radar perspective and reduces the dead detection zone. The millimeter-wave radar detection range is shown in Figure 3.

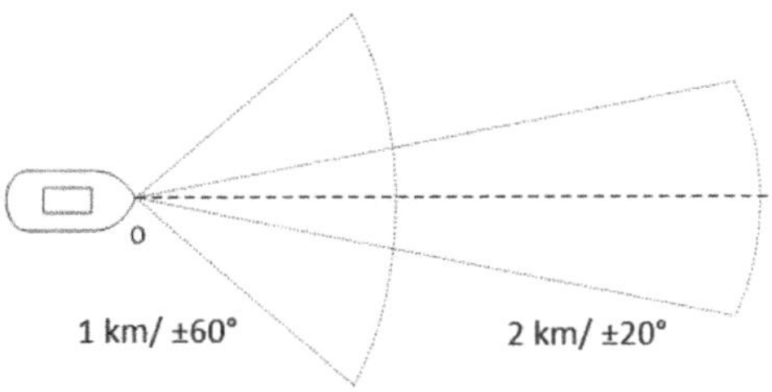

Figure 3. Radar detection range.

Assuming that the homogeneous coordinate of a point on the radar detection plane is $p_r = (x_r, y_r, 1)^T$, and the homogeneous coordinate of a point on the image plane is $p_i = (u, v, 1)^T$, Equation (1) indicates the conversion relationship from the radar coordinate system to the image coordinate system.

$$sp_i = Hp_r \tag{1}$$

where, s is a scalar and H is a 3×3 reversible homography matrix with 8 degrees of freedom. The purpose of calibration is to estimate the homography matrix H. After moving linearly in the common field of vision of the camera and radar during calibration, a series of radar and image points can be captured, and then the image sequence was processed to extract the coordinates of the concentric circles in each image. It should be emphasized here that the concentric circle center coordinates corresponded to the center of the image coordinates of the radar reflector. Finally, the least-squares method was used to fit the image point sequence and the radar point sequence. It was assumed that the connection line of the continuous image center point is l and that of the radar detection point is L, which has the following relationship:

$$l^T p_i = 0,\ L^T p_r = 0 \tag{2}$$

Combined with Equations (1) and (2), the relationship between the straight lines in the image plane and the radar plane is shown in Equation (3):

$$sL = H^T l \tag{3}$$

Solving Equation (3) requires at least four sets of corresponding lines. We found that there was no need to align the time stamp of the radar data and the single frame of the video series, and the linear motion of the calibrated target can be easily captured by the radar due to the motion prediction algorithm adopted when the radar tracks the target. Therefore, the homographic matrix H can be calculated by using the linear-based homographic estimation method.

Figure 4 is the result of calibration using the method proposed in this section, where the blue dot represents the radar detected target and the red dot represents the pixel coordinates, which are consistent with the position of the ship detected in the image to achieve the radar and visual fusion calibration.

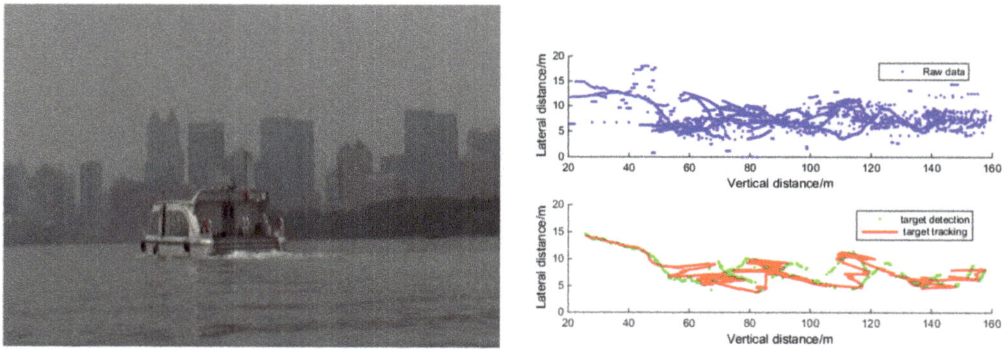

Figure 4. Radar and visual calibration results.

3.2. Object Detection

The purpose of image preprocessing is to capture every frame of the video and detect the moving ships after removing the random noise from the images. At present, the common object detection methods include the frame difference method [21], the motion modeling method [22], and the deep learning method [23]. These methods are simple and fast. However, in contrast, the frame difference and motion modeling methods are susceptible to environmental changes such as light changes and noise.

Through experiments, we found that the You Only Look Once (YOLO) series methods based on deep learning had the best detection results. Therefore, in the following paper, we employed a YOLO v5 network to extract moving ships in images [24]. YOLO v5 detection network is a typical object detection network, which has been widely used in many detection tasks and can meet the requirements of real-time detection of moving ships. Although this method is faced with the challenge of long training times, we can train the network offline in the actual ship detection task, so the training time cost is acceptable for the ship detection task.

This study uses a method based on supervised learning to detect the ship. Specifically, the ship detector based on YOLO is built, and a foreground recognition module is inserted into the detector to ensure that for each detected ship object, the detector would output its specific position in the image coordinate system (usually represented by a rectangular box). In the training process, the YOLO detector model was firstly pre-trained on a public object detection dataset [20], which provided ship-bounding boxes (1000 training samples in total). The parameters of the detector model were fine-tuned by our collected ship dataset (manually annotating ship location) to adapt to the specific scene of intersecting waters.

As shown in Figure 5, the trained detector can successfully detect the moving ship from the images collected by the shore-based camera and give the specific detection bounding box.

Figure 5. Moving ships' detection results.

3.3. Visual Measurement Model of Ship Attitude

The position and speed of ships are physical quantities used to describe a ship's motion state. At the same time, combined with the relationship between static traffic environment and ship position as well as the change of speed, the ship's future motion intention can be effectively predicted. The posture and motion features of ships can be extracted from vision sensors and radar sensors. Compared with the two-dimensional motion features, images often contain more information. Considering that shore-based sensors collect the data in this study, we only extracted the attitude characteristics of the ship from the images collected by a monocular camera. The measurement model of the ship's motion characteristics is shown in Figure 6.

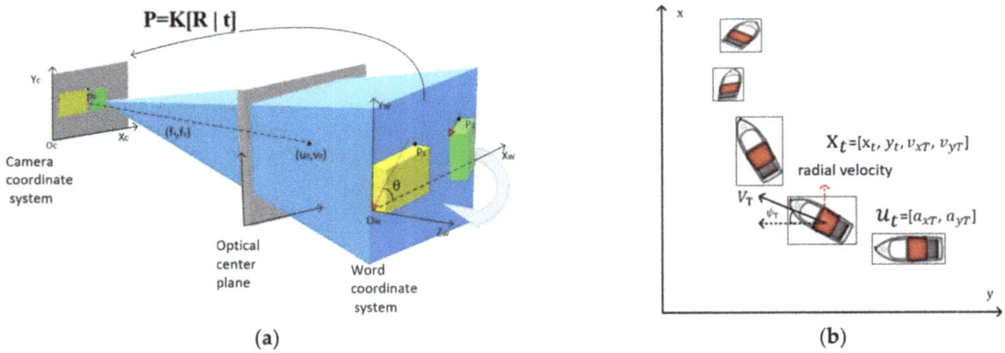

Figure 6. Measurement model of ship motion characteristics. (**a**) Three-dimensional model of ship image coordinate system to world coordinate system transformation. (**b**) The measurement model of ship motion parameters.

Assuming the ship is a cuboid, we built a visual measurement model on this basis. Considering that a cuboid has three visible sides, we defined three coordinate systems, namely a physical coordinate system $oxyz$, a camera coordinate system $o'x'y'z'$, an image coordinate system, XOY. Suppose (x, y, z), (x', y', z') and (x, y) are the coordinates of a point in the object coordinate system, camera coordinate system and image point respectively, then:

$$\begin{bmatrix} XH \\ YH \\ H \end{bmatrix} = \begin{bmatrix} F\,0\,u\,0 \\ 0\,F\,v\,0 \\ 0\,0\,1\,0 \end{bmatrix} \begin{bmatrix} x' \\ y' \\ z' \\ 1 \end{bmatrix} \tag{4}$$

Suppose that the coordinates of the origin of the camera coordinate system in the physical coordinate system are $(X_c,\ Y_c, Z_c)$, then the transformation of coordinate points from the physical coordinate system to the camera coordinate system is:

$$(x'\ y'\ z'\ 1)^T = A(x\ y\ z\ 1)^T \tag{5}$$

where T indicates transpose,

$$\mathrm{A} = R_x(\theta)R_y(\varnothing)R_z(\varphi)D = \begin{bmatrix} a\,b\,c\,p \\ d\,e\,f\,q \\ g\,h\,i\,r \\ 0\,0\,0\,1 \end{bmatrix} \tag{6}$$

$Rx(\theta)$, $Ry(\varnothing)$, and $Rz(\varphi)$ are matrices for rotation around the x, y, and z axes, respectively. While θ, $\varnothing$, and φ are corresponding rotations, D is a translation matrix with a translation of $-(x_c, y_c, z_c)$ and meets the following requirements:

$$-\pi < \theta \le \pi,\ -\frac{\pi}{2} \le \varnothing \le \frac{\pi}{2},\ 0 \le \varphi < 2\pi \tag{7}$$

Substituting Equation (5) into Equation (4), the following relation can be obtained:

$$(XH\ YH\ H)^T = T(x\ y\ z\ 1)^T \tag{8}$$

where,

$$\mathrm{T} = \begin{bmatrix} aF+gu & bF+hu & cF+iu & pF+ru \\ dF+gv & eF+hv & fF+iv & qF+rv \\ g & h & i & r \end{bmatrix} = \begin{bmatrix} t_{11} & t_{12} & t_{13} & t_{14} \\ t_{21} & t_{22} & t_{23} & t_{24} \\ t_{31} & t_{32} & t_{33} & t_{34} \end{bmatrix} \tag{9}$$

In short, the problem to be solved was given a group of known physical coordinates of the object to obtain the camera coordinate parameters: XC, YC, ZC, U, V, and F, and the coordinates of the object were converted from the physical coordinate system to the image coordinate system. Assuming that the height of the ship remained constant, then $z = 0$, $\theta = 0$, $\varnothing = 0$.

3.4. Static Environmental Parameter Measurement Based on Environmental Message

In this study, ships, channel obstacles, and intersection corners were labeled with rectangular boxes to show each object's relative and absolute positions.

The information of the traffic environment layer affects and restricts a ship's path choice, which can be divided into the static environment and dynamic environment features. The static environment features refer to the relative position relationship between ships and static environment space, such as channel structure and static obstacles on the water surface, which mainly affect the long-term trajectory planning of ships. The short-term behavior of ships is also affected by the behavior of dynamic objects in the traffic environment, such as other ships, other moving obstacles, etc., which are called dynamic environment features. Both static environment characteristics and dynamic environment characteristics affect the behavior of ships. When identifying the intention of the ship in the intersection area, it is considered that the static environment feature that affects the intention mainly refers to whether the ship arrives at the intersection area. Only when arriving at the intersection area, the possibility of turning will occur. The minimum relative longitudinal distance between the ship and other ships was selected for a dynamic environment feature, which is used to judge whether the ship is in danger and whether it will choose to turn at this moment. In addition, the relationship between the ship and other static environment features, such as channel structure and static obstacles on the channel, will be taken into account in predicting the ship's future path.

In order to obtain the information from a forward channel structure, ships mainly rely on the electronic chart and an automatic ship identification system to locate. As the research scene selected in this study is a fixed water area, the electronic chart can be used to obtain map information. The shape of the intersection area can be described by map information, and then the relative position relationship between the ship and the intersection area can be obtained by combining the position relationship between the ship and each corner of the intersection area marked by millimeter-wave radar data. That is, the ship's position in the global coordinate system with the intersection center as the origin was determined, and the 2 km range of the channel at the intersection is the intersection area Therefore, the longitudinal distance between the ships entering the intersection area from different directions and the red dotted line in the corresponding intersection area can be used as an observation variable to represent the relative position relationship between the ships and the intersection area. In addition, the areas on the water that are not allowed to pass by ships, such as stationary ships or obstacles in the channel, can also be determined in the established 2D map model.

In this study, only ships were considered as traffic participants in the traffic environment. Therefore, the relationship between the ship and the dynamic traffic environment can be considered as the relative relationship between the target ship and other ships. Here, we considered the minimum distance between the target ship and other ships. that is, the minimum distance between the target ship and other ships continuing at their current speed. According to the relative position relationship between the target ship and other ships, and the speed information of the ship collected by millimeter-wave radar, it was used to represent the danger degree of the current situation, that is, whether there was a potential collision risk between the target ship and other ships.

3.5. Ship Intention Identification Based on Static and Dynamic Parameters

In Section 3.1, we extracted the multiple features of ships and determined the factors that affect the intentions of ships. On this basis, this paper proposes a basic idea to build an intention recognition model framework from the three levels (dynamic environment, static environment, and object factors), constructs a dynamic Bayesian network, and describes the intention inference algorithm in detail, which can be used for the intention recognition of ships in confluence waters of inland rivers.

The Bayesian Network (BN) is also called the Belief Network or Directed Acyclic Graph Model (DAGM). Since ship intention needs to be inferred by combining the factors related to ship intention, each node in the Bayesian network corresponds to intent-related factors, observation quantity, and ship intention, respectively. The probability distribution of each variable was inferred by establishing the probability relationship between nodes to realize the ship intention identification.

Figure 7 shows a dynamic Bayesian network where the rectangular nodes represent discrete variables and hidden variables. The shaded rectangular nodes represent intention variables, and the unshaded rectangular nodes represent intent-related factor variables. The circular nodes are continuous variables, i.e., observed quantities. In this network, the probability distribution values of variables were updated by receiving observations at each moment, and the conditional probability relationship between observation nodes and intent-related factor nodes was obtained by prior knowledge and sample data training.

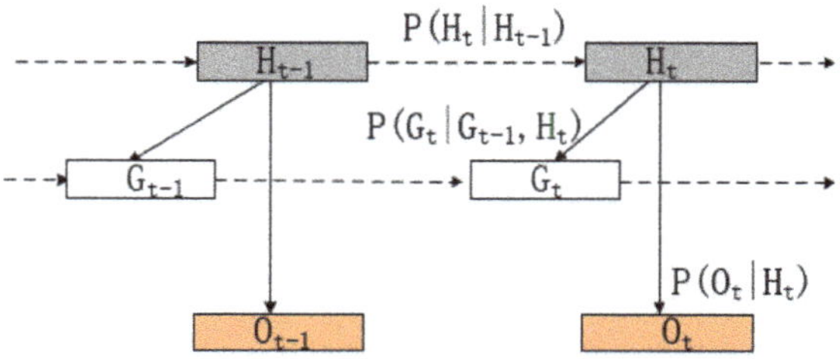

Figure 7. Dynamic Bayesian networks for intention recognition.

Variables in the dynamic Bayesian network mainly included the conditional variable set H and observation variable set O. Conditional variables are all discrete variables that satisfy the Markov hypothesis, and the same node has transition probability relationships at adjacent moments.

The transition probability relation of all nodes in the set of conditional variables can be expressed as:

$$P(H_t \middle| H_{t-1}) = P(H_t{}^{dyn} \middle| H_{t-1}{}^{dyn}) * P(H_t{}^{stat} \middle| H_{t-1}{}^{stat}) * P(H_t{}^{actc} \middle| H_{t-1}{}^{actc}) \quad (10)$$

According to the fitting results of the sample data, the probability distribution of the minimum distance D_{min} between the target ship and other ships conformed to the gamma distribution under the condition of dynamic environmental state H^{dyn}. This is shown in Figure 8 in a static environment state.

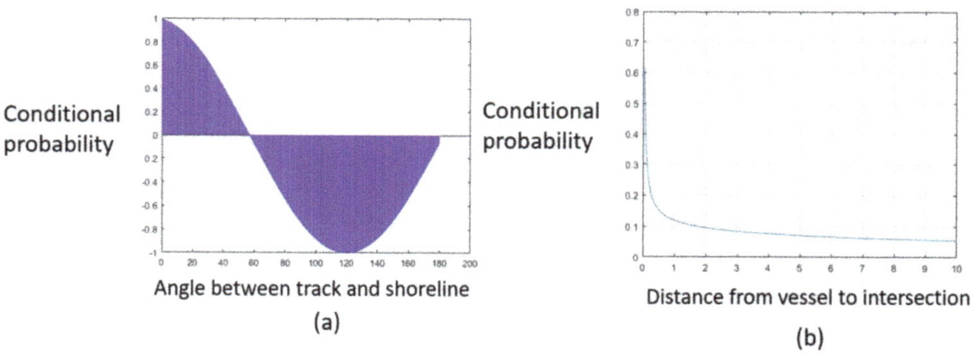

Figure 8. Probability distribution of environmental state quantity. (**a**) Conditional probability distribution of ship turning intention and shoreline angle. Negative probability indicates the probability that the ship turns in the opposite direction. (**b**) Conditional probability distribution of ship turning intention and distance to the intersection.

H^{stat} condition, the probability distribution of the longitudinal distance between the target ship and the intersection area conformed to a Gaussian distribution. Under the condition of continuous object state H^{stat}, the orientation of the target ship θ conformed to the Weibull distribution.

The dynamic Bayesian network can be regarded as a forward-filtering process, and the probability distribution of each variable can be updated when the observed variables are received so as to realize the process of intention inference. In the process of intention inference, the assumed density filtering is adopted as the inference tool. The process of intention inference can be divided into prediction and update.

1. Prediction

The prediction process is to predict the prior distribution of the current moment through the posterior distribution of the previous moment and the fixed transition probability. Based on the joint posterior distribution of the previous moment, the prior joint distribution of the current moment can be calculated according to the transfer probability, and the edge distribution can be obtained by adding, as shown in Equations (11) and (12).

$$\overline{P}(G_t, G_{t-1}, H_t, H_{t-1}) = P(G_t|G_{t-1}, H_t) * P(H_t|H_{t-1}) * \hat{P}_{t-1}(G_{t-1}, H_{t-1}) \tag{11}$$

$$\overline{P}(G_t, H_t) = \sum_{G_{t-1}} \sum_{H_{t-1}} \overline{P}(G_t, G_{t-1}, H_t, H_{t-1}) \tag{12}$$

2. Update

The updating process is to update the posterior distribution of the current moment according to the observed variables of the current moment. Based on the joint prior distribution obtained in the prediction step, the joint posterior distribution at the current moment can be calculated and added according to the observation variables and conditional probability relationship. The posterior distribution of the nodes obtained is shown in Equations (13) and (14).

$$\hat{P}(G_t, H_t) \propto P_t(O_t|H_t) * \overline{P}_t(G_t, H_t) \tag{13}$$

$$\hat{P}_t(G_t) = \sum_{H_t} \hat{P}(G_t, H_t) \tag{14}$$

4. Experiment and Result

This section includes the following parts: (1) experimental scenario design; (2) data set introduction; (3) simulation results; (4) real ships experiments; (5) model evaluation.

4.1. Experimental Scenario Design

In the experiment, the ship traffic scene at the intersection of the Hanjiang River and Yangtze River in Wuhan was selected as the experiment scene. This region covers longitudes 114°260′–114°301′ E and latitudes 30°538′–30°590′ N, as shown in Figure 9.

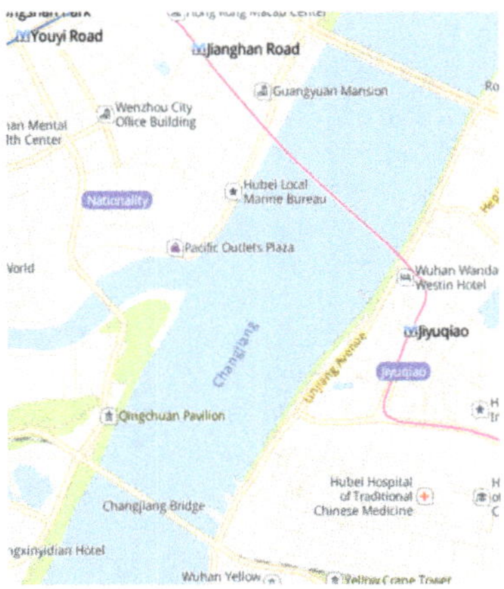

Figure 9. Schematic diagram of experimental data acquisition environment.

In the above inland river intersection scenario, the behaviors of ships were divided into three categories: straight, turn left, and turn right. In addition, combined with the intersection channel structure, it can be divided into up-straight, up-right turn, down-straight, down-left, left-up, and left-down, respectively, corresponding to the situation when ships enter the intersection area from the three different directions of the intersection. According to the above classification method, the collected data can be divided into six scenarios, and the schematic diagram of each scenario is shown in Figure 10. The specific definition of each scenario is as follows:

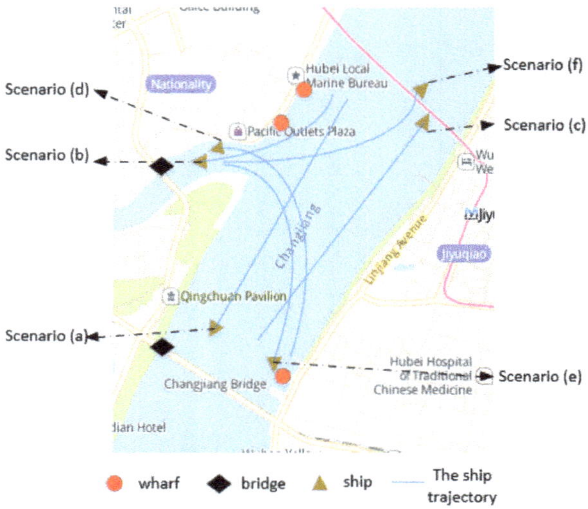

Figure 10. Schematic diagram of ship navigation traffic scene. Scenario (a): The ship enters the intersection area from the lower reaches of the Yangtze River, and the ship goes straight up. Scenario (b): The ship enters the intersection area from the lower reaches of the Yangtze River, and the ship turns right. Scenario (c): The ship enters the intersection area from the upper reaches of the Yangtze River, and the ship goes straight down. Scenario (d): The ship enters the intersection area from the upper reaches of the Yangtze River, and the ship turns left. Scenario (e): The ship enters the intersection area from the Han River, and the ship turns right, across the river, berthing. Scenario (f): The ship enters the intersection area from the Han River, and the ship turns right.

4.2. Data Set Introduction

Currently, the available ship datasets contain fewer scenarios, and the sampling frequency is usually low. The data set of this study is from the surveillance video of the intersection between the Yangtze River and the Han River, and the video data of 86 groups of ships were established. Each set of data recorded a ship that was close to the intersection area and intended to cross the intersection. The longest time of each set of data series was 5 min, and the shortest time was 1 min. After the data was processed by the frame difference method, there were a total of 800 images with a resolution of 1280 × 720. We randomly selected 600 of them as the training set and the remaining 200 as the test set. In the training set, we labeled the ships in it. We used this labeled training set to train the YOLO model and optimize the parameters of the network. When the video was fed into our detection system, the image was first obtained and preprocessed at a certain frame rate. Then, these images were inputted into the YOLO object detection network. Through this detection network, we can extract the target ship from each frame and obtain the position and motion parameters. Thus, the results of the detection and location of the ship in the video were obtained.

4.3. Simulation Results

In order to verify the feasibility of the algorithm proposed in this paper, two scenarios of the ship turning right and the ship going straight were simulated before the actual ship experiment. The simulation results are shown in Figures 11–13.

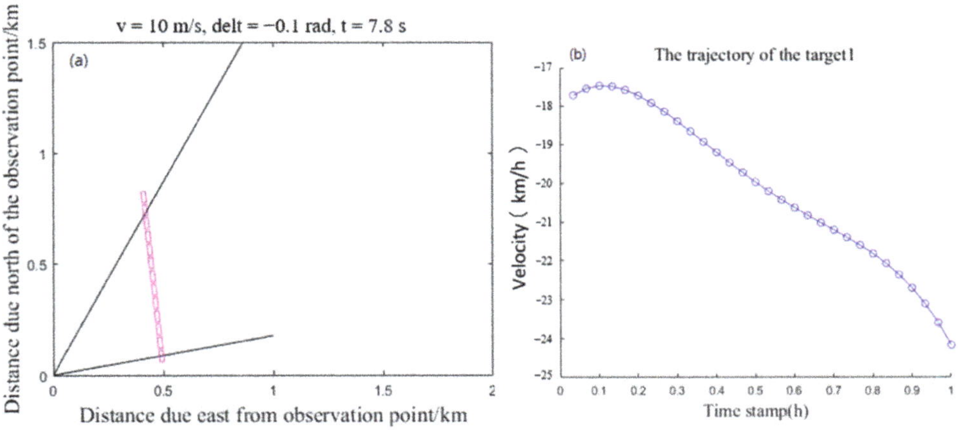

Figure 11. Simulation intention prediction results of rightward ship. (**a**) Simulation of ship movement trajectory. (**b**) Ship speed and direction.

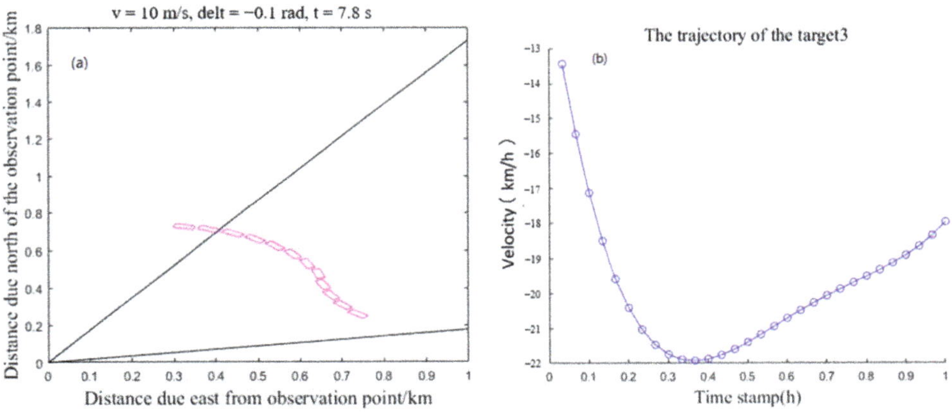

Figure 12. Simulation intention prediction results of direct ship. (**a**) Simulation of ship movement trajectory. (**b**) Ship speed and direction.

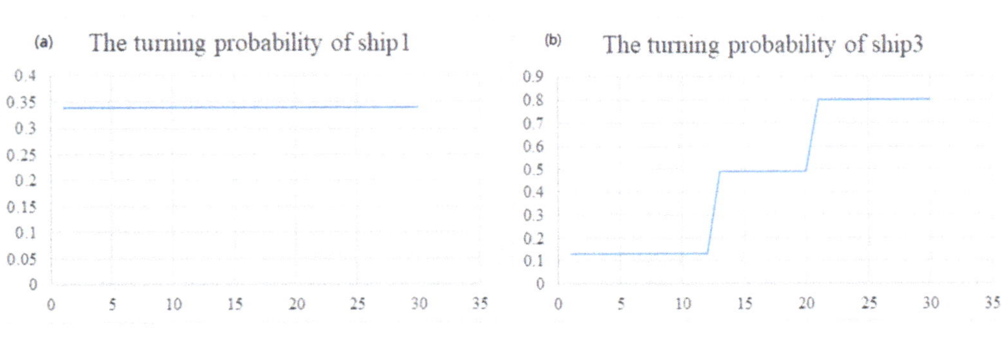

Figure 13. Simulation intention prediction results of direct ship. (**a**) The turning probability of ship1. (**b**) The turning probability of ship3.

Assuming that the speed of the ship remains stable when passing through the intersection area, Figure 11a shows that the ship traveled along a straight line. At this time, the included angle between the ship and the observation point gradually decreased, so the radial speed of the ship in Figure 11b gradually increased. Figure 12a shows that the ship turned right, and the included angle between the ship and the observation point first gradually decreased and then remained unchanged. Therefore, the radial velocity of the ship in Figure 12b first gradually increased and then changed smoothly. As shown in Figure 13, the conclusion obtained by calculating the turning probability of the ship is that the turning probability of the target in Figure 11 is 0.34, and it was determined that the turning will not occur. The turning probability of the target in Figure 12 is 0.81 and it was determined that the turning would occur. Therefore, from the simulation results, the ship intention recognition results were in line with the scene designed by the simulation experiment.

4.4. Real Ship Experiments

In order to more accurately reflect the actual ship navigation, several sets of real ship tests were carried out to evaluate the accuracy of the proposed model for predicting ship intentions in cross channels. We obtained radar and video data collected by monitoring equipment at the Yangtze River and Han River intersections from September 2020 to October 2020. The intersection and dense traffic flow make the study area a highly complex cross-channel, and after data preprocessing and sampling, 500 pieces of data containing moving ships were obtained. We illustrate the prediction results of ship sailing intention in typical scenarios.

The testing process of each trajectory is divided into the following three steps.

1. Target detection and tracking. The ships in the 0–2 km region from the intersection area are detected, and the pixel coordinates and radar coordinates of the ships are output.
2. Track segmentation. The 0–2 km area is divided into 10 track segments, and the length of each track segment is 0.2 km.
3. Predicted intent. The observed trajectory sequences are input to the HMM, LSTM, and the Ours models, and the predicted intent labels and probabilities for each intent class are determined. Return to step 2.

In Figure 14, the black line represents that the ship does not change speed and direction during navigation at the intersection, the green line represents the ship is turning right, and the red line represents the ship is turning left. Next, we analyzed the ship movement process in the scenario. We found that Ship 1 would turn right when entering the monitoring area, and the speed and direction would remain unchanged after turning, while Ship 2 and Ship 3 maintained their original speed and direction. The predicted results of our algorithm are consistent with the actual results (we can obtain the actual sailing results from the surveillance videos in advance by eye).

Figure 15 shows the movement process in this scenario, found that Ship 1 and Ship 2 kept their original speed and direction to move. Ship 3 first maintained speed and direction after entering the monitoring area and then turned right after arriving at the intersection area. After turning right, Ship 3 maintained its speed and direction for the rest of the voyage. The actual results are also consistent with the predicted results of the algorithm.

4.5. Model Evaluation

We conducted numerical experiments on HMM, LSTM, and the Ours ship intention identification method to predict the intent classes of the 500 tested trajectories. To quantitatively evaluate the prediction performance of the three models, we used accuracy and mean square error (MSE) as measures. The mean and variance of the accuracy and MSE at different distances to the precautionary area are listed in Table 2.

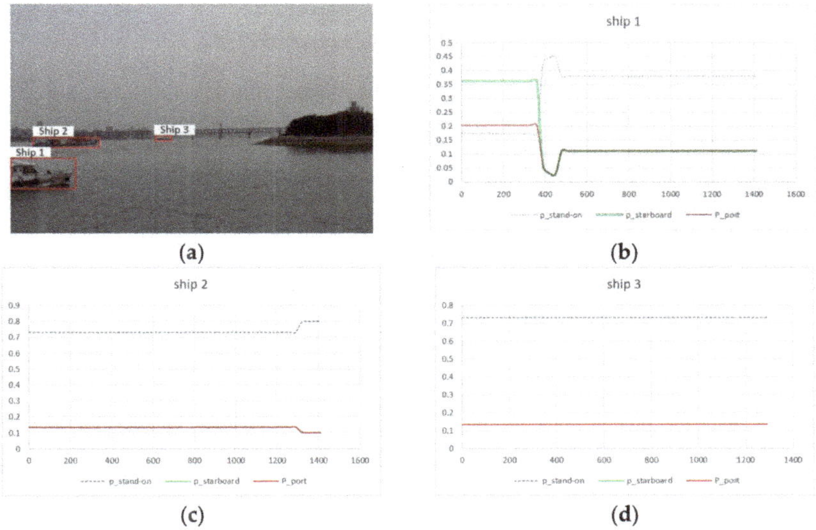

Figure 14. Prediction results of ship navigational intention in typical scenarios of intersection channel. (**a**) Ships detection results. (**b**) Intention identification probability of Ship 1. (**c**) Intention identification probability of Ship 2. (**d**) Intention identification probability of Ship 3.

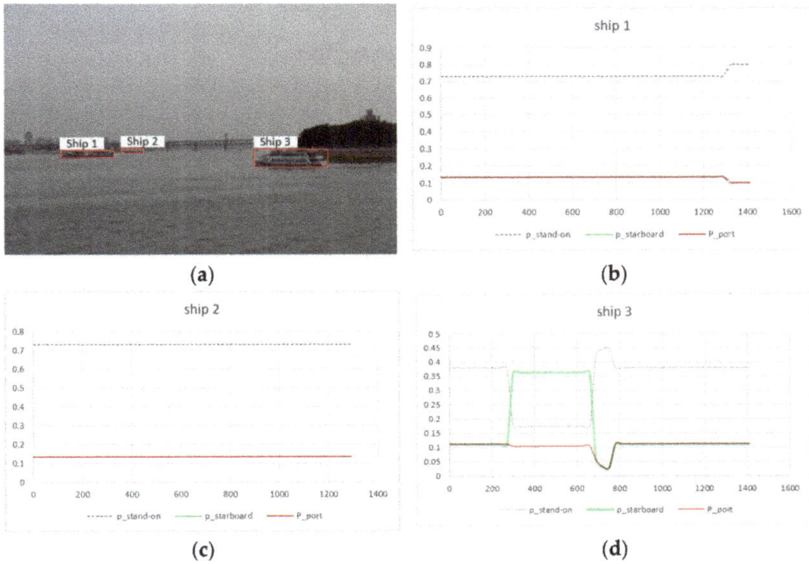

Figure 15. Prediction results of ship navigational intention in typical scenarios of the intersection channel. (**a**) Ships detection results. (**b**) Intention identification probability of Ship 1. (**c**) Intention identification probability of Ship 2. (**d**) Intention identification probability of Ship 3.

Table 2 shows the accuracy of the intention prediction of the three models. Each test takes into account three different distances (0, 1, and 2 km) from the intersection area. The proposed model is better than the LSTM model and HMM model. For example, when d = 0 km, the accuracy of our model is about 30% higher than HMM model and 10% higher than the LSTM model. At d = 1 km, the accuracy of our model is about 12% higher than

HMM and 14% higher than the LSTM model. When the distance from the intersection area increased to 2 km, the preparation rate of the three models decreased. This is an expected result because early prediction is more challenging than later prediction. According to the regulations of the People's Republic of China on river collision prevention, the safety distance of ships with a length of more than 30 m is 2 km, and that of ships with a length of less than 30 m is not less than 1 km. Therefore, our algorithm can effectively avoid collision by predicting the ship's intention within 2 km.

Table 2. Mean (variance) accuracy and MSE of intent prediction models at different distances to precautionary area.

Model	Measure	d = 2.0 km	d = 1.0 km	d = 0.0 km
HMM	Accuracy	0.522	0.619	0.612
	MSE	0.592	0.445	0.421
LSTM	Accuracy	0.446	0.594	0.814
	MSE	0.636	0.508	0.206
Ours	Accuracy	0.652	0.734	0.912
	MSE	0.541	0.343	0.016

5. Conclusions

The intention prediction of ships at intersections can effectively reduce the occurrence of ship collisions. However, the prediction accuracy of ship intention is easily affected by the validity and real-time data. In this paper, we propose an intention prediction model based on the fusion of video and radar data by using the Bayesian framework, and the model is verified on the real channel data at the intersection of the Yangtze River and the Han River. It was found that the ship motion intention is highly correlated with ship motion parameters and environmental factors. In order to effectively utilize this finding, we introduced a Bayesian framework and finally calculated the probability of ship motion intention by reasonably assuming the probability distribution of different factors. Due to the high acquisition frequency of an image and radar monitoring data, ship motion can be stably tracked, which accurately predicts ship intention and improves the real-time decision-making, thus effectively solving the problem of poor real-time prediction of ship intention caused by the data delay of other sensors.

It is undeniable that there are still some shortcomings in the algorithm. For example, it is necessary to integrate the environmental data, video, and radar data of specific areas to identify the ship's intention, which is difficult to be transplanted into the monitoring equipment of mobile ships. In future work, the cross-domain adaptive scene understanding method based on radar and video research will be considered. Then, the ship's intention recognition can be based on the results of dynamic scene understanding, which can be transplanted to mobile ships without static environment data fusion. This will provide a decision-making basis for intelligent ship collision avoidance.

Author Contributions: Conceptualization, Q.C. and C.X.; methodology, Q.C. and C.X.; software, Q.C.; validation, Q.C.; formal analysis, Q.C.; investigation, Q.C.; resources, Y.W.; data curation, Q.C. and M.T.; writing—original draft preparation, Q.C.; writing—review and editing, Q.C.; visualization, Q.C. and M.T.; supervision, Y.W.; project administration, W.Z.; funding acquisition, C.X. and Y.W. All authors have read and agreed to the published version of the manuscript.

Funding: This work was supported by the Natural Science Foundation of Shandong Province under Grant ZR2020KE029; by the National Natural Science Foundation of China under Grant 52001241; by the 111 Project(B21008); by the Zhejiang Key Research Program under Grant 2021C01010.

Institutional Review Board Statement: Not applicable.

Informed Consent Statement: Not applicable.

Data Availability Statement: No public data sets were used in the study.

Conflicts of Interest: The authors declare that they have no known competing financial interests or personal relationships that could have appeared to influence the work reported in this paper.

References

1. Tang, H. *Intelligent Analysis and Research on Behavior Characteristics of Inbound and Outbound Ships*; Dalian Maritime University: Dalian, China, 2020.
2. Steidel, M.; Mentjes, J.; Hahn, A. Context-Sensitive Prediction of Vessel Behavior. *J. Mar. Sci. Eng.* **2020**, *8*, 987. [CrossRef]
3. Pietrzykowski, Z.; Wielgosz, M.; Breitsprecher, M. Navigators' behavior analysis using data mining. *J. Mar. Sci. Eng.* **2020**, *8*, 50. [CrossRef]
4. Zhang, H.; Yang, S.L.; Fan, W.; Shi, H.M.; Yuan, S.L. Spatial analysis of the fishing behaviour of tuna purse seiners in the western and central Pacific based on vessel trajectory date. *J. Mar. Sci. Eng.* **2021**, *9*, 322. [CrossRef]
5. Gao, M.; Shi, G.; Li, S. Online prediction of ship behavior with automatic identification system sensor data using bidirectional long short-term memory recurrent neural network. *Sensors* **2018**, *18*, 4211. [CrossRef]
6. Ma, J.; Li, W.; Jia, C.; Zhang, C.W.; Zhang, Y. Risk prediction for ship encounter situation awareness using long short-term memory based deep learning on intership behaviors. *J. Adv. Transp.* **2020**, *2020*, 8897700. [CrossRef]
7. Tang, H.; Wei, L.; Yin, Y.; Shen, H.; Qi, Y. Detection of abnormal vessel behaviour based on probabilistic directed graph model. *J. Navig.* **2020**, *73*, 1014–1035. [CrossRef]
8. Zissis, D.; Xidias, E.K.; Lekkas, D. A cloud based architecture capable of perceiving and predicting multiple vessel behaviour. *Appl. Soft Comput.* **2015**, *35*, 652–661. [CrossRef]
9. Ma, J.; Jia, C.; Yang, X.; Cheng, X.; Li, W.; Zhang, C. A data-driven approach for collision risk early warning in vessel encounter situations using attention-BiLSTM. *IEEE Access* **2020**, *8*, 188771–188783. [CrossRef]
10. Xue, J.; Chen, Z.; Papadimitriou, E.; Wu, C.; Gelder, P.V. Influence of environmental factors on human-like decision-making for intelligent ship. *Ocean Eng.* **2019**, *186*, 106060. [CrossRef]
11. Alizadeh, D.; Alesheikh, A.A.; Sharif, M. Vessel trajectory prediction using historical automatic identification system data. *J. Navig.* **2021**, *74*, 156–174. [CrossRef]
12. Yu, H.; Fang, Z.; Murray, A.T.; Peng, G. A direction-constrained space-time prism-based approach for quantifying possible multi-ship collision risks. *IEEE Trans. Intell. Transp. Syst.* **2019**, *22*, 131–141. [CrossRef]
13. Suo, Y.; Sun, Z.; Claramunt, C.; Yang, S. A Dynamic Risk Appraisal Model and Its Application in VTS Based on a Cellular Automata Simulation Prediction. *Sensors* **2021**, *21*, 4741. [CrossRef]
14. Alvarellos, A.; Figuero, A.; Sande, J.; Peña, E.; Rabuñal, J. Deep Learning Based Ship Movement Prediction System Architecture. In *Advances in Computational Intelligence*; Springer: Cham, Switzerland, 2019; pp. 844–855.
15. Wang, T.; Wu, Q.; Zhang, J.; Wu, B.; Wang, Y. Autonomous decision-making scheme for multi-ship collision avoidance with iterative observation and inference. *Ocean Eng.* **2020**, *197*, 106873. [CrossRef]
16. Kawamura, K.; Hashimoto, H.; Matsuda, A.; Terada, D. SPH simulation of ship behaviour in severe water-shipping situations. *Ocean Eng.* **2016**, *120*, 220–229. [CrossRef]
17. Praczyk, T. Using evolutionary neural networks to predict spatial orientation of a ship. *Neurocomputing* **2015**, *166*, 229–243. [CrossRef]
18. Murray, B.; Perera, L.P. An AIS-based deep learning framework for regional ship behavior prediction. *Reliab. Eng. Syst. Saf.* **2021**, *215*, 107819. [CrossRef]
19. Giubilato, R.; Chiodini, S.; Pertile, M.; Debei, S. Minivo: Minimalistic range enhanced monocular system for scale correct pose estimation. *IEEE Sens. J.* **2020**, *20*, 11874–11886. [CrossRef]
20. Guo, S.; Zhao, Q.; Cui, G.; Li, S.; Kong, L.; Yang, X. Behind corner targets location using small aperture millimeter wave radar in NLOS urban environment. *IEEE J. Sel. Top. Appl. Earth Obs. Remote Sens.* **2020**, *13*, 460–470. [CrossRef]
21. Ding, X.; Huang, Y.; Li, Y.; He, J. Forgery detection of motion compensation interpolated frames based on discontinuity of optical flow. *Multimed. Tools Appl.* **2020**, *79*, 28729–28754. [CrossRef]
22. Chen, C.; Ma, F.; Xu, X.; Chen, Y.; Wang, J. A novel ship collision avoidance awareness approach for cooperating ships using multi-agent deep reinforcement learning. *J. Mar. Sci. Eng.* **2021**, *9*, 1056. [CrossRef]
23. Ranjan, R.; Patel, V.M.; Chellappa, R. Hyperface: A deep multi-task learning framework for face detection, landmark localization, pose estimation, and gender recognition. *IEEE Trans. Pattern Anal. Mach. Intell.* **2017**, *41*, 121–135. [CrossRef]
24. Wu, J.W.; Cai, W.; Yu, S.M.; Xu, Z.L. Optimized visual recognition algorithm in service robots. *Int. J. Adv. Robot. Syst.* **2020**, *17*, 1729881420925308. [CrossRef]

Journal of
Marine Science and Engineering

MDPI

Article

Multi-Sensor-Based Hierarchical Detection and Tracking Method for Inland Waterway Ship Chimneys

Fumin Wu [1], Qianqian Chen [1,*], Yuanqiao Wen [2,3], Changshi Xiao [1,3,4] and Feier Zeng [1]

1 School of Navigation, Wuhan University of Technology, Wuhan 430063, China; wfm19970@whut.edu.cn (F.W.); cs_xiao@hotmail.com (C.X.); 304716@whut.edu.cn (F.Z.)
2 Intelligent Transportation Systems Research Center, Wuhan University of Technology, Wuhan 430063, China; yqwen@whut.edu.cn
3 National Engineering Research Center for Water Transport Safety, Wuhan 430063, China
4 Institute of Ocean Information Technology, Shandong Jiaotong University, Weihai 264200, China
* Correspondence: chenqq@whut.edu.cn

Abstract: In the field of automatic detection of ship exhaust behavior, a deep learning-based multi-sensor hierarchical detection method for tracking inland river ship chimneys is proposed to locate the ship exhaust behavior detection area quickly and accurately. Firstly, the primary detection uses a target detector based on a convolutional neural network to extract the shipping area in the visible image, and the secondary detection applies the Ostu binarization algorithm and image morphology operation, based on the infrared image and the primary detection results to obtain the chimney target by combining the location and area features; further, the improved DeepSORT algorithm is applied to achieve the ship chimney tracking. The results show that the multi-sensor-based hierarchical detection and tracking method can achieve real-time detection and tracking of ship chimneys, and can provide technical reference for the automatic detection of ship exhaust behavior.

Keywords: ship exhaust behavior; detection and tracking; multi-sensor; deep learning; morphological operation

Citation: Wu, F.; Chen, Q.; Wen, Y.; Xiao, C.; Zeng, F. Multi-Sensor-Based Hierarchical Detection and Tracking Method for Inland Waterway Ship Chimneys. *J. Mar. Sci. Eng.* **2022**, *10*, 809. https://doi.org/10.3390/jmse10060809

Academic Editor: Rafael Morales

Received: 18 March 2022
Accepted: 12 May 2022
Published: 13 June 2022

1. Introduction

The construction of the Yangtze River Economic Belt is one of the key strategies of the national cross-regional coordinated development, and both the "Yangtze River Protection" and the "Yangtze River Green Ecological Corridor" are the top priorities of the construction of the Yangtze River Economic Belt. The International Maritime Organization (IMO) has mandated a gradual reduction of nitrogen oxide and other types of gas emissions [1], and a regulation on sulfur emissions from ships sailing in global waters has been in effect since 1 January 2020 [2]. In addition, the design of ships' intake ports and the exhaust ports of the exhaust gas is being modified in accordance with the requirements of the International Maritime Organization (IMO) [3]. However, the detection of ship exhaust depends on high-sensitivity gas sensors, and it is difficult to obtain evidence. The Pankratova NV study showed that ship exhaust emission data are correlated with ship chimneys [4]; therefore, the method of tracking ship chimney detection based on computer technology is one of the most important tools for scientific and efficient regulation.

Ship chimney detection is the core research content of this paper, and ship detection is the prerequisite and a key technical point of ship chimney detection. Since the ship chimney has small target and inconspicuous features, and the known chimney dataset is very small, it is very difficult to detect the ship chimney directly; on the contrary, the ship has relatively large target and obvious features compared with the chimney, and the dataset is relatively large.

However, currently there are still difficulties and challenges in the field of computer vision for small target detection. In terms of visible images, both traditional manually designed feature operator-based target detection and deep learning-based target detection methods have yet to improve the detection accuracy of small targets. In addition to the characteristics of small targets, the detection of inland river ship chimneys is also affected by the small feature information of ship chimneys. In terms of infrared images, the infrared camera has a small field of view, and the acquired image information is not rich. Although the infrared camera is more sensitive to the target in high temperature regions and the ship chimney is also a high temperature object, the simple use of an infrared camera to detect the chimney is less robust due to the high temperature of the ship itself or the ship's cargo exposed to the sun, as well as the influence of the background buildings and water reflections in the inland river.

Based on the above problems, we found that, on the visible band image, although the ship chimney target is small, the ship target is relatively large and rich in information, and the deep learning technique can be used to detect the ship on the visible band image first, with the aim of narrowing down the detection range of the ship chimney. Then, since the visible band is more sensitive to high temperature regions, the difficulty of detecting the chimney in a small area will be greatly reduced. Therefore, the detection of ship chimneys can eventually be achieved by combining the characteristics of different sensor images, thereby bringing convenience to the subsequent tracking.

The remainder of this paper is organized as follows. Some related works are introduced in Section 2. In Section 3, we will discuss the whole methodology of our algorithm. The experiment and model prediction performance is reported in Section 4. Finally, the work is concluded in Section 5.

2. Related Work

A large number of scholars have also conducted research on ship detection based on computer vision techniques. According to the type of technology used, this research can be divided into traditional-based methods and deep learning-based methods. Most of the traditional methods are designed to detect or recognize a specific scene. Arshad [5] et al. first processed the ship background image using morphological operations, and then used the Sobel operator to perform edge detection of the ship to discriminate it from its background, but it is not effective in the case of complex textures, which have more noise. Zhang X designed a rotated Gaussian mask to model the ship, and, at the same time, contextual information was used to enhance the perception of the ship [6]. Wang Y. [7] et al. proposed a ship detection algorithm based on a background difference method, but the algorithm was aimed at ship detection under a static background, and did not identify, classify, and track targets. Tang Y. [8] et al. adopted the fusion technology of multi-vision to analyze and detect ship targets by monitoring through local entropy and a connected domain, requiring two scans of images, which was inefficient, and the threshold had a great influence on the final effect. Shi W. et al. [9] proposed morphology with multiple structural elements to extract the edge features of ships by using different structural elements, which can fully retain various details of ships while filtering out background noises such as waves, but it is difficult to detect small targets.

In addition to the traditional vision technology-based methods mentioned above, deep learning technology-based methods are the mainstream ship detection methods at present. Excellent target detection methods based on deep learning are the R-CNN series, YOLO series, and SSD series. Cui ZY used a pyramidal structure to connect the convolutional block attention module (CBAM) closely with each feature map connected from top to bottom of the pyramidal network in order to extract rich features containing resolution and semantic information for multi-scale ship detection [10]. Subsequently, Cui ZY proposed a center net-based large SAR image ship detection method for locating the centroid of the target by key point estimation, which can effectively avoid the missed detection of small target ships [11]. Differently, Chen XQ used a convolutional neural network in the YOLO model to extract multi-scale ship features from the input ship images. Then, multiple bounding boxes (i.e., potential ship positions) were generated based on the target confidence, and, finally, the background surround box interference was suppressed to obtain the ship positions in each ship image. Finally, Chen XQ analyzed the spatio–temporal behavior of ships in continuous ocean images based on the ship's kinematic information [12]. Shao ZF used the CNN framework based on depth features, saliency maps, and coastline prior. This work integrated ship discriminative features to detect ship class and location [13]. Yang X proposed a dense feature pyramid network to detect ships in different scenarios, including in the ocean and at ports, in order to solve the problem caused by narrow ship width [14].

In recent years, deep learning methods have been successfully applied to ship detection in synthetic aperture radar (SAR) images. Wei SJ proposed a high-resolution ship detection network based on high-resolution and low-resolution convolutional feature mapping for ship detection in high-resolution SAR images [15]. Similarly, Lin Z, et al. proposed a new fast R-CNN-based network structure based on high-resolution SAR images to further improve ship detection performance by using a squeeze excitation mechanism [16,17]. Jin L., et al. used the SSD model and added a feature fusion module to the shallow feature layer to optimize the feature extraction capability for small objects, and then added the squeeze and excitation network (SE) module to each feature layer to introduce an attention mechanism for the network to achieve small-scale ship detection in remote sensing images [18,19]. Wang Y combined single-shot multibox detector (SSD) with migration learning to solve the ship detection problem in complex environments, such as oceans and islands [20]. Sun J, based on the SSD model, integrated expansion convolution with a multiscale feature fusion to improve small target detection accuracy [21]. Not coincidentally, Chen P, to improve the small target detection accuracy, embedded the elemental pyramid model into the traditional RPN, and then mapped it to a new elemental space for object recognition [22]. The detection of multiscale SAR ships remains a great challenge due to the strong interference and wide variation of scales in the offshore background.

This paper proposes a multi-sensor hierarchical detection tracking algorithm based on deep learning to detect and track ship chimneys. Firstly, the first level detection uses a visible light image input deep-learning target detector to detect the ship target, so as to greatly reduce the target detection range and solve the problem of background interference. Then, in view of the problem that the chimney target is too small to be identified, infrared imaging is adopted for the second-level detection, with the first-level detection result used as the input of the second-level detection. The image is extracted through a two-step Ostu binarization algorithm, image corrosion, and expansion operation. Finally, according to the prior knowledge of the chimney orientation, the candidate area is bisecting to further reduce the detection range and extract the final chimney target, combined with area characteristics. The improved DeepSORT tracking algorithm is used to track the ship chimney, which provides some help for the ship exhaust monitoring.

3. Algorithm Design

3.1. Algorithmic Framework

The framework of the multi-sensor hierarchical detection and tracking algorithm is shown in Figure 1, which is divided into four parts, namely data input, detection, tracking, and data output. Among them, the input data are an infrared camera and a visible camera, and the detection stage is divided into primary detection and secondary detection. The primary detection uses the improved YOLOV3 which was proposed by Joseph Redmon in 2018 as the ship detector, which is improved from two aspects: the design of the a priori frame, and the output of the feature pyramid. The second level detection splits the ship area in the infrared camera according to the first level detection result, and then filters the background by Gaussian filtering and adaptive threshold selection algorithm to obtain the candidate area of the chimney, according to the a priori knowledge. It is known that the chimney detected in this paper is located above the ship area, so the area equalization method is used to narrow the detection range again. Finally, the maximum value of the contour area is calculated as the final detection result.

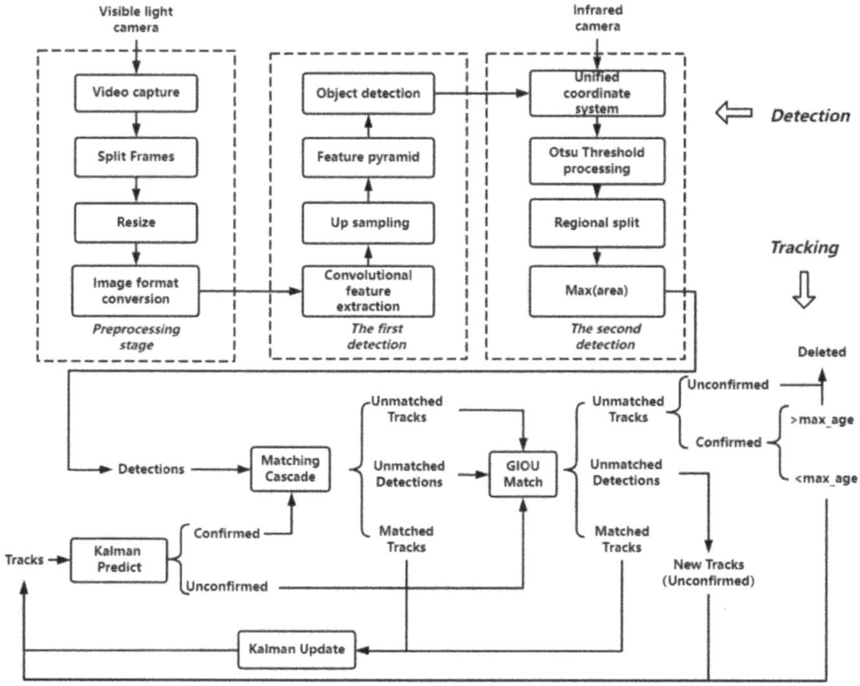

Figure 1. General framework of ship chimney detection and tracking algorithm.

The tracking is performed using the improved DeepSORT [23] algorithm, which mainly consists of a target detection module and a data association module. DeepSORT is used in the real-time target tracking process to first extract the depth features of the target, and then uses Kalman filtering to make predictions, correlate the sequence data, and perform the target matching. Mainly from the calculation of the cost matrix and association, the algorithm is improved. The main steps of the improved DeepSORT tracking are as follows:

(1) Create Tracks according to the results detected in the first frame, and initialize the Kalman filter. Tracks are initially in Unconfirmed state and can be converted to Confirmed state only if they are tracked successfully three times in a row.

(2) Calculate the cost matrix between the tracked target in Tracks and the detected target in the current frame using the improved GIOU.

(3) The cost matrix in Step (2) is input to the improved data association algorithm KM, and three kinds of matching results are obtained: the first category Matched Tracks is the traces matched to the detection results, indicating that the current frame tracks the target in the previous frame, and the values in Tracks are subsequently updated according to Kalman filtering. The second type, Unmatched Detections, is the detection result of unmatched tracks, which means that the target detected in the current frame is a newly appeared target, which is not related to the previous detection result, so a new tracking track needs to be added. The third type, Unmatched Tracks, is the trajectory with unmatched detections, which means that the trajectory existing in the previous frame is lost in the current frame, and if it is an Unconfirmed stable state, the trajectory is deleted directly. If it is a Confirmed stable state, the number of followed traces max_age is increased by 1. When the number of followed traces reaches 30 times, the Confirmed state is converted to an Unconfirmed non-stable state.

(4) For the Confirmed state, Tracks and Detections will use cascade matching to calculate the cost matrix. Cascade matching uses the appearance feature vector to calculate

the cosine similarity, and uses the Marxist distance to exclude the targets between frames that are far away from each other, where the appearance feature vector saves the feature vector of this target in the first 100 frames by default.

(5) There are also three types of cascade matching results: for the Unmatched Tracks and Unmatched Detections states, the algorithm re-calculates these two states together with the Unconfirmed state in Tracks using the GIOU association algorithm. For Matched Tracks states, the variable information in Tracks is updated by Kalman.

(6) The cost matrix in (5) is input into the KM algorithm, and the processing result is similar to step (3).

(7) Loop (4) to (6) steps until the end of the video frame.

3.2. Improved YOLOv3-Based Ship Detection Network

3.2.1. Anchor Improvements

A large number of experiments have shown that the selection and design of Anchor has had a large impact on the results of detection. Through the analysis of our own ship dataset, we know that the ship targets are larger, and the ship lengths and widths are more similar with horizontal orientation. By comparing the characteristics of the COCO dataset, we can see that the default Anchor of YOLOV3 does not meet our actual needs. Based on the above characteristics of the actual ship dataset, we made a specific design for the Anchor of the ship target, aiming to improve the speed and accuracy of ship detection.

In YOLO detection algorithm, the input image is divided into $S \times S$ grids, and each grid is called a Grid Cell. Each Grid Cell is responsible for detecting a target on which the center of the object falls. Each Grid Cell has a prediction box, which we call Anchor, and the number of Anchor for each Grid Cell is different in different versions. In YOLOV1, the image is divided into 7×7 size, and each grid is fixed with only two Anchors with different aspect ratios. Each Grid Cell can predict only one category, so the detection accuracy is low in scenes with dense targets. In YOLOV2, the authors used clustering to cluster the real target aspect ratios of the dataset into five classes by default, thus introducing five Anchors for each Grid Cell, and improving the detection capability for dense objects. In YOLOV3, the authors reduce the number of Anchors for each Grid Cell to three different scales, and introduce the concept of multi-scale feature map fusion to detect targets at different scales with three different scales, so the number of Anchor for each Grid Cell increases to nine.

The targets detected in this paper are ships, which generally have an aspect ratio greater than 1, i.e., the detection frame rectangle is longer than wide, as shown in Figure 2, where the upper left corner shows the distribution of the number of ship types, the upper right corner shows the distribution of the rectangular frame of the ship training set, and the lower left corner shows the distribution of the target center x and y, where the horizontal and vertical coordinates are the ratio of x and y to the actual width and height of the image. The same is true for the lower right corner, where the original width of the image is 1920 pixels and the height is 1080 pixels. From the statistical results, we can see that most of the ship widths are distributed around 0.1~0.3, i.e., 192~576 pixels wide, and the heights are distributed around 0.02~0.1, i.e., 22~108 pixels high. In order to make our designed Anchor aspect ratio closer to the actual ship detection application, the number of each Grid Cell was reduced from three to two, and we kept the default Feature Map with two different scales due to the "small and large" characteristics in inland waters. As a result, the number of Anchors was reduced from the default nine to four. In order to make our designed Anchor aspect ratio closer to the actual ship detection application, we first clustered the aspect of the Bounding Box of the dataset, where there are multiple clustering methods. We borrowed the idea from the YOLOV2 authors, and used k-means algorithm to cluster the data into two classes, and obtained the original dimensions of Anchor for each Grid Cell as (384, 54) and (1152, 216).

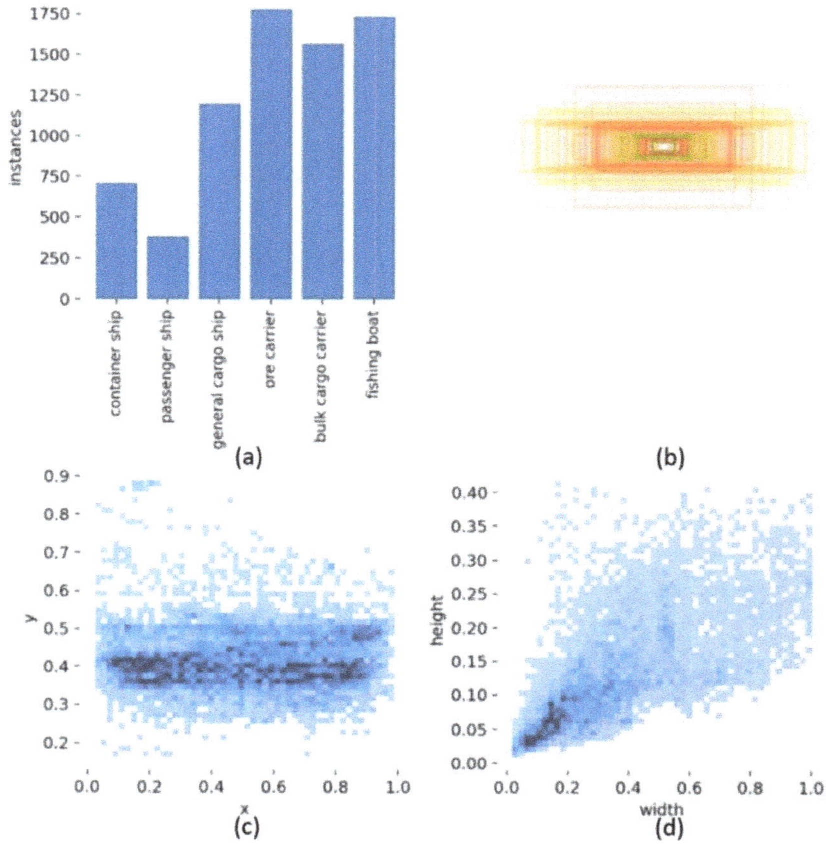

Figure 2. Distribution of Anchor information of a ship dataset. (**a**) Number of ships by class. (**b**) Statistics of Anchor shape. (**c**) Statistics of anchor X y center coordinates (**d**) Statistics of Anchor width height.

3.2.2. Improvement of Feature Pyramids

In YOLOV3, in order to make the detection of objects of different sizes, after the feature extraction network, the features of different feature extraction layers were fused to form new feature maps through Concat and upsampling operations. These different feature maps have the same depth, but different sizes. Fresh feature maps of different sizes, as well as the network structure in YOLOV3, are shown in Figure 3.

The light yellow part of the figure is for the three different scales of 13×13, 26×26, and 52×52. In these different scales, the size of each Cell is inversely proportional to the size of the scale, and for the large scale of 52×52, the corresponding size of each Cell is small, while for the small scale of 13×13, the size of each Cell is large. The small-scale Cell contains less information, and is therefore more suitable for detecting small objects, while the large-scale Cell incorporates more information, and is therefore more suitable for detecting larger objects, as shown in Figure 4. For the large ship in the bottom corner, a 13×13 feature map is generally available, while for the small target ship in the middle, a 26×26 feature map is generally available.

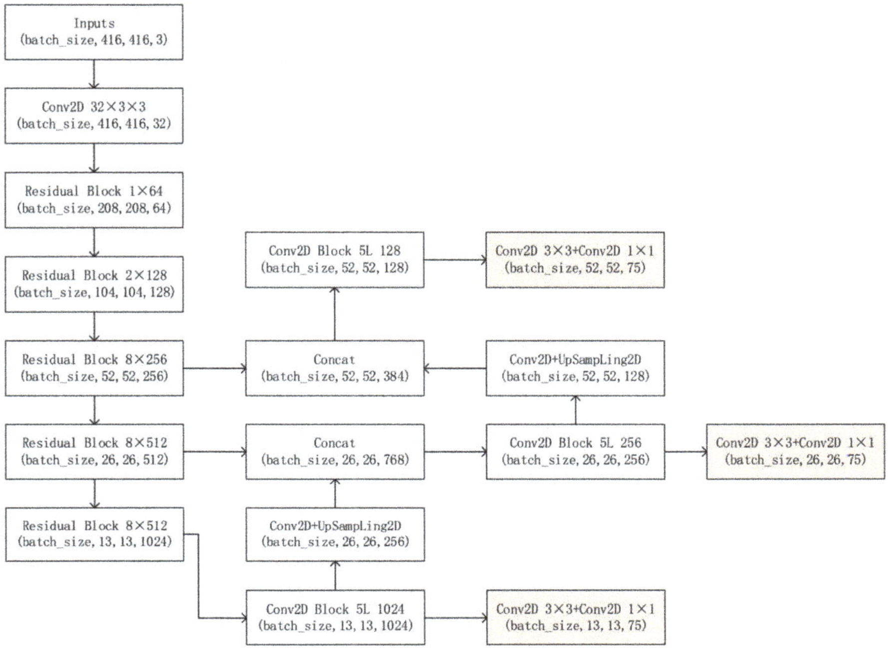

Figure 3. The network structure of YOLOV3.

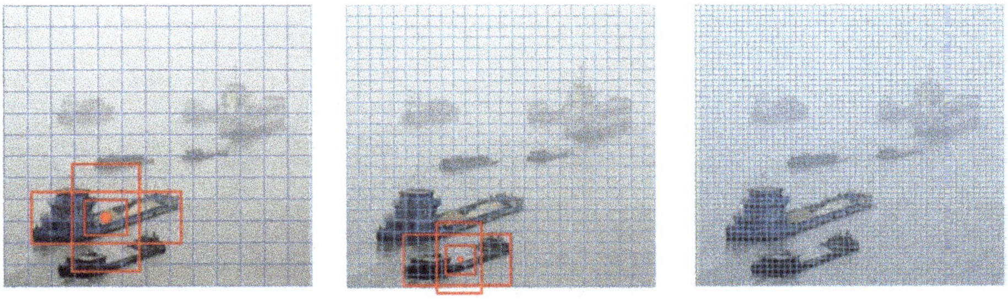

Figure 4. Output diagram of different sizes. (**a**) 13 × 13 grid cell. (**b**) 26 × 26 grid cell. (**c**) 52 × 52 grid cell.

Through analyzing the self-collected ship dataset in this paper, we can see that, in terms of species, the species of ships is much smaller than the open source generic dataset; in terms of scenarios, the river channel in inland waters is limited, and ships can only travel in the area. With the shore camera as the reference point, the width of the river channel greatly limits the size variation of ships, and most of the ships in inland waters are larger in size and belong to large targets, so we can delete the 52 × 52 feature maps used to detect small targets. Just keep the 13 × 13 and 26 × 26 feature maps. This optimization can reduce the parameters for network training, as well as speed up the training of the network. In addition, since the number of feature maps is reduced from three to two, the number of Anchor corresponding to each feature map is also reduced from three to two, so the original 3 × 3 = 9 frames to be detected is reduced to 2 × 2 = 4 frames to be detected when calculating the detection frames. This will greatly reduce the amount of calculation, as well as improve the detection speed of the ship. To sum up, the complete network structure after the improvement designed in this paper is shown in Figure 5.

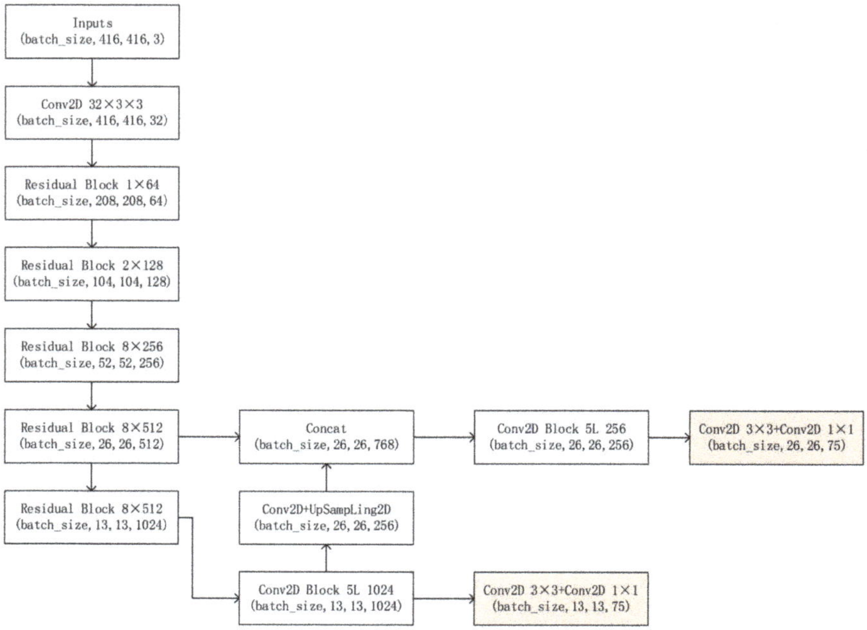

Figure 5. Improved YOLOv3 network.

As shown in Figure 5, the input size of the image is 416 × 416 pixels, and after five down-sampling calculations, a feature layer of size 13 × 13 is obtained, which detects large targets. Ship targets are relatively large targets, so a 52 × 52 feature layer will increase the number of parameters of the model and reduce the detection speed. Therefore, only two output layers of 13 × 13 and 26 × 26 are considered for retention. The goal of reducing the number of parameters and operations is achieved by reducing the number of feature layers to improve the network detection speed.

3.3. *Chimney Detection with Fused Infrared Images*

3.3.1. Threshold Processing

The video saved by the infrared heat-sensing camera used in this paper was later processed and saved locally as an RGB three-channel image as well. In order to facilitate the subsequent thresholding, the RGB image needed to be converted to a grayscale image. The conversion of RGB to gray scale image is represented by Equation (1)

$$\mathrm{Gray} = \mathrm{R} \times 0.299 + \mathrm{G} \times 0.578 + \mathrm{B} \times 0.114 \quad (1)$$

After the grayscale processing, a bimodal image can be obtained by counting the individual grayscale values, and due to the processing of bimodal images, this subsection uses Otsu's algorithm, which attempts to find a threshold that minimizes the weighted intra-class variance given by the relation:

$$\sigma^2 = \omega_1 \cdot (\mu_1 - \mu_0)^2 + \omega_2 \cdot (\mu_2 - \mu_0)^2 \quad (2)$$

where σ^2 is the interclass variance of foreground and background, ω_1 and ω_2 represent the proportion of background and foreground pixels in the total image, μ_1 and μ_2 represent the average grayscale of background and foreground, respectively, and μ_0 represents the average grayscale of the whole image. Expanding Equation (2) yields:

$$\begin{aligned}\sigma^2 &= \omega_1 \cdot \mu_1^2 + \omega_2 \cdot \mu_2^2 \\ &-2(\omega_1 \cdot \mu_1 + \omega_2 \cdot \mu_2) \cdot \mu_0 + \mu_0^2\end{aligned} \tag{3}$$

According to the mathematical definition formula of expectation $E(X) = \sum_{k=1}^{\infty} x_k \cdot p_k$, we can deduce that:

$$\mu_0 = \omega_1 \cdot \mu_1 + \omega_2 \cdot \mu_2 \tag{4}$$

Bringing (4) into (3), $\sigma^2 = \omega_1 \cdot \mu_1^2 + \omega_2 \cdot \mu_2^2 - \mu_0^2$ is again replaced using the relationship between Equation (4) and $\omega_2 = 1 - \omega_1$:

$$\begin{aligned}\sigma^2 &= \omega_1 \cdot \mu_1^2 + \frac{\omega_2^2 \cdot \mu_2^2}{1-\omega_1} - \mu_0^2 \\ &= \omega_1 \cdot \mu_1^2 + \frac{(\mu_0 - \omega_1 \cdot \mu_1)^2}{1-\omega_1} - \mu_0^2 \\ &= \frac{\omega_1}{(1-\omega_1)} \cdot (\mu_1 - \mu_0)^2\end{aligned} \tag{5}$$

Using Equation (5), we only need to count the pixels before the current iteration of grayscale, which greatly improves the efficiency of the program.

As can be seen in Figure 6, some background noise points can be effectively removed after Gaussian filtering. Compared with a fixed threshold, the Ostu algorithm is more likely to try to find a threshold to reasonably separate the foreground and background.

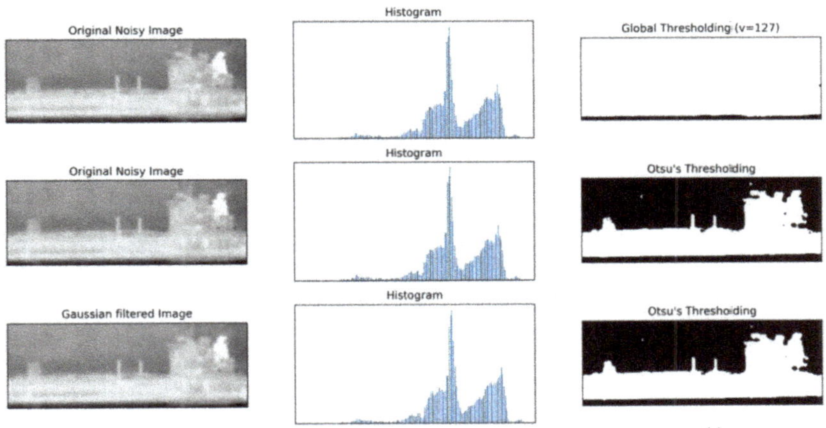

(a) (b) (c)

Figure 6. Comparison of infrared image binarization. (**a**) input images. (**b**) histogram. (**c**) Processing results of thresholds.

3.3.2. Coordinate Fusion

In order to collect experimental data, we have independently developed a set of experimental systems, which consists of a visible camera, an infrared camera and a gimbal that can be rotated coaxially, which can locate and track the target in real time. The visible camera has a resolution of 1920 × 1080, and the thermal imaging camera is a custom thermal imaging camera from Golder Infrared with a resolution of 640 × 512 resolution and a rotation angle of −120–120° for the gimbal. The multi-sensor coaxial rotation system is shown in Figure 7.

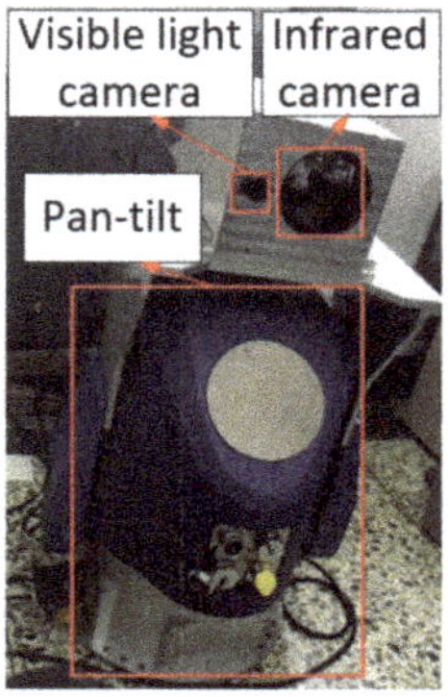

Figure 7. Multi-sensor coaxial rotation system.

Therefore, the coordinates of the same object in different cameras in the same frame are represented differently, as shown in Figure 8.

$$x_2 = \frac{x_1}{1920} \times 640 \tag{6}$$

$$y_2 = \frac{y_1}{1080} \times 512 \tag{7}$$

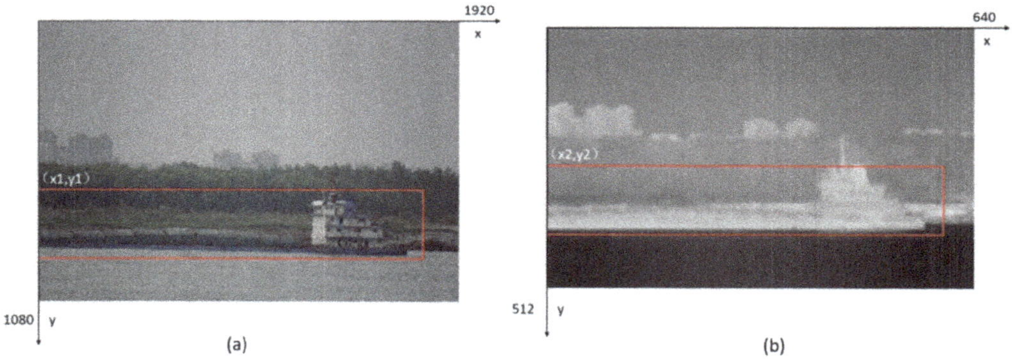

Figure 8. Schematic diagram of the same ship position in different coordinate systems. (**a**) Detection results in the visible light camera coordinate system. (**b**) Detection results in infrared camera coordinate system.

Therefore, this paper needs to convert the coordinates to ensure the accuracy of the search area of the ship's chimney. For different size images, the size is different, but the position of each coordinate point relative to the upper left corner (zero point) is the same after conversion to a right-angle coordinate system, so the coordinates can be converted using the scale relationship. For the image, the coordinate system is two-dimensional, so it needs to be converted separately in the x and y directions. Supposing that the coordinates of the same object P (x_1, y_1) on the 1920 × 1080 resolution image and (x_2, y_2), the specific value of x, y of (x_2, y_2) should be as shown in the operation of Equation (6), according to the image scale. By using the above-mentioned coordinate conversion equation after corresponding the visible image to the infrared band image, it aims to ensure that the location of the ship's chimney is found accurately, rather than deviations due to coordinate conversion.

3.4. Improving DeepSORT Algorithm

3.4.1. GIOU Loss Function

IOU (Intersection over Union), also known as intersection and merge ratio, is a measure of the accuracy of detecting the corresponding object in a given dataset. DeepSORT (Deep Simple Online and Realtime Tracking) uses the IOU of the detection frame and tracking frame as the loss matrix in the correlation algorithm. The input IOU ranges between [0, 1] with scale invariance, and the equation is shown in Equation (8):

$$\text{IOU} = \frac{S_A \cap S_B}{S_A \cup S_B} \tag{8}$$

where S_A is the area of the predicted box, and S_B is the area of the real box. If IOU is used as a measure of the overlap between boxes, the following problems will occur

(1) IOU is always 0 when there is no overlap between the prediction box and the real box, as shown in Figure 9, state 1, where the red prediction box and the blue real box have no intersection, and the value of IOU is 0.

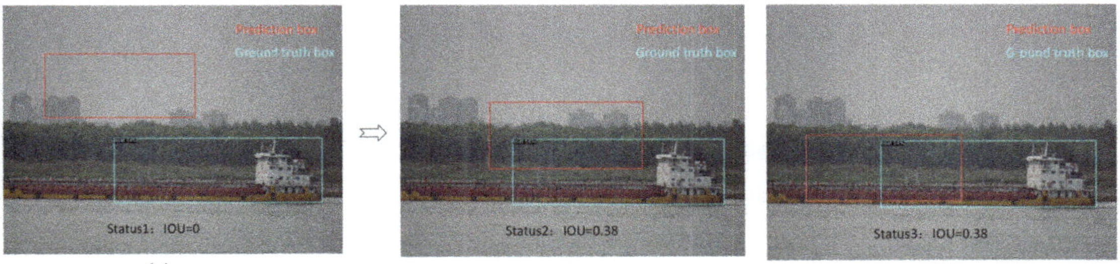

(a) (b) (c)

Figure 9. Schematic diagram of different overlapping shapes of IOU. (**a**) status1: IOU = 0. (**b**) status2: IOU = 0.38. (**c**) status3: IOU = 0.38.

(2) When the IOUs intersect and have the same value, it is impossible to distinguish the various cases of IOUs. There can be many kinds of overlapping shapes for the same IOU value, and they are different in effect. As shown in Figure 9 state 2 and state 3, the IOUs of both the prediction box and the real box in state 2 and state 3 are equal to 0.38, but state 2 is an up-and-down intersection, and state 3 is a horizontal intersection.

In order to solve the above problem, this paper uses GIOU (Generalized Intersection over Union) to replace IOU in the DeepSORT algorithm. GIOU loss focuses not only on overlapping regions, but also on non-overlapping regions, which distinguishes the cases with the same IOU but different forms of overlap, and solves the problem that there can be no gap between non overlapping frames. The value range of GIOU is $[-1, 1]$ with the following formula.

$$\text{GIOU} = \text{IOU} - \frac{|C - (A \cup B)|}{|C|} \tag{9}$$

where C is the smallest outer rectangle of the prediction frame and the target frame, as shown in the left of Figure 10. In Equation (8) is the difference set, as shown in the blue part in Figure 10.

As shown in Figure 11, suppose A is the ship target at frame n and B and C are the ship targets at frame n + 1, where the IOUs of A and B are $4/28 \approx 0.14$, and the IOUs of A and C are also $4/28 \approx 0.14$. Since their IOU values are equal, the difference cannot be measured if IOUs are used. However, the GIOU of A and B is $4/28 - (36 - 28)/36 \approx -0.08$, while the GIOU of A and C is $4/28 - (28 - 28)/28 \approx 0.14$, which shows that the correlation between A and C will be greater than that between A and B. Therefore, in the ship tracking task, we prefer to consider ship C as the target position of ship A in the next frame, which is also consistent with the fact that inland ships travel slowly and have low deformation in the video sequence.

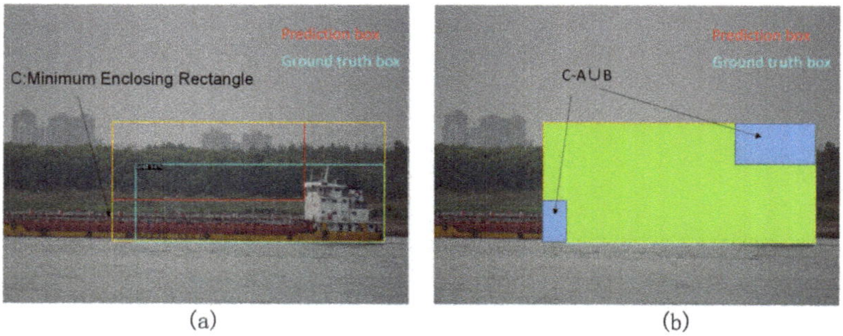

Figure 10. Schematic diagram of GIOU. (**a**) C is Minimum Enclosing Rectangle. (**b**) The areas of C-A∪B.

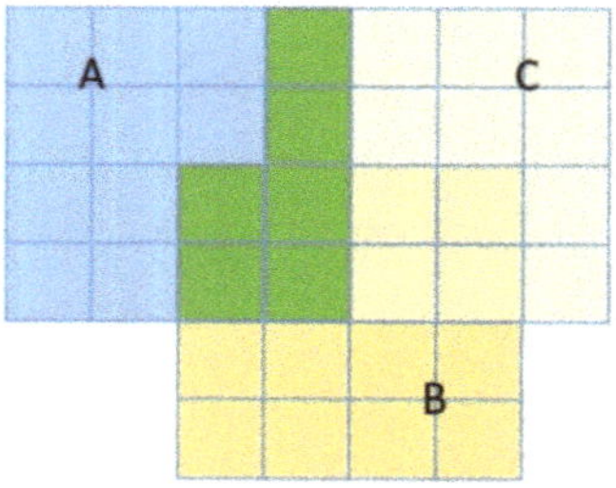

Figure 11. Schematic diagram of ship intersection between simulated frames. A is detection of n frame. B and C are detections of n + 1 frame.

3.4.2. KM Association Algorithm

In multi-target tracking tasks, the main purpose of data association is to perform matching of multiple targets between frames, including emerging targets, the disappearance of old targets, and the ID matching problem between the previous frame and the current frame. The DeepSORT default data association algorithm uses the Hungarian algorithm, and the core idea is to find the maximum matching algorithm of the augmented path for the bipartite graph. As shown in Table 1.

Table 1. Inter-frame target matching.

n + 1 Frame / n Frame / GIOU	N1	N2	N3	N4
M1	0.8	0.6	0	0
M2	0	0.6	0.9	0
M3	0.9	0.8	0.5	0
M4	0	0	0	0

M1~M4 are the four tracked targets in the nth frame, N1~N4 are the four newly detected targets in the nth + 1 frame, and the association degree index between the targets is measured by the GIOU loss function in the previous section. Since M4 is not associated with any detected targets in the new frames, M4 is the old target tracking loss case; N4 is the detected targets in the new frames, which belong to the new target emergence case and will be assigned new IDs to track. The association algorithm discussed in this section then solves the matching problem between M1~M3 and N1~N3. If the Hungarian algorithm with no weight value is used, the matching results are generally: M1 matches with N1, M2 matches with N2, and M3 matches with N3 when the threshold value is taken as 0.5. The Hungarian algorithm considers both to be correlated as long as it is greater than the specified threshold, that is, it considers that M2 and N2 and N3 are matched while ignoring the fact that M2 is more correlated with N3. It is this matching method, which is regarded as leveling, that leads to low tracking accuracy.

The KM algorithm is an improvement of the Hungarian algorithm, in which the weights of the edge values are increased to achieve optimal weight matching based on the Hungarian algorithm. The steps to solve the target tracking problem involve using the KM algorithm [11]. The results detected in the *n*th and *n*th + 1 frames are used as vertices to form the point set M and the point set N, respectively, the GIOU of the detection frame and the prediction frame is used as the edge value connected between the two points, with the ID of each vertex in M set to M_i, the initial weight set to the maximum edge value W of the edge connected to that point, and each point in the point set N set to N_i, with the initial weight set 0. If the point set M is satisfied $M_i + N_i = W_{ij}$, then the M_i and N_i will be matched; if not satisfied, then the point set M in the conflict will be minus *d*, and the point set N in the conflict will be plus *d*, here set to 0.1. The specific process is shown in Figure 12.

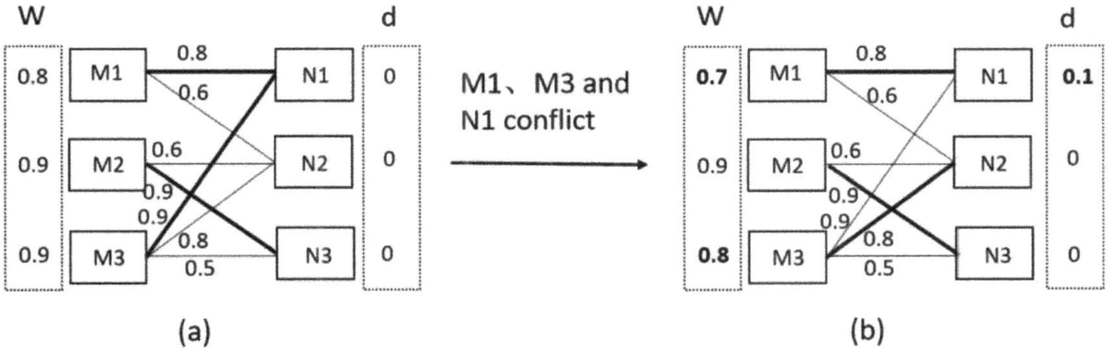

Figure 12. Schematic diagram of KM algorithm. (**a**) Initialize W, d. (**b**) Resolve conflict.

In Figure 11a, the KM algorithm assigns the initial value of W in the target M1~M3 from the maximum weight edge, and the initial value of d in the target N1~N3 is 0. After initialization, it is found that M1 and M3 are matched with N1, and try to change the edge weights of M1 and M3 to other values, but they do not satisfy $M_i + N_i = W_{ij}$. As such, a conflict arises. In order to resolve the conflict, the KM algorithm subtracts 0.1 from the W value of M1 and M3, and adds 0.1 to the d value of N1. At this point, M3 and N2 satisfy 0.8 + 0 = 0.8, and M1 and N1 also satisfy 0.7 + 0.1 = 0.8. The matching results obtained using the KM algorithm are: M1 matches N1, M2 matches N3, and M3 matches N2. The KM algorithm (total weight 0.8 + 0.9 + 0.8 = 2.5) is better than the Hungarian algorithm (total weight 0.8 + 0.6 + 0.5 = 1.9). 0.5 = 1.9) at matching yields with greater correlations.

4. Experimental Results and Analysis

4.1. Network Training Experiments

Due to the complex conditions of inland waters, the changeable weather, and the diversity of inland vessel types, datasets also require a large number of data sources. There are four main ways through which data sources were collected in this section: (1) ship images were collected and screened through search engines such as Baidu, Google, and Bing [24]; (2) high-definition surveillance cameras were built at fixed locations next to both banks of the Wuhan basin of the Yangtze River, which captured images cropped from the videos of ship navigation between 11 June 2019 and 17 November 2019; (3) the image of the ship was captured with a digital camera in the Changjiang River Basin of Wuhan City, such as: erqi River Bank, Tianxingzhou Ferry Port, Hankou River Bank at a frequency of one per second, from 21 July 2020 to 26 November 2020. In this paper, data were collected from many locations and over a large time span, so the collected ship dataset meets the requirements of large data volume and sample types. The number of data sources is summarized as shown in Table 2. The ship types are divided into six categories as shown in Figure 13, and the statistical information of ship image data is shown in Table 3.

Table 2. Statistics of the number of data sources.

Source	Quantity (Sheets)	Percentage (%)
Search Engine	416	5.9
Field shooting	2061	29.4
Surveillance video data	4523	64.6

(a) (b) (c) (d) (e) (f)

Figure 13. Vessel classification. (**a**) ore carrier. (**b**) bulk cargo carrier. (**c**) general cargo ship. (**d**) container ship. (**e**) fishing boat. (**f**) passenger ship.

Table 3. Number and proportion of ship types.

Ship Type	Quantity (Sheets)	Percentage (%)
Ore Ships	1751	25.0
Bulk Carrier	1498	21.4
Miscellaneous Cargo Ships	1149	16.4
Container ship	702	10.0
Fishing boats	1597	22.8
Passenger Ship	303	4.3

The ratio of the training set, validation set, and test set was 16:3:1. In order to ensure the objectivity of the experimental results, the hyper-parameter settings were consistent for different models, and some of the hyper-parameter settings related to the experiments are shown in Table 4.

Table 4. Hyper-parameter settings.

Hyper-Parameter Name	Numerical Value
Batch Size	64
Weight Decay	0.0005
Momentum	0.937
IOU Threshold	0.2
Loss Gain	21.35
Epoch	100
Learning Rate	0.002324

4.2. Chimney Inspection Experiments

The graded detection results are shown in Figure 14 below, where each column represents a set of data. The first and second rows show the raw data from the visible and infrared cameras, respectively; the third row shows the first-level detection, i.e., the result of detection by the deep learning detector; the fourth row passes the first-level detection result and uses the two-step Ostu binarization algorithm to obtain the region with higher temperature, and further filters the non-chimney highlighted region by leveling the upper and lower regions; the fifth row filters the noise points through the image erosion operation, and expands the chimney candidate region through the expansion operation to expand the chimney candidate area; and the sixth row calculates the maximum value of the area of the chimney candidate area, and draws the contour of the maximum value as the final chimney detection area. After the first-level detection to narrow the range, the background interference can be reduced, and the accuracy of ship chimney detection can be improved.

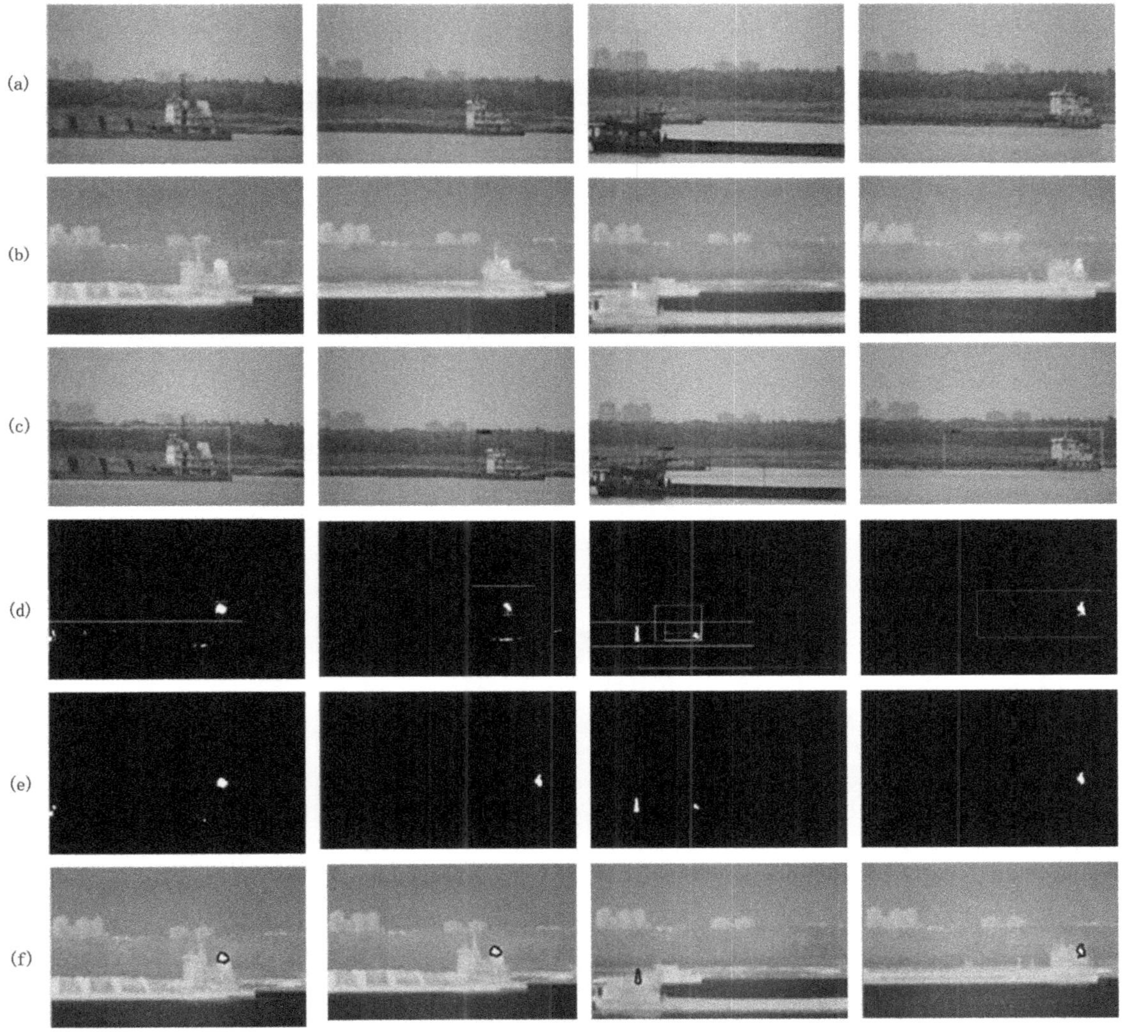

Figure 14. Chimney inspection. (**a**) Visible light input diagram. (**b**) Infrared camera input diagram. (**c**) First-level test results. (**d**) Optimization of detection range. (**e**) Secondary test results. (**f**) The results are displayed in the original image.

4.3. Model Evaluation

4.3.1. Evaluation Index of Ship Detection Model

We conducted numerical experiments on YOLOV3 [10] and our ship detection method. To evaluate the performance of the two models, we used the evaluation metrics, mainly Precision (accuracy), Recall (recall), mAP (mean average precision), and F1 (F1-Measure), and the calculation formula is as in Equation (9), where AP is the area value under the curve calculated by integration after the P-R curve is smoothed, and mAP is the mean value of AP for all categories.

$$\begin{cases} \text{precision} = \frac{TP}{TP+FP} \\ \text{recall} = \frac{TP}{TP+FN} \\ \text{F1} = \frac{2 \times \text{precision} \times \text{recall}}{\text{precision} + \text{recall}} \end{cases} \tag{10}$$

where TP means the sample is marked as positive, FP means the sample is marked as positive by error, TN means the sample is marked as negative by correct, and FN means the sample is marked as negative by error.

After 100 rounds of iterative training using the migration study methodology, the training results are shown in Figure 15. The abscissa represents the number of training iterations, and the ordinate indicates accuracy, average recall, average accuracy and F1, respectively. As can be seen from the result graph, after the number of iterations reaches 40 rounds, the four basic parameters are stable at about 92%.

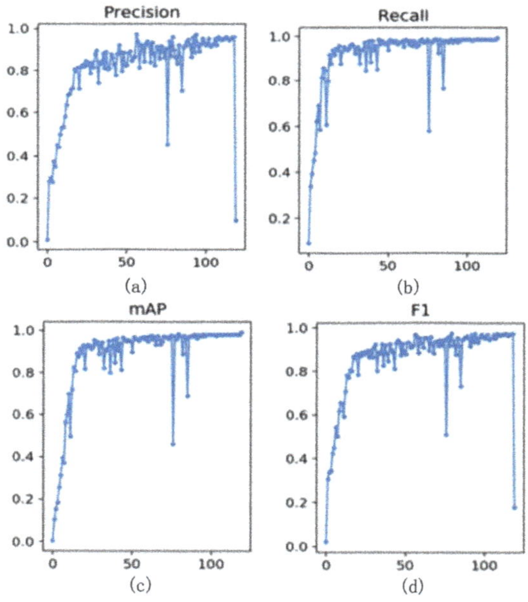

Figure 15. Results after 100 iterations. (**a**) Accuracy results. (**b**) Recall results. (**c**) mAP results. (**d**) F1 results.

In order to illustrate the effectiveness of our model, we carried out experiments in the same software and hardware environment, and the specific parameters are shown in Table 5. The calculation times of YOLOV3 model and our model were counted. To ensure the fairness of time cost, our calculation time was divided into two parts: training time, and verification time. The time consumed by each epoch is calculated in Table 4.

Table 5. Computational cost (SECOND).

Method	YOLOV3		OURS	
Category	Train	Val	Train	Val
Time	9.46	5.48	6.83	3.52

From the data in Table 4, we can see that the average calculation time of each epoch of YOLOV3 is 9.46, while the time of our model is lower, at 6.83. Compared with the calculation time of the validation model, our model is also faster. Therefore, our model has a good effect on real-time tasks, such as the detection and tracking of ship chimneys in inland rivers.

Under the same hyper-parameters, dataset, and the same experimental environment, the improved network and the original YOLOV3 network were compared experimentally, and the experimental results for each category are shown in Figure 16. The horizontal coordinates are the confidence values, and the vertical coordinates are the values of each metric at the current confidence level, and the overall values are shown in Table 6.

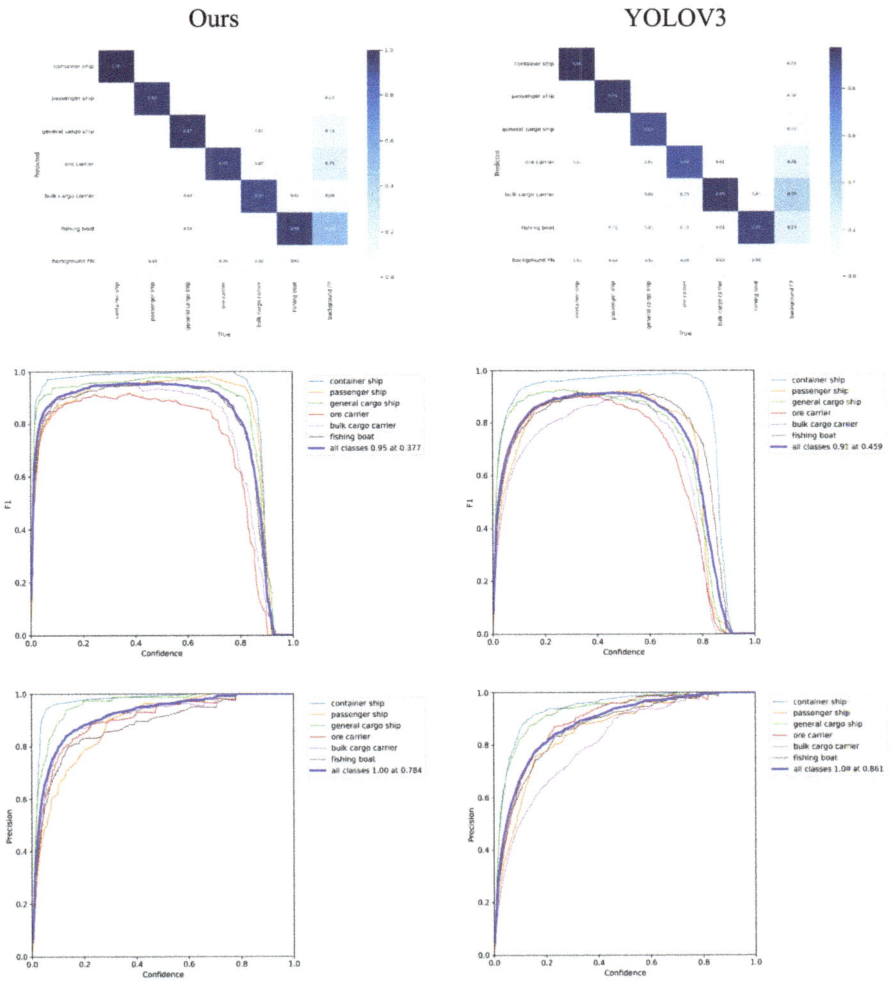

Figure 16. *Cont.*

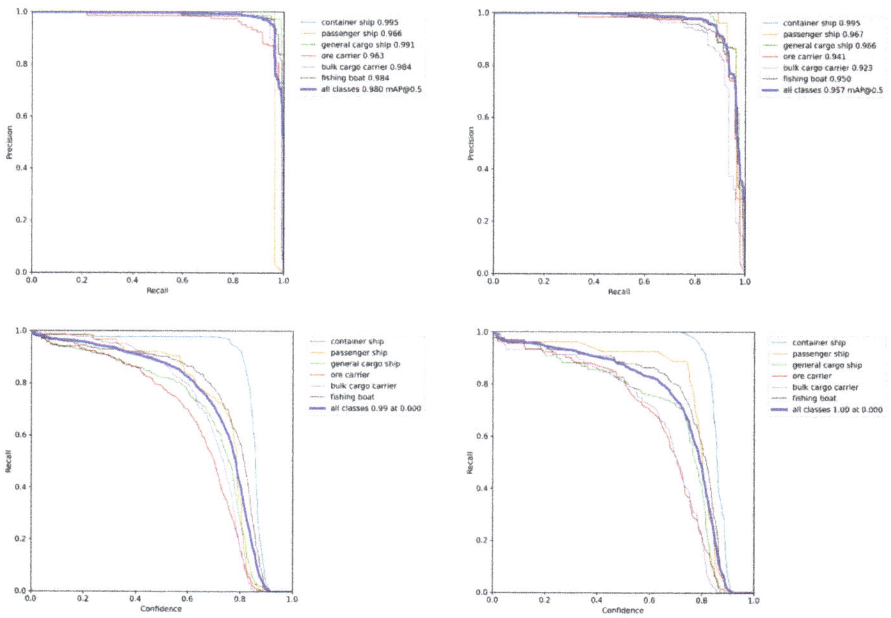

Figure 16. Graph of the results of various ship indicators on the test set.

Table 6. Overall index results on the test set.

Evaluation Indicators	Ours	YOLOV3
Precise	0.95	0.93
recall	1.00	0.99
mAP	0.980	0.957
F1	0.95	0.91
FPS	36	28

From the experimental results, our model can reach an accuracy of 1 when the confidence level is taken as 0.784, while the original model needs to be taken as 0.861 to reach 1. We can also see from the confusion matrix that the improved model has significantly less false detections than the original model, and the detector in this paper outperforms the YOLOV3 model in terms of accuracy, recall, mAP, and F1 indexes, especially in terms of detection speed, which is significantly higher than in the original model. The visualization of detection results is shown in Figure 17.

4.3.2. Ship Tracking Model Evaluation Index

In this paper, three metrics were chosen to evaluate the effectiveness of multiple object tracking: (1) ID switch indicates the number of times the target label is changed in a tracking track, and the smaller the value, the better; (2) multiple object tracking accuracy (MOTA) mainly considers the matching errors of all objects in the tracking process, mainly the FP, FN, and ID switch. MOTA gives a very intuitive measure of the performance of the tracking algorithm in detecting objects and maintaining the trajectory, independent of the progress of target detection. A larger MOTA value indicates a better performance of the model. MOTA is calculated as:

$$M_{OTA} = 1 - \frac{\sum (A_{FP} + A_{FN} + A_{ID})}{\sum A_{GT}} \tag{11}$$

where A_{FP} is the number of false positive cases, A_{FN} is the number of false negative cases, M_{OTA} is the multi-target tracking accuracy, A_{ID} is the number of ID switches, and A_{GT} is the number of labeled targets; (3) FPS, the number of image frames per second processed by the model—the larger the value, the better the processing effect. To verify the performance of the method in this paper in chimney tracking, the test was conducted on the video surveillance data of the Yangtze River Bridge, and the test results are shown in Figure 18. The results are also shown in Table 5.

Figure 17. Visualization of test results.

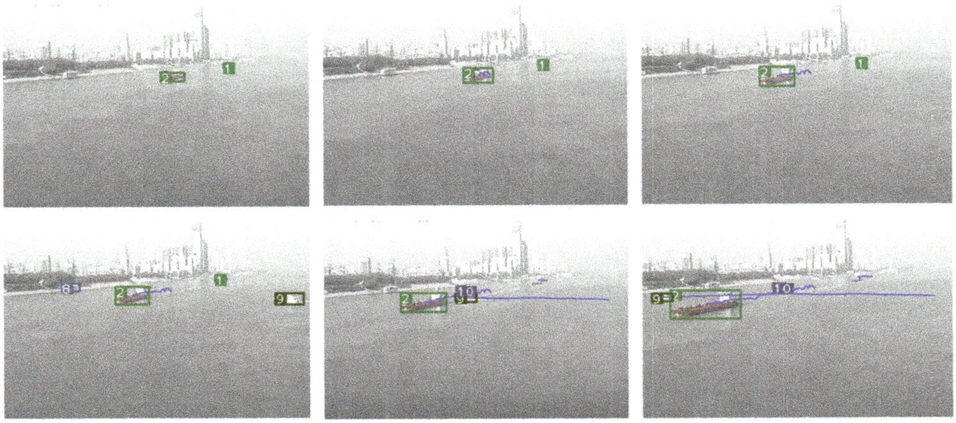

Figure 18. Vessel tracking visualization.

Among them, the blue box is the box detected by the deep learning detector, and the yellow box is the final box after the Kalman filter update. From the experimental results, it can be seen that the ID jump frequency decreases when the target is occluded, and the accuracy rate increases by 0.04. The specific experimental results evaluation index is shown in the Table 7.

Table 7. Results of tracking metrics in the test set.

Evaluation Indicators	Ours	DeepSORT
Multi-target tracking accuracy	0.75	0.71
Number of target marker changes	4	6
Frame rate	6.1	7.4

5. Conclusions

In this paper, we propose a deep learning-based multi-sensor hierarchical detection and tracking method for inland river ship chimneys, which makes full use of the image characteristics of different sensors, and combines the hierarchical idea to solve the problems encountered in practical engineering problems. The method uses visible images with rich feature information, combining deep neural networks to detect inland river ships, filtering irrelevant background information, and using the infrared camera's sensitivity to temperature to locate ship chimneys to ensure high accuracy of detection results under inland river waters with complex backgrounds. The reliability and practicality of the method are proved by field experiments. It makes a certain contribution to assisting the monitoring of automatic air pollution.

Author Contributions: Conceptualization, F.W. and Q.C.; methodology, F.W.; software, F.W.; validation, F.W.; formal analysis, Q.C.; investigation, F.W.; resources, Y.W and C.X.; data curation, F.W. and F.Z.; writing—original draft preparation, F.W. F.Z. and Q.C.; writing—review and editing, Q.C.; visualization, Y.W.; supervision, Y.W.; project administration, Y.W; funding acquisition, Y.W. and C.X. All authors have read and agreed to the published version of the manuscript.

Funding: This work was supported by the Natural Science Foundation of Shandong Province under Grant ZR2020KE029; by the National Natural Science Foundation of China under Grant 52001241; by the 111 Project (B21008); by the Zhejiang Key Research Program under Grant 2021C01010.

Institutional Review Board Statement: Not applicable.

Informed Consent Statement: Not applicable.

Data Availability Statement: The data presented in this study are openly available in [SeaShips] at [doi:10.1109/TMM.2018.2865686], reference number [24].

Conflicts of Interest: The authors declare that they have no known competing financial interests or personal relationships that could have appeared to influence the work reported in this paper.

References

1. Píštěk, V.; Kučera, P.; Fomin, O.; Lovska, A. Effective mistuning identification method of integrated bladed discs of marine engine turbochargers. *J. Mar. Sci. Eng.* **2020**, *8*, 379. [CrossRef]
2. Choi, Y.S.; Lim, T.W. Numerical simulation and validation in scrubber wash water discharge from ships. *J. Mar. Sci. Eng.* **2020**, *8*, 272. [CrossRef]
3. Kim, K.H.; Kong, K.J. 1D-3D Coupling for gas flow analysis of the air-intake system in a compression ignition engine. *J. Mar. Sci. Eng.* **2021**, *9*, 553. [CrossRef]
4. Pankratova, N.V.; Belikov, I.B.; Belousov, V.A.; Kopeikin, V.M.; Skorokhod, A.I.; Shtabkin, Y.A.; Malafeev, G.V.; Flint, M.V. Concentration and isotopic composition of methane, associated gases, and black carbon over Russian Arctic seas (shipborne measurements). *Oceanology* **2020**, *60*, 593–602. [CrossRef]
5. Arshad, N.; Moon, K.-S.; Kim, J.-N. Multiple Ship Detection and Tracking Using Background Registration and Morphological Operations. In *Signal Processing and Multimedia, International Conferences on Signal Processing, Image Processing and Pattern Recognition/Multimedia, Computer Graphics and Broadcasting*; Springer: Berlin/Heidelberg, Germany, 2010; pp. 121–126.
6. Zhang, X.; Wang, G.; Zhu, P.; Zhang, T.; Li, C.; Jiao, L. GRS-Det: An anchor-free rotation ship detector based on Gaussian-mask in remote sensing images. *IEEE Trans. Geosci. Remote Sens.* **2020**, *59*, 3518–3531. [CrossRef]
7. Wang, Y.; Mao, L. Ship Target Detection and Traffic Statistics Algorithm for Waterway. *J. Wuxi Polytech.* **2015**, *14*, 37–40.
8. Tang, Y.; Liu, S.; Zhou, C.; Yan, H.; Jin, H.; Zhu, Y. Intelligent monitoring system of inland waterway based on multi-vision information fusion. *J. Image Graph.* **2008**, *8*, 1608–1616.

9. Shi, W.; An, B. Port Ship Detection Method Based on Multi-Structure Morphology. *J. Comput. Syst. Appl.* **2016**, *25*, 283–287.
10. Cui, Z.; Li, Q.; Cao, Z.; Liu, N. Dense attention pyramid networks for multi-scale ship detection in SAR images. *IEEE Trans. Geosci. Remote Sens.* **2019**, *57*, 8983–8997. [CrossRef]
11. Cui, Z.; Wang, X.; Liu, N.; Cao, Z.; Yang, J. Ship detection in large-scale SAR images via spatial shuffle-group enhance attention. *IEEE Trans. Geosci. Remote Sens.* **2020**, *59*, 379–391. [CrossRef]
12. Chen, X.; Qi, L.; Yang, Y.; Luo, Q.; Postolache, O.; Tang, J.; Wu, H. Video-based detection infrastructure enhancement for automated ship recognition and behavior analysis. *J. Adv. Transp.* **2020**, *2020*, 7194342. [CrossRef]
13. Shao, Z.; Wang, L.; Wang, Z.; Du, W.; Wu, W. Saliency-aware convolution neural network for ship detection in surveillance video. *IEEE Trans. Circuits Syst. Video Technol.* **2019**, *30*, 781–794. [CrossRef]
14. Yang, X.; Sun, H.; Fu, K.; Yang, J.; Sun, X.; Yan, M.; Guo, Z. Automatic ship detection in remote sensing images from google earth of complex scenes based on multiscale rotation dense feature pyramid networks. *Remote Sens.* **2018**, *10*, 132. [CrossRef]
15. Wei, S.; Su, H.; Ming, J.; Wang, C.; Yan, M.; Kumar, D.; Shi, J.; Zhang, X. Precise and robust ship detection for high-resolution SAR imagery based on HR-SDNet. *Remote Sens.* **2020**, *12*, 167. [CrossRef]
16. Lin, Z.; Ji, K.; Leng, X.; Kuang, G. Squeeze and excitation rank faster R-CNN for ship detection in SAR images. *IEEE Geosci. Remote Sens. Lett.* **2018**, *16*, 751–755. [CrossRef]
17. Fu, J.; Sun, X.; Wang, Z.; Fu, K. An anchor-free method based on feature balancing and refinement network for multiscale ship detection in SAR images. *IEEE Trans. Geosci. Remote Sens.* **2020**, *59*, 1331–1344. [CrossRef]
18. Chang, Y.L.; Anagaw, A.; Chang, L.; Wang, Y.C.; Hsiao, C.Y.; Lee, W.H. Ship detection based on YOLOv2 for SAR imagery. *Remote Sens.* **2019**, *11*, 786. [CrossRef]
19. Jin, L.; Liu, G. An approach on image processing of deep learning based on improved SSD. *Symmetry* **2021**, *13*, 495. [CrossRef]
20. Wang, Y.; Wang, C.; Zhang, H. Combining a single shot multibox detector with transfer learning for ship detection using sentinel-1 SAR images. *Remote Sens. Lett.* **2018**, *9*, 780–788. [CrossRef]
21. Sun, J.; Xu, Z.; Liang, S. NSD-SSD: A Novel Real-Time Ship Detector Based on Convolutional Neural Network in Surveillance Video. *Comput. Intell. Neurosci.* **2021**, *2021*, 7018035. [CrossRef] [PubMed]
22. Chen, P.; Li, Y.; Zhou, H.; Liu, B.; Liu, P. Detection of small ship objects using anchor boxes cluster and feature pyramid network model for SAR imagery. *J. Mar. Sci. Eng.* **2020**, *8*, 112. [CrossRef]
23. Wojke, N.; Bewley, A.; Paulus, D. Simple online and realtime tracking with a deep association metric. In Proceedings of the 2017 IEEE International Conference on Image Processing (ICIP), Beijing, China, 17–20 September 2017; pp. 3645–3649.
24. Shao, Z.; Wu, W.; Wang, Z.; Du, W.; Li, C. SeaShips: A Large-Scale Precisely Annotated Dataset for Ship Detection. *IEEE Trans. Multimed.* **2018**, *20*, 2593–2604. [CrossRef]

Journal of
Marine Science and Engineering

MDPI

Article

Use of Hybrid Causal Logic Method for Preliminary Hazard Analysis of Maritime Autonomous Surface Ships

Di Zhang [1,2,3], Zhepeng Han [1,4], Kai Zhang [4,5], Jinfen Zhang [2,4], Mingyang Zhang [6] and Fan Zhang [2,7,*]

1 School of Transportation and Logistics Engineering, Wuhan University of Technology, 1040 Heping Avenue, Wuhan 430063, China; zhangdi@whut.edu.cn (D.Z.); hanzhepeng@whut.edu.cn (Z.H.)
2 National Engineering Research Center for Water Transport Safety, Wuhan University of Technology, 1040 Heping Avenue, Wuhan 430063, China; jinfen.zhang@whut.edu.cn
3 Inland Port and Shipping Industry Research Co., Ltd., Shaoguan 512100, China
4 Intelligent Transportation Systems Research Center, Wuhan University of Technology, 1040 Heping Avenue, Wuhan 430063, China; zhang_kai03@126.com
5 Chinese Flight Test Establishment, 8 Shifeiyuan Road, Xi'an 710089, China
6 Department of Mechanical Engineering, Aalto University, 02150 Espoo, Finland; mingyang.0.zhang@aalto.fi
7 School of Navigation, Wuhan University of Technology, 1040 Heping Avenue, Wuhan 430063, China
* Correspondence: michael_zf@whut.edu.cn

Abstract: Recently, the safety issue of maritime autonomous surface ships (MASS) has become a hot topic. Preliminary hazard analysis of MASS can assist autonomous ship design and ensure safe and reliable operation. However, since MASS technology is still at its early stage, there are not enough data for comprehensive hazard analysis. Hence, this paper attempts to combine conventional ship data and MASS experiments to conduct a preliminary hazard analysis for autonomy level III MASS using the hybrid causal logic (HCL) method. Firstly, the hazardous scenario of autonomy level III MASS is developed using the event sequence diagram (ESD). Furthermore, the fault tree (FT) method is utilized to analyze mechanical events in ESD. The events involving human factors and related to MASS in the ESD are analyzed using Bayesian Belief Network (BBN). Finally, the accident probability of autonomy level III MASS is calculated in practice through historical data and a test ship with both an autonomous and a remote navigation mode in Wuhan and Nanjing, China. Moreover, the key influence factors are found, and the accident-causing event chains are identified, thus providing a reference for MASS design and safety assessment process. This process is applied to the preliminary hazard analysis of the test ship.

Keywords: maritime autonomous surface ships; hybrid causal logic; preliminary hazard analysis; risk assessment; hazard identification

Citation: Zhang, D.; Han, Z.; Zhang, K.; Zhang, J.; Zhang, M.; Zhang, F. Use of Hybrid Causal Logic Method for Preliminary Hazard Analysis of Maritime Autonomous Surface Ships. *J. Mar. Sci. Eng.* **2022**, *10*, 725. https://doi.org/10.3390/jmse10060725

Academic Editor: Alessandro Ridolfi

Received: 20 April 2022
Accepted: 23 May 2022
Published: 25 May 2022

1. Introduction

Thanks to the rapid development of the artificial intelligence and 5G technology, autonomous ships will become one of the key transportation vehicles in the future [1–3]. Nowadays, several companies and organizations have performed research on MASS. The vehicle ferry Falco successfully navigated autonomously during its voyage between Parainen and Nauvo, and its return journey was conducted under remote control [4]. Wärtsilä successfully tested such innovative technology into a voyage, during which a vessel was automatically controlled by a software, while manual intervention and control was still possible at any time [5]. YARA and Kongsberg are building a ship named "YARA Birkeland", which will be the world's first fully electric and autonomous container vessel upon completion [6]. DNV GL built a 1:20 scale model of MASS to investigate sensor fusion and collision avoidance [7]. The AAWA project aimed to produce the preliminary specifications for the next generation of advanced ship solutions [8]. Finally, the MUNIN research project developed a technical concept for the operation of an unmanned merchant ship and assessed its technical, economic and legal feasibility [9]. In order to clarify the definition of autonomous ships, the International Maritime Organization (IMO) defines autonomous ships as maritime autonomous surface ships (MASS). MASS is classified into four degrees according to their autonomy level, as follows [10,11]. Note that, during the navigation of MASS, the MASS can change the autonomy level according to the scenario:

- Autonomy level I: Ship with automated processes and decision support: Seafarers are on board to operate and control shipboard systems and functions. Some operations may be automated;
- Autonomy level II: Remotely controlled ship with seafarers on board: The ship is controlled and operated from another location, but seafarers are on board;
- Autonomy level III: Remotely controlled ship without seafarers on board: The ship is controlled and operated from another location. There are no seafarers on board;
- Autonomy level IV: Fully autonomous ship: The operating system of the ship is able to make decisions and determine actions by itself.

The safety of MASS will become a key issue for autonomous ship operations. MASS should have the desired level of safety, i.e., at least the same safety level as conventional ships [12]. Researchers believe that, compared to conventional ships, MASS are more economical and safer due to the reduction in crew on board [13,14]. Moreover, changed technologies, systems and procedures also bring new influence factors [15–17]. Thus, there is an urgent need for a risk assessment of MASS to assist MASS design.

Maritime risk assessments are considered a hotspot for MASS [18–20]. Due to complexity and novelty of MASS, several studies were performed for hazard identification, which is the basis for risk assessment. Fan et al. proposed a framework for the identification of factors that influence the navigational risk of remotely controlled MASS without crew on board [21]. It classifies a total of 55 influence factors into ship-related, human-related, environment-related and technology-related factors. More in detail, failure of onboard equipment may result in the degradation or failure of functions related to propulsion. At the same time, the results show that the majority of these influence factors are related to human error. Kretschmann et al. [22] found 23 identified hazards with acceptable risk based on a formal safety assessment (FSA). These hazards are related to various influence factors such as weather, equipment and cyber security. Human errors may be related to remote monitoring, control and maintenance. At the same time, this study shows that a failure of the power and propulsion system will lead to unacceptable consequences. Wróbel et al. [19] reviewed a hundred maritime accident reports, analyzing various safety hazards that lead to accidents for conventional ships based on what-if and human factors analysis and classification system for marine accident (HFACS-MA) methods, and considering the impact of these safety hazards on MASS. The results show the existence of the human factor in unmanned systems' operation, as long as people are involved in operation. In summary, almost all studies on MASS hazard identification mentioned the complexity and diversity of MASS influencing factors, as well as the significant influence of mechanical failure and human error.

Based on hazard identification, some studies have been conducted to analyze equipment failure and human error. In relation to the human error in SCC, Ramos et al. [23] divided the possible human error process into four stages, and established an event tree model of the MASS. Moreover, they classified the influencing factors, describing their differences across various human factor reliability analysis methods and the shortcomings of the current behavior influencing factor set, simulating the human–machine interaction process and proposing an avoidance based on hierarchical task analysis. Man et al. [24] invited six participants to conduct a scenario-based simulation as proposed operators in the SCC. Their conclusions suggest that human factor issues, such as psychophysical and perceptual limitations of operators, decision-making latencies and automation bias, may remain in systems assembled by assumed reliable technological components. Zhang et al. [11] presented a model based on the Technique for Human Error Rate Prediction (THERP) and on Bayesian Network, which can depict the causal relationship focused on human–autonomy collaboration and perform a quantitative assessment. Unlike for human errors, research on equipment failure focuses mainly on power and propulsion systems. Bolbot et al. [25] analyzed the hazards related to the electric propulsion system based on the System theoretic process analysis (STPA) method. In addition, Bolbot et al. [26] combined event tree analysis (ETA), fault tree and STPA method to analyze a simplified diesel electric propulsion system and identify the hazardous scenarios leading to a blackout. Wang et al. [27] determined the weakness of the ship power system and put forward a design of the ship power plant. These studies provided a reference for MASS designers in case of human error or equipment failure. However, they overlooked the influence of individual factors on the safety of the entire MASS, and often neglected the mutual influence of different factors.

The hazard scenario of MASS usually gradually evolves from a hazard event. Different outputs of safety barriers in this process will lead to different end states. The interaction among influence factors needs to be taken into consideration in this complex process [28]. Thieme et al. [29] formulated nine criteria and used them to assess 64 relevant ship risk models since 2005. The results show that none of them are suitable to be directly used for MASS risk assessment. In fact, MASS risk assessment should comprehensively include various influence factors, instead of only analyzing specific factors. Accordingly, new methods have been applied for MASS risk assessment. The STPA method has been applied to MASS, as it can analyze the interactions between its components. Valdez Banda et al. [30] applied the STPA to analyze the safety hazards in the foreseen functioning of two concepts of autonomous ferries operating in urban waterways in, and near, the city of Turku in Finland. Employing the STPA, a safety-controlled structure and hazard list has been created for the system to ensure that remotely controlled ships do not have a negative impact on maritime safety [18]. Wróbel et al. [31] applied the STPA to identify the hazards, formulate hazard mitigation and improve the safety performance of autonomous ships. In addition, Utne et al. [16] proposed a framework combining STPA and Bayesian Belief Networks to establish an online risk model for autonomous ships. In parallel, Ramos et al. [32] proposed the human–system interaction in autonomy (H-SIA) method, which consists of an event sequence diagram (ESD) and concurrent task analysis (CoTA), to analyze the system as a whole and focus on the interactions between sub-systems. At the same time, Ramos et al. [28] extended the H-SIA to include the paths to failure through the Fault Tree (FT). However, these approaches can only be used in qualitative analyses, and are not suitable to perform quantitative analyses. The relationship of potential hazards of MASS can be easily described by these qualitative methods. However, the failure probability and sensitivity of potential hazards cannot be obtained. The results have limited contribution to the safety design of MASS.

Since MASS is still in the experimental stage and concept stage rather than the operation stage, there are insufficient data to quantitatively analyze the risk of MASS. A preliminary risk analysis should take place to evaluate the ability of the MASS to operate safely and reliably during the concept and experimental stages [12]. In this study, we want to develop a model which can perform a preliminary hazard analysis of MASS. For the function during concept stage, the historical data such as failure rate are used for qualitative analysis. For the function during the experimental stage, the experimental data of the MASS model are used to develop the quantitative model. This result will be used to further improve the performance of MASS experiment. At the same time, the data can assist in judging whether these concepts of MASS are suitable or not and help develop the function which is still in the concept stage.

The shift from conventional ships to autonomous ships is a gradual process [21]. Compared with conventional ships, the MASS will be equipped with an autonomous system (AS) that may help or replace human decision-making and action. At the highest level of autonomy, MASS can be controlled by AS completely. Given the current development of MASS technology, in the near future, MASS will have a constrained autonomy, and their operation will be supervised or controlled by a shore control center (SCC) [33]. Autonomy level III MASS will be an important stage with the participation of AS and operators in

SCC. According to the elaboration of autonomy level III, MASS are equipped with AS, an advanced sensor module, a SCC, a satellite communication equipment, alarm devices, other facilities and without anyone onboard. Various sensors will provide sufficient data for AS system and SCC to identify the navigation status and environment. The AS system can control navigation according to the surrounding environment and ship condition; in case of hazardous events, it will propose strategies to guarantee the safety of MASS. At the same, the operator in SCC will supervise the operation of MASS, including the operating environment, decision proposed by AS, etc. The remote operator has the highest right to take over the control of MASS at any time. In case the AS system cannot propose effective measures or a situation develops in a particularly difficult direction, the SCC can take over the control of the MASS and dispatch a professional team to deal with problems [34]. Above all, the autonomy level III MASS is a suitable object to conduct a preliminary hazards analysis for MASS.

The hybrid causal logic (HCL) methodology provides a vehicle for the identification and communication of cause–effect relations including those associated with human, organization and system hardware and software, and the physical and regulatory environment of the system [35]. The HCL method uses ESD as the first layer to describe system behavior, and then provides a more detailed picture of the contributing causes by using FTs. Fault tree analysis is the one of the popular techniques used for reliability studies for a complicated system [36]. Fault trees are widely used in mechanical systems with obvious structure and causal logic such as the aviation industry and offshore systems. Mohaghegh et al. [37] applied the HCL method to include the organizational roots of risk. Groth et al. [38] introduced a software platform for the HCL method and applied it to analyze a type of aviation accidents. Røed et al. [39] discussed the applicability of HCL to the offshore industry and its relationships with the barrier and operational risk analysis project (BORA). Sklet et al. [40] applied the HCL to analyze the installation-specific factors with respect to technical systems, operational conditions, and human and organizational factors. Thus, the HCL method is a suitable tool to analyze MASS, as it includes various influence factors.

Based on these considerations, this article hopes to introduce the HCL method into MASS to assist the early design of MASS. Taking contact hazards as an example, this paper applies the experimental MASS model and historical data to conduct hazard analysis on MASS. The ESD was applied to define the hazard scenario, focusing on the interaction between AS and operators in the SCC. For non-human-related events (such as mechanical failure) that can be decomposed into the equipment level, we applied the FT to develop a branch model to analyze in detail the influence factors. The concept of the mechanical system of MASS and the failure data of conventional ships were used to conduct a preliminary analysis. As for human- and organization-related events, due to their uncertainties, we applied the Bayesian Belief Network (BBN) to analyze in detail the influence factors based on the experimental statistics This process was applied to demonstrate a case study of a test ship, equipped with an autonomous navigation mode and a remote navigation mode in Wuhan and Nanjing, China.

The rest of this paper is organized as follows. Section 2 describes the HCL methodology used to develop the model. Section 3 presents the MASS hazard scenarios. Section 4 introduces the quantitative case study of contact scenario. Finally, Section 5 presents the conclusions of this study and the future work.

2. Methodology

HCL methodology is a powerful modeling tool for developing hazards scenarios and search the more detail potential hazards. Figure 1 presents the main framework and the flowchart of the HCL method. The application of HCL can be divided into 4 steps and described in detail below.

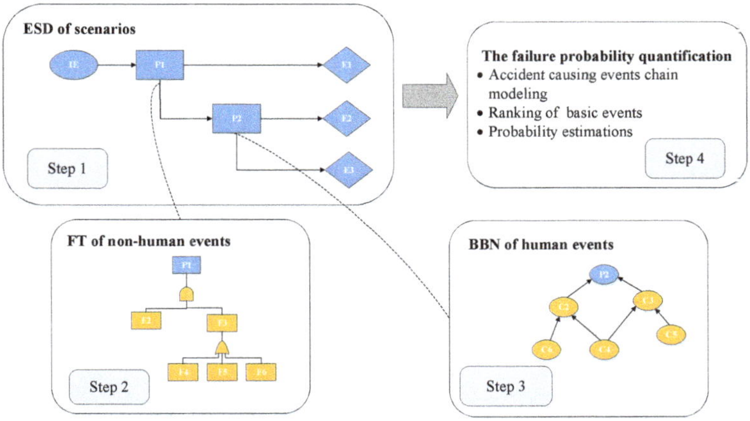

Figure 1. Framework and flowchart of the HCL method.

Step 1: Development of a MASS hazard scenario using ESD. ESDs are used to define the system hazard scenarios. The ESD presents a temporal sequence of events, from an initiating event to various end states. The initiating event (IE) is commonly a hazardous event or a source of risk. Once a hazardous event occurs, some safety barriers, regarded as pivotal events in ESD, should be adopted to prevent or mitigate the hazard. The output of safety barriers (i.e., normal or failure of operation) determines whether or not the hazardous event evolves into an accident. Different pivotal events and their output will lead to different end states, such as safe or accident states. In order to determine the probability of each end state, the probability of each pivotal event output (i.e., normal or failure of operation) must be obtained. According to the characteristics of pivotal events, their detailed influence factors can be analyzed using FT and BBN. In this study, the equipment events were analyzed using FT, as shown in Step 2. The events involving human factors were analyzed using BBN, as shown in Step 3.

Step 2: Analysis of mechanical events using FT. The FT is used to develop a branch model to quantitative analyze mechanical events in ESD. Fault tree analysis is the one of popular techniques used for reliability studies for a complicated system. The system failure event is regarded as top event. The subsystem failure events which may cause the top events are identified and linked to top event through logical connective function (such as AND/OR gate) [36]. Fault trees are widely used in mechanical systems with obvious structure and causal logic such as the aviation industry and offshore systems. The quantitative analysis of the fault tree first needs to convert the logical structure established by it into an equivalent probability expression. Once the failure rate and operation time are obtained, the failure probability of the basic event can be calculated. Thus, according to the equivalent probability expression, the failure probability of the top event can also be obtained.

Step 3: Analysis of events related to human factors using BBN. Unlike for mechanical events, the events related to human factors are non-deterministic and uncertain, and can be effectively analyzed using BBN. The BBN network consists of nodes and directed arcs. The events involving human factors in ESD are regarded as target nodes in the BBN network. The detailed influence factors of the events involving human factors are regarded as sub-nodes. The nodes are divided into various states according to their characteristics and requirements, while the arcs between nodes represent the direct influences. Similar to FT, the BBN also allows us to quantify the probability of events in the ESD when the probability of root nodes and conditional probability table are obtained (see further details in Section 3.3).

Step 4: Quantification of the failure probability. The probability of events in ESD are calculated in Steps 2 and 3. This way, we obtain the occurrence of various end states by logics. At the same time, the hazard scenario can be expressed by the accident-causing events. These chains of events can be ranked according to their probabilities. In addition, important measures are adopted to provide information about the criticality of basic events according to their contribution to the overall system performance (see Section 4 for further details).

3. HCL Model for the Hazard Scenario for Mass

The preliminary hazards analysis for MASS should at least cover the relevant hazards such as collision/contact, grounding, unable to detect, etc. [12,41–43]. In this section, we take contact with foreign objects/obstacles (non-detected and detected) as an example. Contact refers to ships striking or being struck by an external object include floating object, fixed object or flying object. According to the definition of the contact scenario, several experiments were carried out in the Tangxun Lake in Wuhan and in the Qinhuai River in Nanjing, China [44]. Through the experiments and historical database, the hazard model for contact scenarios of MASS is developed.

3.1. Develop a MASS Hazard Scenario Using ESD

It is important to understand the entire process of MASS contact scenarios. Once an external object occurs, the AS and the operators in the SCC have a responsibility to detect it and avoid [28]. The MASS will strike or be struck by an external object if the course/speed of the vessels does not change.

To assist in the analysis of the contact scenarios for MASS, the ESD is used to develop a model. *IE* usually refers to potentially hazardous events that may lead to accidents. In the contact scenario, the initiating event (IE) is commonly an external object appears on the planned sailing route. For a better description, several pivotal events and end states of the contact scenario are classified into three stages: (1) hazardous event perception; (2) decision-making; and (3) execution based on the experimental situation combined with experts' knowledge [21,28,45]. They are described as follows:

- λPerception stage: In this stage, the external object is perceived by the MASS; accordingly, information should be acquired based on sensors and human perceptions. Through the analysis of information, the MASS can detect the external object in two ways [12]. The first way mainly relies on sensing devices and AS and is labeled as 'detection by AS' (P1). Accordingly, the MASS is equipped with various sensing devices that ensure a timely perception of hazardous events. The second way is labeled as 'detection by SCC' (P2), where operators in SCC should monitor the MASS in case the external object is not perceived by the sensing devices. A failure in the perception stage will directly lead to an accident.
- λDecision-making stage: In this stage, an agent (either the AS or the operators in the SCC) should propose an effective strategy to prevent contact with external objects according to the data and information gathered at the perception stage. This covers situation assessment, diagnosis and response planning [28]. In this stage, the AS should control the ship and propose a strategy to avoid the external events, an occurrence labeled as 'control by AS' (P3). If the AS cannot propose an effective strategy, the operators in the SCC should take over the control of MASS, an occurrence labeled as 'remote control by the SCC' (P4).
- λExecution stage: In this stage, the MASS should successfully execute the strategy selected at the decision-making stage. More in detail, the actuators will operate a control system to change the course/speed according to the strategy [12]. In this study, the execution system mainly includes the 'steering system' (P5) and the 'power and propulsion system' (P6).

The normality or failure of operation of pivotal events will lead to different end states. In this study, four end states were determined. In the 'normal navigation' (E1) end state, the MASS successfully avoid the objects and has the ability to continue navigation. In 'accident due to perception failure' (E2), the MASS does not recognize external objects and struck with them. In 'accident due to decision-making failure' (E3), the MASS does not propose effective strategies to avoid the external object. Finally, in 'accident due to execution failure' (E4), the MASS does not adjust the speed and course lead in a timely manner due to a mechanical failure resulting in a contact accident.

The description of the pivotal events and of the end states in the contact scenario is presented in Table 1. At the same time, the ESD model for the MASS contact scenario was elaborated and is shown in Figure 2.

Table 1. Description of the nodes in the proposed ESD model.

Stage	Label	Event	Description	Reference
Perception stage	P1	Detection by AS	During the navigation of MASS, equipment such as sensors, laser and range finder should detect navigational hazards or abnormal operational conditions all the time.	[34]
	P2	Detection by SCC	During navigation, the MASS should transmit images and sounds to the SCC, so that the operators may detect the hazardous event.	[28]
Decision-making stage	P3	Control by AS	The AS should choose the optimal maneuver to stop the hazardous event according to the information gathered.	[28]
	P4	Remote control by SCC	When the situation requires navigational operation from the SCC, the operators in the SCC will go into the situation handling room to handle the risk.	[46]
Execution stage	P5	Steer system	The steer system has the responsibility to actuate ship motion. The MASS should control the direction to avoid the hazardous event.	[16]
	P6	Power and propulsion system	The power and propulsion system has the responsibility to actuate ship motion. The MASS should control the speed to avoid the hazardous event.	[47]
End state	E1	Normal navigation	The MASS successfully handles the hazardous event and continues navigation.	[28]
	E2	Accident due to perception failure	MASS does not recognize external objects and struck with it.	[11]
	E3	Accident due to decision-making failure	MASS does not propose an effective strategy to avoid the object.	[11]
	E4	Accident due to execution failure	MASS does not timely adjust the speed and course lead due to a mechanical failure resulting in a contact accident.	[48]

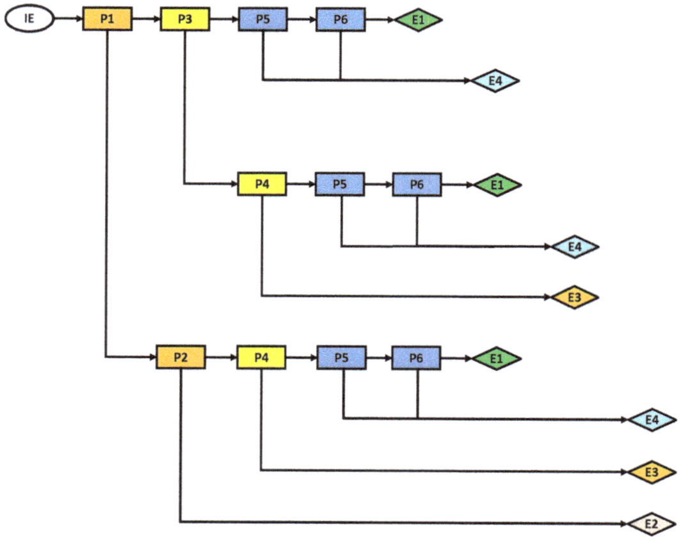

Figure 2. ESD model of the MASS contact scenario.

3.2. Analysis of Mechanical Events Using FT

In order to prevent the contact accident, MASS needs to adjust the course and speed which mainly relied on steer system and power and propulsion system. In this study, we developed a model for the MASS power and propulsion system using the FT method as an example.

Since there is no MASS in operation, its mechanical system structure and failure data cannot be obtained. In the current study, the researchers usually use the failure data of conventional ships to continue the research about the MASS [49]. Thus, in this section, we will develop a FT for the mechanical events of MASS based on the MUNIN report and DNV GL guideline.

In conventional ships, machinery problems have a very high frequency of causing minor incidents which, however, will be more severe in MASS without maintenance [22]. The power and propulsion system of a conventional ship, which includes the main engine, the propeller and the auxiliary system, is considered to be the cause of major ship technical failures. Thus, the normal operation of the mechanical system is key for MASS navigation. There are different opinions about the MASS power and propulsion system. Some projects, such as the AAWA project and the ReVolt project, selected batteries as power source because they have a good efficiency and can ensure zero emissions [8]. In the MUNIN project, the diesel engine propulsion line was selected as the propulsion system [22]. Although the forms of power and propulsion are different, it is commonly accepted that MASS should be purposely built with redundant energy propulsion systems. In this study, we adopted the requirement that MASS should be arranged with a minimum of two independent propulsion lines, as proposed by DNV GL. In parallel, each propulsion line should have a sufficient capacity to meet the specifications for normal operation [12]. This arrangement has two advantages: (i) the two propulsion lines are redundant; and (ii) two independent propulsion lines can prevent common cause failures. In this study, considering that the energy provided by the battery is not enough to support long-term sailing, the diesel electric propulsion was selected as the power and propulsion system. The equipment in the power and propulsion system is shown in Table 2.

Table 2. Description of the components of the power and propulsion system.

Event	Description	References
Bus bar	The bus dispatching power according to the load.	[50]
Transformer	The transformer has the responsibility to obtain different voltage levels and sometimes also to phase shift.	[50]
Converter	The frequency converter has the responsibility to control the shaft line speed.	[51]
Electric motor	The electrical motor is the commonly used device for the conversion of electrical power into mechanical power.	[50]
Diesel generator	Diesel engines supply power to the electric generator shaft.	[52]
Propeller	The electric propulsion motor drives the propeller to provide propulsion.	[51]

According to the FT logic and to the equipment of diesel electric propulsion, we established the FT of diesel electric propulsion systems for MASS. The failure of operation of the 'power and propulsion system' (P6) was regarded as the top event and was labeled as F1. A failure of both the first diesel electric propulsion line (F2) and the second diesel electric propulsion line (F3) will lead to propulsion loss (F1). The second diesel electric propulsion line (F2) has the same arrangement as, and is independent from, the first diesel electric propulsion line (F1). We took the first propulsion line as an example. The single diesel electric propulsion line can be decomposed into three elements: power plant, distribution and loads. The power plant (F5) includes three diesel generators (F16, F17 and F18), two of which can provide sufficient power. Multiple diesel generator sets feed a fixed-frequency high-voltage electrical bus (F6), upon which the distribution depends to dispatch power according to the load. In this section, we only consider the load of the propulsion. This bus feeds the electrical propulsion motor drive, in most cases through a transformer (F7). The electric propulsion motor (F9) drives a frequency converter (F8) to control the shaft line speed and the propeller (F10) to provide propulsion to the MASS [51]. The propulsion system failure was modeled by using FT, as shown in Figure 3. The nodes in the FT are shown in Table 3.

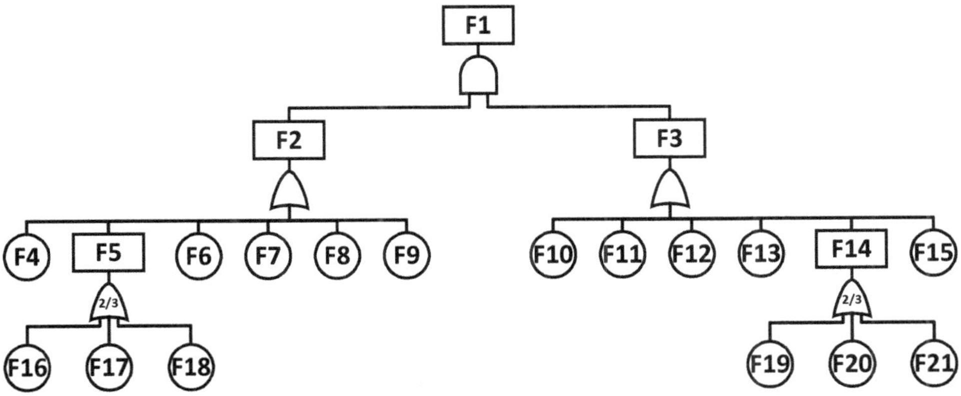

Figure 3. FT of the power and propulsion system.

Table 3. Nodes in the FT of the power and propulsion system.

Node	Event
F1	Power and propulsion system
F2	Diesel electric propulsion 1st line
F3	Diesel electric propulsion 2nd line
F4, F15	Propeller
F5, F14	Power plants
F6, F13	Bus bar
F7, F12	Transformer
F8, F11	Converter
F9, F10	Electric motor
F16, F17, F18, F19, F20, F21	Diesel generator

3.3. Analysis of Events Related to Human Factors Using BBN

Although autonomy level III MASS have no crew on board, the human error in the SCC can still lead to contact, especially in the remote driving mode. In this step, we used BBN to develop a branch model for the 'remote control by the SCC' (P4), which was defined as the target node of the BBN model (C1). The influence of the detailed variables on the 'remote control by the SCC' (C1) is mainly reflected in the form of the various nodes in the network. We first investigate the historical literature to obtain potential influence factors with their associated definitions and descriptions. After that, develop and apply contact scenarios in Tanxun Lake and Qinhuai River, and remotely control MASS ships to conduct contact avoidance experiments. After experimentation and expert judgment, 15 influence factors that influence the 'remote control by SCC' (C1) are regarded as sub-nodes, as shown in Table 4. The process employed is as follows:

Table 4. The influence factors of remote control.

Label	Node	Description	References
C1	Remote control by SCC	The MASS switches into remote control. The operators in the SCC should send updated route information or directly control the MASS and propose a strategy to handle hazardous events.	[32]
C2	Operators' performance	Operators' performance during remote control.	[53]
C3	Ship condition	The various technical aspects of the ship condition, including, but not limited to, communication and engine conditions.	[54]
C4	Operating environment	Weather conditions and traffic density.	[13]
C5	Fatigue	Although more advanced technology can reduce operators' fatigue, a long work schedule may still lead to fatigue.	[11]
C6	Situational awareness	The operators should ensure an appropriate situation awareness of the MASS, despite the physical distance with the crew and vessel.	[9]
C7	Experience	The operators should have the theoretical knowledge of and experience in remote control in a virtual environment.	[11]
C8	Communication and collaboration	The SCC is articulated into specific roles (e.g., supervisor, captain and engineer), which need to communicate and collaborate with each other to handle hazardous events.	[46]

Table 4. *Cont.*

Label	Node	Description	References
C9	Ship's feedback	The MASS should establish a two-way communication with the SCC. The ship's feedback means that the information is transferred from the ship to the SCC.	[31]
C10	Software performance	The onboard software decides the operation of the MASS and the communication with the SCC.	[55]
C11	SCC's feedback	The MASS should establish a two-way communication with the SCC. The ship's feedback means that the information is transferred from the SCC to the ship.	[31]
C12	Communication quality	Quality of the communication between the SCC and the ship.	[33]
C13	Communication bandwidth	Available communication bandwidth during operations.	[12]
C14	Weather conditions	Heavy weather conditions may push the ability to control the ship to the limit, while at the same time affecting communication.	[30]
C15	Traffic density	Traffic density could be specified based on the relevance of potential accident risks in the area.	[13]

(1) Determination of BBN nodes

The 'remote control by the SCC' (C1) is influenced not only by the operators' performance, but also by the ship condition and operating environment. Different from the 'remote control by SCC' (C1), which is a binary node, these influence factors have multiple states. The sub-nodes are classified into multiple states according to the criteria presented in Table 5.

Table 5. Multiple states and description of the sub-nodes in BBN.

Label	Node	State	Description
C2	Operators' performance	Good (a)	The operators are able to operate the ship and to handle hazardous events.
		Medium (b)	The operators are able to fulfil basic requirements for ship operation.
		Bad (c)	Failure by the operators to operate the ship as required.
C3	Ship's condition	Good (a)	The automated function can assist the operator to drive well.
		Medium (b)	The automated function can meet the basic driving requirements.
		Bad (c)	The automated function cannot meet the driving requirements.
C4	Operating environment	Good (a)	The environment has no impact on remote driving.
		Medium (b)	The environment has a slight impact on remote driving.
		Bad (c)	The environment has a serious impact on remote driving.
C5	Fatigue	Good (a)	The operator does not feel tired at all.
		Medium (b)	The operator is slightly tired.
		Bad (c)	The operator feels tired after a long time of operation.
C6	Situational awareness	Good (a)	The operator can clearly judge the situation.
		Medium (b)	The operator can basically judge the situation.
		Bad (c)	The operator cannot accurately judge the situation.
C7	Experience	Good (a)	Operators have sufficient remote driving experience.
		Medium (b)	Operators have some remote driving experience.
		Bad (c)	Operators do not have remote driving experience.

Table 5. *Cont.*

Label	Node	State	Description
C8	Communication and collaboration	Good (a) Medium (b) Bad (c)	Communication is performed as required, obtaining useful results. Communication meets basic information delivery requirements. Failure to communicate and collaborate as required.
C9	Ship's feedback	Good (a) Medium (b) Bad (c)	The ship can feedback sufficient information. The ship can feedback information related to driving. The ship gives no feedback on the ship's situation.
C10	Software performance	Good (a) Medium (b) Bad (c)	The software can meet the driving requirements well. The software can meet the basic driving requirements. The software cannot meet the driving requirements.
C11	SCC's feedback	Good (a) Medium (b) Bad (c)	The SCC has a timely response to the ship. The SCC can feedback information related to driving. The SCC does not respond to the ship in time.
C12	Communication quality	Good (a) Medium (b) Bad (c)	Information can be transmitted well between the ship and the SCC. Incomplete information is transmitted, but sufficient to drive. Sufficient information cannot be transmitted.
C13	Communication bandwidth	Good (a) Medium (b) Bad (c)	Approaching the maximum value. Within the normal range of communication equipment. Only a small amount of information can transmit.
C14	Weather conditions	Good (a) Medium (b) Bad (c)	Have almost no effect on ships. Have a certain impact on ships. Have a greater impact on ship control.
C15	Traffic density	Good (a) Medium (b) Bad (c)	More than three ships. The ship encounters other ships. No ships around when the ship is sailing.

a, b, c represent the abbreviations for the good, medium and bad states, respectively.

(2) Analysis of BBN nodes

The label C1 refers to a situation where the operators in the SCC remotely control the ship and handle the hazardous events. This node is mainly related to three aspects: 'operators' performance' (C2); 'ship's condition' (C3); and 'operating environment' (C4).

The label C2 refers to the operators' performance during the remote control of MASS in the contact scenario. During the remote driving mode, the SCC will assign a group of people including a supervisor, an engineer and a captain to remotely drive the MASS. After a long work schedule, the operators may be in a state of 'fatigue' (C5). 'Situational awareness' (C6) refers to operators' awareness of the current emergency situation of MASS. 'experience' (C7), 'communication and collaboration' (C8) and 'ship's feedback (C9) influence the 'situational awareness' (C6). In terms of 'experience' (C7), the crew group should not only master the ability of remote driving, but also have experience in handling various hazardous events. 'Communication and collaboration' (C8) means that the crew group needs to exchange information and collaborate to propose effective strategies.

The SCC operators cannot handle hazardous events without the support of ship's function. The label C3 refers to whether or not the ship can capture and deliver the necessary information needed by the SCC, which depends on 'software performance' (C10), 'SCC's feedback' (C11) and 'operating environment' (C4). 'Ship's feedback' (C9) and 'SCC's feedback' (C11) refer to the quality of the data and information transferred between the ship and the SCC, which depends on 'software performance' (C10) and 'communication quality' (C12). In turn, 'communication quality' (C12) is related to 'communication bandwidth' (C13) and "operating environment" (C4), and determines the sufficient and timely delivery of information. In case of insufficient communication between the ship and the SCC, 'software performance' (C10) should give priority to providing the urgently needed information, which affects both the 'ship's feedback' (C9) and 'SCC's feedback' (C11).

The label C4 refers to the surrounding environment of MASS, it includes 'weather conditions' (C14) and 'traffic density' (C15), which will affect the difficulty of remote driving. After determining the nodes, and according to the relationship between them, a model of remote control error was developed, as shown in Figure 4.

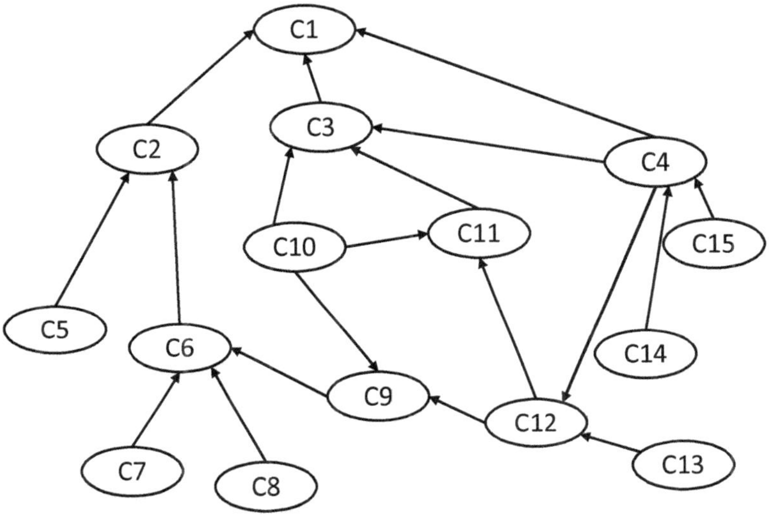

Figure 4. BBN model for MASS remote control.

4. Case Study

A case study of preliminary hazard analysis of MASS contact scenario based on experimental data, historical data and experts' judgement is presented. According to the definition of the contact scenario, several experiments were carried out in the Tangxun Lake in Wuhan and in the Qinhuai River in Nanjing, China [44]. The experimental ship employed is a 1:7 scale MASS model with three operation modes, namely remote driving, crew maneuvering and autonomous driving [45,48,56]. It weighs 5.5 ton and is about 7.2 m in length. Its profile and propeller rudder are consistent with MASS. The ship is equipped with various sensors, a laser radar, cameras and other hardware, which allow us to obtain the surrounding weather, traffic and other navigation environment information in a timely manner. In this quantitative analysis, the events related to the autonomous navigation are all obtained from experimental data. The mechanical systems of MASS are determined through the quantitative analysis of historical data.

4.1. Quantification of the Nodes of the FT Model

Quantitative analysis of the FT consists of transforming its logical structure into an equivalent probability expression by "minimal cut set" method at first [36]. Take the F2 FT as an example, the logical structure of F2 is transformed into equivalent probability expression in Equation (1).

$$P(\text{F2}) = P(\text{F4}) + P(\text{F6}) + P(\text{F7}) + P(\text{F8}) + P(\text{F9}) + P(\text{F16}) \times P(\text{F17}) + P(\text{F16}) \times P(\text{F18}) + P(\text{F17}) \times P(\text{F18}) \quad (1)$$

In order to quantify the failure of top events, the failure probability of basic events (equipment) in FT had to be obtained. Because the MASS power and propulsion line is the same as conventional ships, the existing failure data on the power and propulsion system of conventional ships and other industries could be used to estimate the failure probability of the power and propulsion system. The failure rate of each component in FT is shown in Table 6.

Table 6. Equipment failure rate data in the FT model.

Label	Equipment	Failure Rate (hour)	References
F16, F17, F18, F19, F20, F21	Diesel generator	7.59×10^{-5}	[57]
F6, F13	Electric bus	1.50×10^{-6}	[58]
F9, F10	Electric motor	6.74×10^{-5}	[57]
F7, F12	Main transformer	6.47×10^{-7}	[58]
F8, F11	Converter	2.66×10^{-5}	[58]
F4, F15	Propeller	5.00×10^{-6}	[59]

In this study, the following assumptions were made in the development of the FT to calculate the failure probability of the propulsion system:

- λThe failure rate of each component is a constant value.
- λDuring MASS navigation, no maintenance and repair activities are performed. In this study, the voyage of the MASS was considered to last 30 days, or 720 h.
- λWhile the MASS is in the port, the SCC should dispatch engineers to repair and maintain the system. This is a perfect-repair process, which means that the power and propulsion system can be the same as the new equipment.
- λFailure processes are modeled with an exponential distribution.

The probability of failure P was calculated based on the fact that the equipment's failure rate λ and the period per hour t were known, as follows [60]:

$$P(t) = 1 - e^{-\lambda t} \tag{2}$$

Using Equation (2), failure probabilities of the basic events (the failure probability of equipment) in the power and propulsion system could be obtained. Based on the equivalent probability expression and failure probability of basic events, the failure probability of the diesel electric propulsion 1st line (F2) was calculated as equal to 7.36×10^{-2}, as shown in in Equation (3).

$$\begin{aligned} P(\mathrm{F2}) &= P(\mathrm{F4}) + P(\mathrm{F6}) + P(\mathrm{F7}) + P(\mathrm{F8}) + P(\mathrm{F9}) + P(\mathrm{F16}) \times P(\mathrm{F17}) + P(\mathrm{F16}) \times P(\mathrm{F18}) + P(\mathrm{F17}) \times P(\mathrm{F18}) \\ &= \left(1 - e^{-1.50\times10^{-6}\times24\times30}\right) + \left(1 - e^{-6.74\times10^{-5}\times24\times30}\right) + \left(1 - e^{-6.47\times10^{-7}\times24\times30}\right) + \left(1 - e^{-2.66\times10^{-5}\times24\times30}\right) \\ &+ \left(1 - e^{-5.00\times10^{-6}\times24\times30}\right) + 3 \times \left(1 - e^{-5.00\times10^{-6}\times24\times30}\right) \times \left(1 - e^{-5.00\times10^{-6}\times24\times30}\right) \\ &= 0.0736 \end{aligned} \tag{3}$$

Similar to the diesel electric propulsion 1st line (*F2*), the failure probability of the propulsion system (*F1*) is 5.41×10^{-3}. Compared to conventional ships, which only have one propulsion line, the failure probability of the MASS propulsion system is lower. Moreover, the value of the normal operation of the 'power and propulsion system' (*P6*) is 0.9946.

4.2. Quantification of the Nodes in BBN Model

Based on the proposed BBN model and on the multiple states of nodes, experiments were conducted from October 2019 to November 2019 in a section of the Qinhuai River in Nanjing, to simulate the MASS contact scenario. Conventional ships include mainly passenger cruise ships, cleaning boats, patrol boats and others. Ferries and docks are present on both sides of the riverbank; there are several bridges above the water area, and the river channel is narrow. The experimental MASS model and the surrounding environment are shown in Figure 5. We selected some representative risk scenarios in the experiment, simultaneously recording all the information on the MASS model. In parallel, we determined the current states of related risk factors and the conditional probability distribution of the intermediate variables according to interviews and observations.

Figure 5. The Shore Control Center, several models of ships and the MASS model.

4.2.1. Prior Probability Determination of Each Root Node

By analyzing the record of the experiments, we regarded the frequency of occurrence of each root node as the prior probability. For objective factors, such as communication bandwidth, the communication bandwidth is recorded and classified in every experiment. In the experiment, the percentage of the number of times in which the communication bandwidth state is good, medium or bad is regarded as the prior probability. Subjective data that reflect the operators' performance, such as experience and fatigue, were evaluated through interviews. Taking the experience node as an example, an operator who has no remote control experience, has undergone remote control training and has sufficient remote control experience will be the experimental personnel. The percentage of the total number of experiments performed by these three types of people is regarded as the prior probability. The prior probability of each root node is shown in Table 7.

Table 7. Prior probability of each root node.

Node	Name	Good (a)	Medium (b)	Bad (c)
C5	Fatigue	0.8802	0.1030	0.0168
C7	Experience	0.6593	0.2201	0.1206
C8	Communication and collaboration	0.8264	0.1220	0.0516
C10	Software performance	0.3407	0.4396	0.2198
C13	Communication bandwidth	0.4286	0.1538	0.4176
C14	Weather conditions	0.8021	0.1429	0.0549
C15	Traffic density	0.7033	0.1978	0.0989

a, b, c represent the abbreviations for the good, medium and bad states, respectively.

4.2.2. Conditional Probability Table (CPT) Estimation

Both the arcs and the CPTs in the BBN reflect the causal relationship between the nodes. For the BBN, there are large number of CPTs that need to be determined. At the same time, it is difficult to accurately quantify the limited experimental data. Therefore, we adopted the method proposed by Røed et al. [39] to allocate CPTs. This method provides a structured way to derive the CPTs, thereby making it relatively less time-consuming. It is structured as follows. At the same time, this article provides a suitable way to convert experimental statistics into CPT:

- Determination of the relative importance weights between parent nodes and child node.

First, different parent nodes affecting the same child node have different degrees of importance, which can be addressed by assigning a weight w_i for each parent i through expert judgement. The sum of the weight of all parent nodes should be equal to 1. To this end, we adopted the interval type-2 fuzzy analytic hierarchy process (IT2FAHP) method proposed by Hu et al. [61]. The linguistic terms for importance as shown in Table 8. Based on the experimental certainty of the MASS model in the Qinhuai River in Nanjing, China, and the previously established BBN, a questionnaire on the importance of the parent nodes was designed and used to query three experts. To achieve a single view on the importance of parent nodes, we used the TIT2-WAA operation to aggregate the fuzzy judgment proposed by three experts. After TIT2-WAA operation, the fuzzy weight of each parent node was obtained. Finally, the fuzzy weights were defuzzified and normalized to obtain the relative weights of each parent node. The rationality of the result was further corrected through expert opinions. Taking the "operator performance" node as an example, the hierarchical structure is shown in the Figure 6. Three MASS remote operators gave a judgment on the relative importance of the two nodes, as shown in Table 9. After TIT2-WAA operation and defuzzification, the relative weights of fatigue 'C5′ and situation awareness 'C6′ are 0.4 and 0.6. The more detailed method and equation are in Hu et al. [61]. The relative weights of all nodes in the BBN are shown in Table 10.

Table 8. Linguistic terms for importance.

Linguistic Variable	Trapezoidal Interval Type-2 Fuzzy Sets
Absolutely Strong (AS)	((7,8,9,9;1,1),(7.2,8.2,8.8,9.0;0.8,0.8))
Very Strong (VS)	((5,6,8,9;1,1),(5.2,6.2,7.8,8.8;0.8,0.8))
Fairly Strong (FS)	((3,4,6,7;1,1),(3.2,4.2,5.8,6.8;0.8,0.8))
Slightly Strong (SS)	((1,2,4,5;1,1),(1.2,2.2,3.8,4.8;0.8,0.8))
Equal (E)	((1,1,1,1;1,1),(1,1.1,1,1;1,1))
If candidate I has one of the above linguistic variables assigned to it when compared with candidate j, then j has reciprocal value when compared with i.	Reciprocals of above

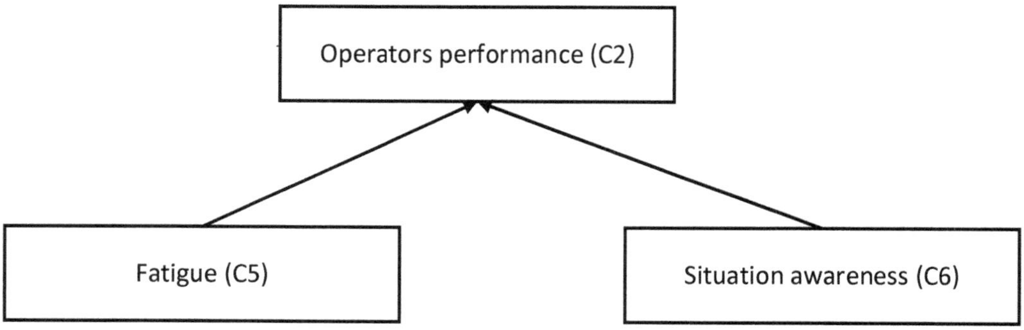

Figure 6. The hierarchical structure of C2 node.

Table 9. The relative importance of C5 and C6 nodes.

	Operator 1st		Operator 2nd		Operator 3rd	
	C5	C6	C5	C6	C5	C6
C5	E	E	E	1/SS	E	E
C6	1/E	E	SS	E	E	E

Table 10. Relative weight of the parent nodes.

Parent Node	Child Node	Relative Weight
Weather conditions (C14) Traffic density (C15)	Operating environment (C4)	0.6 0.4
Communication bandwidth (C13) Operating environment (C4)	Communication quality (C12)	0.64 0.36
Communication quality (C12) Ship's feedback (C9) Software performance (C10)	Ship's condition (C3)	0.42 0.16 0.42
Communication quality (C12) Software performance (C10)	SCC's feedback (C11)	0.48 0.52
Software performance (C10) Communication quality (C12)	Ship's feedback (C9)	0.55 0.45
Ship's feedback (C9) Experience (C7) Communication and collaboration (C8)	Situation awareness (C6)	0.43 0.21 0.36
Fatigue (C5) Situation awareness (C6)	Operators' performance (C2)	0.4 0.6
Operators' performance (C2) Ship's condition (C3) Operating environment (C4)	Remote control by SCC (C1)	0.215 0.215 0.57

- Determination of the weight distance between the parent node state and the child state.

After that, the distance between the parent node state and the child node state should be determined. The distance represents the difference between the parent node state and the child node state. The probability of a state of a child node is close to or equal to the state of its parent node. Therefore, if the parent node is in a 'good' state, the probability that the child node is in a good state should be greater than a medium state than a bad state. Taking the node 'communication quality' (C12) as an example, if 'operating environment' (C4) and 'communication bandwidth' (C13) are in a good state, the probability that 'communication quality' will be in a good state is bigger than that in a medium and bad state. Røed et al. [39] argued that no matter how large the difference between the state of the parent node and the child node, the relative distance can be reflected by obtaining the absolute value of distance. However, Li et al. [62] contended that the state of the child node, i.e., whether it is better or worse than the state of the parent node, influences the distance. The change in a different direction should be recorded with different importance. This means that the positive distance and the negative distance can be weighted, and then they cancel each other. In this study, we adopted the method proposed by Li et al. [62]. The good, medium and bad states of each node are marked as a, b and c, respectively. The formula to calculate the weighted distance is shown in Equation (4):

$$D_j = \left| \sum_{i=1}^{n} D_{ij} \times w_i \right|, D_j \in [0, 2] \tag{4}$$

where $i, j \in \{a, b, c\}$ and D_{ij} refers to the distance between the state of the parent node i and the state of the child node j. If the parent node is in a "good (a)" state, and the child node is in a "medium (b)" state, then the corresponding distance value is 1. n is the number of parent nodes corresponding to the child node and w_i represents the relative weight value of the corresponding parent nodes.

We took the node 'communication quality' (C12) as an example, as shown in Figure 7. C12 has two parent nodes, i.e., 'operating environment' (C4) and 'communication bandwidth' (C13). We assumed that the parent nodes C4 and C13 are in a "good (a)" and "medium (b)" state, respectively. At the same time, assuming that C12 node is in a "good (a)" state, the distance between C12 and C4 is 0; correspondingly, the distance between C12 and C13 is -1. As shown in Table 8, the weights of C4 and C13 are $w_{C4} = 0.36$ and $w_{C13} = 0.64$, respectively, and its weighted distance is. $D_a = |w_{c4} \times 0 + w_{c13} \times -1| = |0.36 \times 0 + 0.64 \times -1| = 0.64$ Similarly, $D_b = 0.36$ and $D_c = 1.36$.

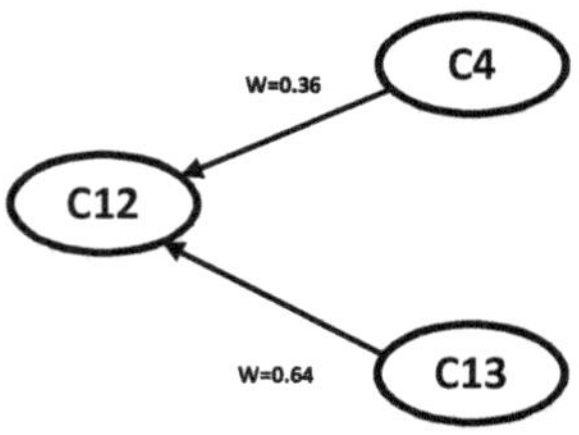

Figure 7. Relationship between C4, C13 and C12.

- Determination of the CPTs of the child nodes.

The CPTs of the child nodes were determined based on experimental statistics, following Røed et al. [39], who calculated it using Equation (5). The good, medium and bad states of each node were marked as a, b and c, respectively.

$$P_j = \frac{e^{-RD_j}}{\sum_{j=a}^{c} e^{-RD_j}}, P_j \in [0,1] \tag{5}$$

In Equation (5), the numerator represents the probability distribution in each state, where $j \in \{a, b, c\}$ and R refers to the modified index value. The higher the R index, the lower the probability that the child node in focus is in a state derived from its parents' states.

The R value was determined using the statistical data of the MASS model experiment. First, we selected representative statistical data in the record as the basis. For example, when the C12 is obtained, C4 is in a "good (a)" state and C13 is in a "medium (b)" state. Second, the upper limits and the "medium (b)" state of the data's probability distribution were used to calculate the value of R When C4 is in a "good (a)" state, C13 is in a "medium (b)" state; in this case, there are 33 sets of data selected by the experiment, 9 of which are for C4 in a "good (a)" state, and the other 24 for C4 in a "medium (b)" state, with 0 groups for C4 in a "bad (c)" state. Therefore, the upper limit probability value of 0.27 and the intermediate state probability value of 0.73 could be used for calculation. The calculation process of the R value of the C12 node is shown in Equations (6) and (7) as follows:

$$P_a/P_b = \frac{\frac{e^{-RD_{a1}}}{\sum_{j=1}^{3} e^{-RD_j}}}{\frac{e^{-RD_{b1}}}{\sum_{j=1}^{3} e^{-RD_j}}} = \frac{0.27}{0.73} = 0.37 \tag{6}$$

$$P_a/P_b = e^{-0.64R}/e^{-0.36R} = 0.37 \Rightarrow R = 3.55 \tag{7}$$

The values of D_a, D_b, D_c were calculated according to Equation (4). For example, when the parent node C13 is in a "good (a)" state, C4 is in a "good (a)" state and the weighted distance among the "good (a)", "moderate (b)" and "bad (c)" states of the C12 node are $D_a = 0, D_b = 1, D_c = 2$, respectively. After obtaining the D and R values, the conditional probability distribution of this child node could be obtained as shown in Equations (8)–(10):

$$P_a = \frac{e^{-RD_a}}{\sum_{j=1}^{3} e^{-RD_j}} = \frac{e^{-0R}}{e^{-0R} + e^{-1R} + e^{-2R}} = \frac{e^{-0\times 3.55}}{e^{-0\times 3.55} + e^{-1\times 3.55} + e^{-2\times 3.55}} = 0.9713 \quad (8)$$

$$P_b = \frac{e^{-RD_b}}{\sum_{j=1}^{3} e^{-RD_j}} = \frac{e^{-1R}}{e^{-0R} + e^{-1R} + e^{-2R}} = \frac{e^{-1\times 3.55}}{e^{-0\times 3.55} + e^{-1\times 3.55} + e^{-2\times 3.55}} = 0.02790 \quad (9)$$

$$P_c = \frac{e^{-RD_c}}{\sum_{j=1}^{3} e^{-RD_j}} = \frac{e^{-2R}}{e^{-0R} + e^{-1R} + e^{-2R}} = \frac{e^{-2\times 3.55}}{e^{-0\times 3.55} + e^{-1\times 3.55} + e^{-2\times 3.55}} = 0.0008 \quad (10)$$

The CPT of "communication quality" (C12) is shown in Table 11. Similarly, we obtained other weighted distances for each combination of any state of the parent that pushes the child node in different states. The BBN model can be quantified by inputting the obtained CPTs and the prior probability of the collected root node.

Table 11. CPT of 'communication quality' (C12).

Node		State and Probability								
C13		a			b			c		
C4		a	b	c	a	b	c	a	b	c
C12	a	0.9714	0.7244	0.1692	0.2645	0.0204	0.0204	0.02315	0.0077	0.0008
	b	0.0278	0.2679	0.8076	0.7150	0.7150	0.7150	0.8076	0.2679	0.0278
	c	0.0008	0.0077	0.0231	0.0204	0.2644	0.2645	0.1692	0.7243	0.9714

a, b, c represent the abbreviations for the good, medium and bad states, respectively.

4.2.3. Failure Probability Quantification of Remote Control Errors

The 'remote control by the SCC' (C1) is a binary node (success, failure), as such, it is completely different from the other nodes, which have multiple states. Thus, the 'remote control by the SCC' (C1) cannot be calculated using the aforementioned method. Røed et al. [39] proposed applying the barrier and operational risk analysis (BCRA) method to calculate the probability of a binary node. This method is articulated in three steps.

First, the basic probability of the event in focus is assigned through the use of historical genetic data combined with a model. Then, the maximum deviation from the basic error probability of the target node, by considering the worst and best states of its parent node, is determined. The values of the adjustment factors proposed by Røed were adopted [39], as shown in Table 12.

Table 12. Adjustment factors for the basis error probabilities.

State of the Parent Node	Adjustment Factor Q
Good (a)	0.1
Medium (b)	1.0
Bad (c)	10

a, b, c represent the abbreviations for the good, medium and bad states, respectively.

Finally, the conditional probability of the target node is determined. Accordingly, the CPTs were calculated based on the parent node states and the adjustment factors Q_i as follows:

$$P_j = P_{basis} \sum_{i=1}^{n} w_i \sum_{k=a}^{c} P_{ik} Q_{ik} P_j \in [0, 1] \quad (11)$$

where P_{ik} is the probability of each parent i to be in each state $k = a, b, c$; Q_{ik} is the corresponding adjustment factor according to Table 10; and w_i is the weight of the parent nodes i, whose sum is 1. The index j indicates the possible states of the event we are considering (i.e., success or failure).

According to experiment statistics and literature review, the basic probability of the remote control error is 8.58×10^{-3} [11]. The 'remote control by the SCC' (C1) has three parent nodes, i.e., 'operators' performance' (C2), 'ship's condition' (C3) and 'operating environment' (C4). When the weights and the probability distributions of three parent nodes are known, the 'remote control by the SCC' (C1) can be calculated, as shown in Table 13. After calculation, the failure probability of the 'remote control by the SCC' (C1) is 7.722×10^{-3}. Therefore, the success probability of 'remote control by SCC' is 0.9923.

Table 13. Probability of the 'remote control by the SCC' (C1) and its parent node.

Node	State	Probability
Operators' performance (C2)	Good (a)	0.5206
	Medium (b)	0.4489
	Bad (c)	0.0304
Ship's condition (C3)	Good (a)	0.3416
	Medium (b)	0.5629
	Bad (c)	0.0955
Operating environment (C4)	Good (a)	0.7000
	Medium (b)	0.2585
	Bad (c)	0.0415
Remote control by the SCC (C1)	Success	0.9923
	Failure	0.0077

a, b, c represent the abbreviations for the good, medium and bad states, respectively.

4.3. Failure Probability Quantification of the MASS Contact Scenario

Once the normal operation and failure probability of pivotal events are calculated, several end states probability in the MASS contact scenario are obtained. As shown in Sections 4.2 and 4.3, the probability of several events in ESD was calculated. The probability that the 'power and propulsion system' (P6) works normally, calculated by using the FT model in Section 4.2, is 0.9946. The probability of success of the 'remote control by the SCC' (P4), calculated by using the BBN model in Section 4.3, is 0.9923. In the same way, the normal operation and failure probability of other pivotal events was calculated according to the experiment and historical data. Different outputs of pivotal events will lead to different end states, such as safe or accident states, with different probabilities. After calculating the probability of each pivotal event in ESD, we could obtain the probability of each end state in the MASS hazard scenarios, according to the following steps:

- Calculation of the end states' probability of the MASS contact scenario.

The probability of each end state was obtained according to the HCL quantitative calculation method. The probability values of all end states are listed in Table 14.

Table 14. Failure probability of the end states.

End State	End State Type	Probability
E1	Normal navigation	9.45×10^{-1}
E2	Contact due to perception stage failure	3.67×10^{-2}
E3	Contact due to decision-making stage failure	1.68×10^{-3}
E4	Contact due to execution stage failure	1.66×10^{-2}

As shown in Table 14, the probability of MASS avoiding the external events and continuing operation is 9.45×10^{-1}. According to the Table 14, the failure of perception stage and execution stage is the main cause of contact accidents. Thus, the perception stage of MASS is the first safety barrier of hazard scenario. It is necessary to ensure that the sensor equipment and the perception of the operator can perceive the risk and ensure that the risk will be detected immediately. For the execution stage of MASS, although the MASS is equipped with a redundant system, it is still very likely to cause an accident. The probability of the contact scenario can be mitigated by shortening the sailing time.

- Calculation of the accident-causing event chains.

In the HCL method, through the combination of the ESD model, the FT model and the BBN model, the events in the ESD model were extended to the FT and the BBN, and then different accident-causing event chains and their probability could be obtained. We selected the five accident-causing event chains with the highest risk and they are shown in Table 15.

Table 15. Five accident-causing event chains with the highest risk.

No.	Accident-Causing Event Chains *	Probability
1	IE-P1(0)-P2(0)-End 2	3.676×10^{-2}
2	IE-P1(0)-P2(1)-P4(0)-End 3	1.671×10^{-3}
3	IE-P1(1)-P3(1)-P5(1)-P6(0): F8,F11-End 4	1.795×10^{-4}
4	IE-P1(1)-P3(1)-P5(0)-End 4	1.234×10^{-4}
5	IE-P1(1)-P3(1)-P5(1)-P6(0): F9,F11-End 4	8.421×10^{-5}

* The normal functioning of the pivotal event is marked as 1; its failure is marked as 0.

As shown in Table 15, the accident-causing event chain with the highest risk is the one that leads to accident end state, due to the failure to perceive the danger (E2). This shows that the perception stage is the most important stage in the MASS hazard scenarios. Secondly, the second main cause of accident-causing event chains is that the operators in the SCC did not propose an effective strategy which leads to accident end state (E3). Thus, it is necessary to train remote operators and maintain the equipment, while at the same time MASS should avoid sailing in bad environmental conditions. Thirdly, most of the occurrences in all accident-causing event chains relate to the failure of the mechanical system (E4), which is the last guarantee for the safe navigation of the MASS. Before the voyage, detailed planning and preparation work should be carried out. Reasonable remedial measures are an important way to effectively improve the safety of the MASS. Finally, the third, fourth and fifth accident-causing event chains involved the failure of operation of the steering system and of the propulsion system. Therefore, in order to guarantee the safety of MASS, it is necessary to design a redundant steering and propulsion system, as well as to propose a maintenance plan for the mechanical system. Through appropriate technical solutions, the MASS risk can be reduced to an acceptable level.

- Identification of the influence factors in the power and propulsion system leading to a failure of the MASS emergency response process (E4).

The reliability of the propulsion system has relatively the largest impact on MASS navigation accidents. In order to support the future design of the MASS power and propulsion system, it is necessary to identify the most influencing equipment in the power and propulsion system. Using the existing evaluation indicators comprehensively, the basic events or risk factors with the highest impact on risk can be identified for improvement. The Fussell–Vesely (VF) importance measure is an evaluation criterion that represents the impact of components on the total failure probability of a system [63]:

$$VF(S,e) = P(e|S) = \frac{P(S \cdot e)}{P(S)} = \frac{P(S|e)P(e)}{P(S)} \tag{12}$$

When the MASS has an accident, we selected E4 to measure the importance factors. As shown in Table 16, the failure of the converter, failure of the diesel generator and failure of the electric motor are the most important factors. Therefore, priority should be given to the maintenance of this equipment. In the future design, a more reasonable redundancy design and maintenance plan will improve propulsion reliability, especially of the converter, the diesel generator and the electric motor.

Table 16. The VF of the power and propulsion system equipment across accident end states.

Label	Component	VF
F8, F11	Converter	0.02651
F16, F17, F18, F19, F20, F21	Diesel generator	0.01445
F9, F10	Electric motor	0.01243
F6, F13	Bus bar	0.00151
F7, F12	Transformer	0.00104
F4, F15	Propeller	0.00046

- Identification of the influence factors in the remote driving mode.

In order to analyze the influence of each factor contributing to the failure of remote driving, the sensitivity of the BN model of remote driving is analyzed in this section. First, the probability of each parent node is assigned the value of one. Then, the probability variation table of target node is obtained. Take the weather condition (C14) as an example, set the probability of being in a "good" state to 100%, obtain the probability of C2, C3 and C4. Based on the Equation (11), the failure probability of "remote driving" is 0.00619. Similarly, the other nodes in BN are assessed. Figure 8 shows the probability change in "remote driving" after adjusting each node. The sensitivity of the nodes affecting remote driving is ranked as follows: C10 > C14 > C7 > C15 > C13 > C8 > C5.

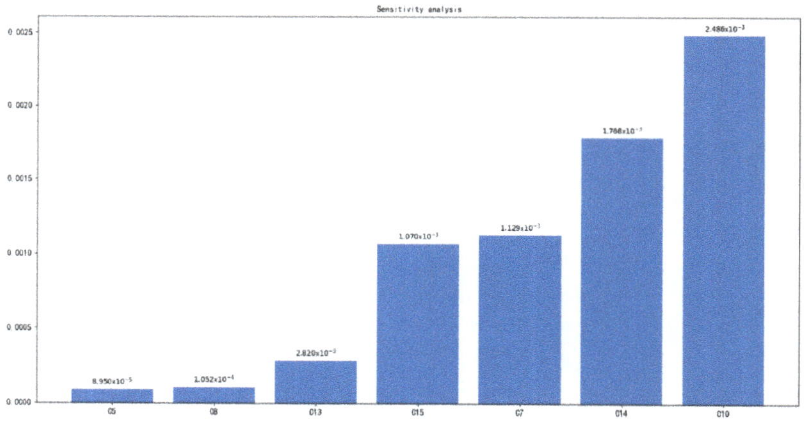

Figure 8. Sensitivity analysis of remote driving mode.

Based on the result, "software performance" (C10) is the most sensitivity node. During the remote driving, the software should be more attention. At the same time, the external influence factors such as "weather condition" (C14) and "traffic density" (C15) will significantly affect the failure probability of remote driving. Among the influence factors related to the operator, "experience" (C7) is the most important factor. In summary, the software in SCC should be updated in time to ensure high availability and quality. The SCC should strengthen the training about contact scenarios in case the operator is unfamiliar with remote driving or does not understand external object avoidance rules.

5. Conclusions and Future Work

With the increase in the use of automation technology in the maritime industry, MASS risk influence factors are increasingly various and complex. This paper is an attempt to conduct a preliminary hazard analysis of MASS in the design and experimental stages based on the conceptual design of MASS, historical data and experiments of conventional ships. The applicability of the HCL method to MASS was demonstrated through a case study of a contact scenario for a MASS model ship. Key conclusions can be summarized as follows:

- The use of the HCL method allows a clear classification of the pivotal events of the hazard scenarios.
- The paper established a branch model to analyze the events in the ESD and used FT and BBN to analyze influence factors in a more detailed way according to their characteristics.
- The importance of more detailed influencing factors is quantified based on the FT and BBN method.
- The HCL method provides a quantitative calculation result of the MASS hazardous scenario and presents a way to verify whether the conceptual design of MASS is reasonable and can help find the weak links in the MASS experiment.

Based on the analysis and test ship, redundant design for MASS is necessary. For example, the operators in the SCC can perceive the risk in case of AS system failure. In relation to the power and propulsion system, at least two independent power and propulsion lines can mitigate the failure probability. However, the development of MASS is still in an early phase. With the development of technology, more risk influence factors will arise and the cooperation between AS and the operators in the SCC will be further discussed. For example, the control priority between the operators in the SCC and AS may change with the development of technology. Moreover, this paper analyzed in detail both mechanical and human events, while overlooking software events. In the future, an important problem to address is how to include software events in risk assessments. The failure probability and the conclusions of the present study can be used as references for the design of MASS.

Author Contributions: Writing—original draft preparation, Z.H., K.Z.; writing—review and editing, D.Z., J.Z. and F.Z.; methodology: Z.H., K.Z. and M.Z.; conceptualization: D.Z., J.Z. and F.Z.; funding acquisition, D.Z. All authors have read and agreed to the published version of the manuscript.

Funding: The research was supported by the Hubei Provincial Natural Science Foundation of China (2019CFA039), the Natural Science Foundation of China (No.52071247) and the innovation and entrepreneurship team import project of Shaoguan city (201208176230693).

Institutional Review Board Statement: Not applicable.

Informed Consent Statement: Not applicable.

Data Availability Statement: Not applicable.

Conflicts of Interest: We declare that we have no financial and personal relationships with people or organizations that can inappropriately influence our work. There is no professional or other personal interest of any nature or kind in any product, service and/or company that could be construed as influencing the position presented in, or in the review of, the manuscript entitled, 'Use of the Hybrid Causal Logic Method for Hazard Identification of Maritime Autonomous Surface Ship Operation'.

References

1. Lyu, H.; Yin, Y. COLREGS-constrained real-time path planning for autonomous ships using modified artificial potential fields. *J. Navig.* **2019**, *72*, 588–608. [CrossRef]
2. Chen, Z.; Chen, D.; Zhang, Y.; Cheng, X.; Zhang, M.; Wu, C. Deep learning for autonomous ship-oriented small ship detection. *Saf. Sci.* **2020**, *130*, 104812. [CrossRef]
3. Yuan, X.; Zhang, D.; Zhang, J.; Zhang, M.; Guedes Soares, C. A novel real-time collision risk awareness method based on velocity obstacle considering uncertainties in ship dynamics. *Ocean. Eng.* **2021**, *220*, 108436. [CrossRef]

4. Rolls-Royce. Rolls-Royce and Finferries Demonstrate World's First Fully Autonomous Ferry. Available online: https://www.rolls-royce.com/media/press-releases/2018/03-12-2018-rr-and-finferries-demonstrate-worlds-first-fully-autonomous-ferry.aspx (accessed on 11 April 2022).
5. Wärtsilä. World's First Autodocking Installation Successfully Tested by Wärtsilä. Available online: https://www.wartsila.com/media/news/26-04-2018-world-s-first-autodocking-installation-successfully-tested-by-wartsila-2169290 (accessed on 11 April 2022).
6. Kongsberg.com. Autonomous Ship Project, Key Facts about YARA Birkeland. Available online: https://www.kongsberg.com/maritime/support/themes/autonomous-ship-project-key-facts-about-yara-birkeland (accessed on 11 April 2022).
7. DNV. The ReVolt—DNV. Available online: https://www.dnvgl.com/technology-innovation/revolt/index.html (accessed on 11 April 2022).
8. Jokioinen, E.; Poikonen, J.; Jalonen, R.; Saarni, J. *Remote and Autonomous Ships-The Next Steps*; Rolls Royce Plc: London, UK, 2016.
9. Burmeister, H.C.; Bruhn, W.; Rødseth, Ø.J.; Porathe, T. Autonomous unmanned merchant vessel and its contribution towards the e-Navigation implementation: The MUNIN perspective. *Int. J. e-Navig. Marit. Econ.* **2014**, *1*, 1–13. [CrossRef]
10. International Maritime Organization (IMO). IMO Takes First Steps to Address Autonomous Ships. Available online: http://www.imo.org/en/MediaCentre/PressBriefings/Pages/08-MSC-99-MASS-scoping.aspx (accessed on 12 September 2020).
11. Zhang, M.; Zhang, D.; Yao, H.; Zhang, K. A probabilistic model of human error assessment for autonomous cargo ships focusing on human–autonomy collaboration. *Saf. Sci.* **2020**, *130*, 104838. [CrossRef]
12. *DNVGL-CG-0264*; Class Guideline—Autonomous and Remotely Operated Ships. DNV GL: Bærum, Norway, 2018. Available online: https://rules.dnv.com/docs/pdf/DNV/cg/2018-09/dnvgl-cg-0264.pdf (accessed on 19 April 2022).
13. Utne, I.B.; Schjølberg, I.; Roe, E. High reliability management and control operator risks in autonomous marine systems and operations. *Ocean. Eng.* **2019**, *171*, 399–416. [CrossRef]
14. Burmeister, H.C.; Bruhn, W.C.; Rødseth, Ø.J.; Porathe, T. Can unmanned ships improve navigational safety? In Proceedings of the Transport Research Arena, Paris, France, 14–17 April 2014.
15. Goerlandt, F. Maritime Autonomous Surface Ships from a risk governance perspective: Interpretation and implications. *Saf. Sci.* **2020**, *128*, 104758. [CrossRef]
16. Utne, I.B.; Rokseth, B.; Sørensen, A.J.; Vinnem, J.E. Towards supervisory risk control of autonomous ships. *Reliab. Eng. Syst. Saf.* **2020**, *196*, 106757. [CrossRef]
17. Zhou, X.Y.; Liu, Z.J.; Wang, F.W.; Wu, Z.L.; Cui, R.D. Towards applicability evaluation of hazard analysis methods for autonomous ships. *Ocean. Eng.* **2020**, *214*, 107773. [CrossRef]
18. Wróbel, K.; Montewka, J.; Kujala, P. System-theoretic approach to safety of remotely-controlled merchant vessel. *Ocean. Eng.* **2018**, *152*, 334–345. [CrossRef]
19. Wróbel, K.; Montewka, J.; Kujala, P. Towards the assessment of potential impact of unmanned vessels on maritime transportation safety. *Reliab. Eng. Syst. Saf.* **2017**, *165*, 155–169. [CrossRef]
20. Bačkalov, I. Safety of autonomous inland vessels: An analysis of regulatory barriers in the present technical standards in Europe. *Saf. Sci.* **2020**, *128*, 104763. [CrossRef]
21. Fan, C.L.; Wróbel, K.; Montewka, J.; Gil, M.; Wan, C.P.; Zhang, D. A framework to identify factors influencing navigational risk for Maritime Autonomous Surface Ships. *Ocean. Eng.* **2020**, *202*, 107188. [CrossRef]
22. Kretschmann, L.; Rodseth, O.; Tjora, A. *Report D9. 2: Qualitative Assessment. Maritime Unmanned Navigation through Intelligence in Networks (MUNIN)*; MUNIN: Munich, Germany, 2015.
23. Ramos, M.A.; Utne, I.B.; Mosleh, A. On factors affecting autonomous ships operators performance in a Shore Control Center. In Proceedings of the Probabilistic Safety Assessment and Management 4, PSAM, Los Angeles, CA, USA, 14 September 2018.
24. Man, Y.; Weber, R.; Cimbritz, J.; Lundh, M.; MacKinnon, S.N. Human factor issues during remote ship monitoring tasks: An ecological lesson for system design in a distributed context. *Int. J. Ind. Ergon.* **2018**, *68*, 231–244. [CrossRef]
25. Bolbot, V.; Theotokatos, G.; Boulougouris, E.; Vassalos, D. Comparison of diesel-electric with hybrid-electric propulsion system safety using System-Theoretic Process Analysis. In *Royal Institution of Naval Architects—International Conference on Power and Propulsion Alternatives for Ships, London, UK, 22–23 January 2019*; RINA: London, UK, 2019.
26. Bolbot, V.; Theotokatos, G.; Vassalos, D. Using system-theoretic process analysis and event tree analysis for creation of a fault tree of blackout in the Diesel-Electric Propulsion system of a cruise ship. In Proceedings of the 13th International Marine Design Conference, IMDC, Espoo, Finland, 10–14 June 2018.
27. Wang, T.; Liu, J.; Zeng, F. Application of QFD and FMEA in ship power plant design. In Proceedings of the 10th International Symposium on Computational Intelligence and Design, ISCID, Hangzhou, China, 9–10 December 2017.
28. Ramos, M.A.; Thieme, C.A.; Utne, I.B.; Mosleh, A. Human-system concurrent task analysis for maritime autonomous surface ship operation and safety. *Reliab. Eng. Syst. Saf.* **2020**, *195*, 106697. [CrossRef]
29. Thieme, C.A.; Utne, I.B.; Haugen, S. Assessing ship risk model applicability to Marine Autonomous Surface Ships. *Ocean. Eng.* **2018**, *165*, 140–154. [CrossRef]
30. Banda, O.A.V.; Kannos, S.; Goerlandt, F.; van Gelder, P.H.; Bergström, M.; Kujala, P. A systemic hazard analysis and management process for the concept design phase of an autonomous vessel. *Reliab. Eng. Syst. Saf.* **2019**, *191*, 106584. [CrossRef]
31. Wróbel, K.; Montewka, J.; Kujala, P. Towards the development of a system-theoretic model for safety assessment of autonomous merchant vessels. *Reliab. Eng. Syst. Saf.* **2018**, *178*, 209–224. [CrossRef]

32. Ramos, M.A.; Utne, I.B.; Mosleh, A. Collision avoidance on maritime autonomous surface ships: Operators' tasks and human failure events. *Saf. Sci.* **2019**, *116*, 33–44. [CrossRef]
33. Porathe, T.; Hoem, Å.; Rødseth, Ø.J.; Fjørtoft, K.; Johnsen, S.O. At least as safe as manned shipping? Autonomous shipping, safety and "human error". Safety and Reliability–Safe Societies in a Changing World. In Proceedings of the European Safety and Reliability Conference, ESREL, Trondheim, Norway, 17–21 June 2018.
34. Wróbel, K.; Krata, P.; Montewka, J.; Hinz, T. Towards the development of a risk model for unmanned vessels design and operations. *Trans. Nav. Int. J. Mar. Navig. Saf. Sea Transp.* **2016**, *10*, 267–274. [CrossRef]
35. Mosleh, A.; Dias, A.; Eghbali, G.; Fazen, K. An integrated framework for identification, classification, and assessment of aviation systems hazards. In *Probabilistic Safety Assessment and Management*; PSAM: London, UK, 2004.
36. Wang, C. *Hybrid Causal Logic Methodology for Risk Assessment*; University of Maryland: College Park, MD, USA, 2007.
37. Mohaghegh, Z.; Kazemi, R.; Mosleh, A. Incorporating organizational factors into Probabilistic Risk Assessment (PRA) of complex socio-technical systems: A hybrid technique formalization. *Reliab. Eng. Syst. Saf.* **2009**, *94*, 1000–1018. [CrossRef]
38. Groth, K.; Wang, C.; Mosleh, A. Hybrid causal methodology and software platform for probabilistic risk assessment and safety monitoring of socio-technical systems. *Reliab. Eng. Syst. Saf.* **2010**, *95*, 1276–1285. [CrossRef]
39. Røed, W.; Mosleh, A.; Vinnem, J.E.; Aven, T. On the use of the hybrid causal logic method in offshore risk analysis. *Reliab. Eng. Syst. Saf.* **2009**, *94*, 445–455. [CrossRef]
40. Sklet, S.; Aven, T.; Hauge, S.; Vinnem, J.E. Incorporating human and organizational factors in risk analysis for offshore installations. In Proceedings of the 16th European Safety and Reliability Conference, ESREL, Gdynia-Sopot-Gdansk, Poland, 27–30 June 2005.
41. Zhang, M.; Zhang, D.; Goerlandt, F.; Yan, X.; Kujala, P. Use of HFACS and fault tree model for collision risk factors analysis of icebreaker assistance in ice-covered waters. *Saf. Sci.* **2019**, *111*, 128–143. [CrossRef]
42. Zhang, M.; Conti, F.; Le Sourne, H.; Vassalos, D.; Kujala, P.; Lindroth, D.; Hirdaris, S. A method for the direct assessment of ship collision damage and flooding risk in real conditions. *Ocean. Eng.* **2021**, *237*, 109605. [CrossRef]
43. Zhang, M.; Zhang, D.; Fu, S.; Kujala, P.; Hirdaris, S. A Predictive Analytics Method for Maritime Traffic Flow Complexity Estimation in Inland Waterways. *Reliab. Eng. Syst. Saf.* **2022**, *220*, 108317. [CrossRef]
44. WUT. Autonomous Ship Test. Available online: http://wts.whut.edu.cn/zxxw/201910/t20191024_414554.shtml (accessed on 3 September 2020).
45. Yan, X.; Ma, F.; Liu, J.; Wang, X. Applying the navigation brain system to inland ferries. In Proceedings of the 18th Conference on Computer and IT Applications in the Maritime Industries, COMPIT, Tullamore, Ireland, 25–27 March 2019.
46. Man, Y.; Lundh, M.; Porathe, T.; MacKinnon, S. From desk to field-Human factor issues in remote monitoring and controlling of autonomous unmanned vessels. *Procedia Manuf.* **2015**, *3*, 2674–2681. [CrossRef]
47. Chen, P.; Huang, Y.; Mou, J.; van Gelder, P.H.A.J.M. Probabilistic risk analysis for ship-ship collision: State-of-the-art. *Saf. Sci.* **2019**, *117*, 108–122. [CrossRef]
48. Wang, T.; Wu, Q.; Zhang, J.; Wu, B.; Wang, Y. Autonomous decision-making scheme for multi-ship collision avoidance with iterative observation and inference. *Ocean. Eng.* **2020**, *197*, 106873. [CrossRef]
49. Eriksen, S.; Utne, I.B.; Ltzen, M. An rcm approach for assessing reliability challenges and maintenance needs of unmanned cargo ships. *Reliab. Eng. Syst. Saf.* **2021**, *210*, 107550. [CrossRef]
50. Ådnanes, A.K. *Maritime Electrical Installations and Diesel Electric Propulsion*; ABB AS Marine: Oslo, Norway, 2003.
51. Geertsma, R.D.; Negenborn, R.R.; Visser, K.; Hopman, J.J. Design and control of hybrid power and propulsion systems for smart ships: A review of developments. *Appl. Energy* **2017**, *194*, 30–54. [CrossRef]
52. Bolvashenkov, I.; Kammermann, J.; Herzog, H.G. Reliability assessment of a fault tolerant propulsion system for an electrical helicopter. In Proceedings of the 12th International Conference and Exhibition on Ecological Vehicles and Renewable Energies, EVER, Monte Carlo, Monaco, 11–13 April 2017.
53. Porathe, T.; Prison, J.; Man, Y. Situation awareness in remote control centres for unmanned ships. In Proceedings of the Human Factors in Ship Design & Operation, London, UK, 26–27 February 2014.
54. Wahlström, M.; Hakulinen, J.; Karvonen, H.; Lindborg, I. Human factors challenges in unmanned ship operations–insights from other domains. *Procedia Manuf.* **2015**, *3*, 1038–1045. [CrossRef]
55. Hogg, T.; Ghosh, S. Autonomous merchant vessels: Examination of factors that impact the effective implementation of unmanned ships. *Aust. J. Marit. Ocean. Aff.* **2016**, *8*, 206–222. [CrossRef]
56. Yang, F.; Liu, J.; Li, S.; Ma, F. Virtual-Real Interaction Tests for Functional Testing of Smart Ships. In Proceedings of the 30th International Ocean and Polar Engineering Conference. International Society of Offshore and Polar Engineers, Shanghai, China, 11–16 October 2020.
57. OREDA. *Offshore Reliability Data Handbook*, 4th ed.; DNV: Oslo, Norway, 2003.
58. IAEA-TECDOC-478. *Component Reliability Data for use in Probabilistic Safety Analysis*; IAEA: Vienna, Austria, 1988.
59. Markus, H.; Knut, E.K.; Jose, I.A. Towards a Design Framework for Maritime Asset Servitization. In Proceedings of the 30th European Safety and Reliability Conference and the 15th Probabilistic Safety Assessment and Management Conference, Venice, Italy, 1–5 November 2020.
60. Leoni, L.; BahooToroody, A.; De Carlo, F.; Paltrinieri, N. Developing a risk-based maintenance model for a Natural Gas Regulating and Metering Station using Bayesian Network. *J. Loss Prev. Process Ind.* **2019**, *57*, 17–24. [CrossRef]

61. Hu, J.; Zhang, Y.; Chen, X.; Liu, Y. Multi-criteria decision making method based on possibility degree of interval type-2 fuzzy number. *Knowl. Based Syst.* **2013**, *43*, 21–29. [CrossRef]
62. Li, P.C.; Zhang, L.; Dai, L.C.; Li, X.F. Study on operator's SA reliability in digital NPPs. Part 3: A quantitative assessment method. *Ann. Nucl. Energy* **2017**, *109*, 82–91. [CrossRef]
63. Meng, F.C. Relationships of Fussell–Vesely and Birnbaum importance to structural importance in coherent systems. *Reliab. Eng. Syst. Saf.* **2000**, *67*, 55–60. [CrossRef]

Journal of
Marine Science and Engineering

MDPI

Article

Available-Maneuvering-Margins-Based Ship Collision Alert System

Lei Du [1], Osiris A. Valdez Banda [2] and Zhongyi Sui [1,*]

[1] School of Navigation, Wuhan University of Technology, Wuhan 430063, China
[2] Department of Mechanical Engineering, School of Engineering, Aalto University, 02150 Espoo, Finland
* Correspondence: suizy@whut.edu.cn

Abstract: The timing of a ship taking evasive maneuvers is crucial for the success of collision avoidance, which is affected by the perceived risk by the navigator. Therefore, we propose a collision alert system (CAS) based on the perceived risk by the navigator to trigger a ship's evasive maneuvers in a timely manner to avoid close-quarters situations. The available maneuvering margins (AMM) with ship stability guarantees are selected as a proxy to reflect the perceived risk of a navigator; hence, the proposed CAS is referred to as an AMM-based CAS. Considering the dynamic nature of ship operations, the non-linear velocity obstacle method is utilized to identify the presence of collision risk to further activate this AMM-based CAS. The AMM of a ship are measured based on ship maneuverability and stability models, and the degree to which they violate the risk-perception-based ship domain determines the level of collision alert. Several typical encounter scenarios are selected from AIS data to demonstrate the feasibility of this AMM-based CAS. The promising results suggest that this proposed AMM-based CAS is applicable in both ship pair encounter and multi-vessel encounter scenarios. Collision risk can be accurately detected, and then a collision alert consistent with the risk severity is issued. This proposed AMM-based CAS has the potential to assist autonomous ships in understanding the risk level of the encounter situation and determining the timing for evasive maneuvers. The advantages and limitation of this proposed method are discussed.

Keywords: collision alert system (CAS); available maneuvering margins (AMM); ship domain; ship stability; maritime safety

Citation: Du, L.; Valdez Banda, O.A.; Sui, Z. Available-Maneuvering-Margins-Based Ship Collision Alert System. *J. Mar. Sci. Eng.* **2022**, *10*, 1123. https://doi.org/10.3390/jmse10081123

Academic Editor: Dracos Vassalos

Received: 20 June 2022
Accepted: 11 August 2022
Published: 15 August 2022
Corrected: 15 December 2022

1. Introduction

Although many advanced methods and technologies have been applied in the maritime field, ship collisions still occur frequently, posing a threat to maritime transportation safety [1–5]. Ship collision alert systems (CAS) are widely applied to prevent ship collision by alerting the navigators to take evasive maneuvers in a timely manner to eliminate the existing collision risk [6–15]. Some limitations or simplifications of these methods make it challenging to put them into practical use. One is the inadequate consideration paid to the dynamic nature of ship maneuvers. This can be proven by the most widely used blind sailing hypothesis that when there is a risk of collision, the ship is assumed to sail in a straight line with a constant speed. As a matter of fact, the ship will normally take evasive maneuvers for collision avoidance [16,17]. The second limitation is the neglect of the risk resolution. The existing research mainly utilized the danger level of approaching ships as a basis to quantify risk severity. The risk resolution of a ship reflects her capability to eliminate the existing danger, which is critical to the success of collision avoidance. Under the same circumstances, a ship with a higher risk resolution is more likely to eliminate the risk, so the risk is relatively low. Therefore, risk measures independent of conflict resolution may lead to inaccurate detection of actual danger [18]. Third, these methods are mainly designed for ship pair encounters without considering traffic conditions [10]. The traffic complexity increases the likelihood of serious encounters [19–21], so the applicability of these methods in complicated multi-vessel encounters has not been confirmed.

Many methods have been proposed to alert the navigator of imminent danger, including the last time to maneuver [22], the minimum distance to collision [23], the last line of defense [16] and critical safety area [24]. A projected collision can still be avoided if the navigator takes evasive maneuvers before reaching the critical condition. "Action too late" is the primary cause of collisions [25]. However, these methods cannot provide an optimal solution for starting evasive maneuvers to avoid serious encounters, such as close-quarters situations and imminent danger.

Our previous work proposes a risk-perception-based ship domain [26]. This ship domain reveals the general strategy of a ship determining the timing for taking evasive maneuvers. The boundary of the risk-perception-based ship domain is quantified by statistically analyzing the perceived risk of a navigator when the first evasive maneuvers started over a large sample of vessel encounters taken from AIS data. A non-linear velocity obstacle (NLVO) algorithm is adopted to detect collision risk with the dynamic nature of ship maneuvers considered. Available maneuvering margins (AMM) are utilized as a proxy to measure the perceived collision risk by the navigator, so the risk resolution of a ship is considered. Although this risk-perception-based ship domain considers the dynamic nature of ship operations and risk resolution, this risk-perception-based ship domain cannot be directly used to define when a ship should maneuver for collision avoidance in practical applications. The constraint of ship stability is not considered. Some drastic maneuvers leading to the success of collision avoidance can create risk of the ship capsizing. The neglect of ship stability leads to an inaccurate estimation of ship's capability to eliminate the collision risk.

Therefore, the principal aim of this work is to construct a CAS based on the perceived risk by the navigator, which is applicable in encounter scenarios with various traffic complexities, including both ship pair encounters and multi-vessel encounters. The available maneuvering margins (AMM) with ship stability guarantees are selected as a proxy to reflect the perceived risk of a navigator, hence this proposed CAS is referred to as AMM-based CAS. The existence of collision risk activates this AMM-based CAS, whereas the degree of the violation of this risk-perception-based ship domain determines the level of collision alert. To be clear, our work alerts the ship of a collision in a timely manner and supports in determining the timing for taking evasive maneuvers rather than directly proposing collision avoidance maneuvers. This ship collision alert system intends to further contribute to the development of maritime autonomous surface ships (MASS), particularly in assisting their strategizing for collision avoidance.

The remainder of this paper is organized as follows. Section 2 reviews the related literature. Section 3 focuses on the proposal of the AMM-based CAS, including framework construction and methodology development. Section 4 presents the case study to demonstrate the feasibility of this proposed method. Discussion and conclusions are addressed in Section 5.

2. Related Work

The CAS proposed in this work alerts users to the presence of hazards with specific risk levels and reminds them to be prepared for the response rather than providing them with solutions. Two important components in establishing CAS are collision risk identification and risk level quantification. A large amount of research work has been conducted on these two topics [27].

For ship collision risk identification, distance to closest point of approach (DCPA) and time to closest point of approach (TCPA) are two typical risk indicators [28]. Collision risk is evaluated based on the combination of DCPA and TCPA to assess the collision risk in the Yangtze River [29] and in the Madura Strait [30]. Another popular approach is based on ship domain theory, originally intended to determine the capacity of waterways and further developed to support collision avoidance [31]. The shapes and dimensions of the ship domain have been significantly modified and developed to suit various application scenarios, see a detailed review in [32]. Additonally, the concepts of probability of collision [33], collision threat parameter area (CTPA) [34], velocity obstacle (VO) [35] and fuzzy collision danger domain are introduced to detect collision risk [36]. With the development of methods for collision risk identification, more and more scholars have realized that ignoring the dynamic nature of ship behavior reduces the accuracy of risk detection [37]. For instance, the non-linear velocity of obstacle method has been introduced and developed to accurately detect the presence of collision risk [18].

For risk level quantification, Zhang et al. developed vessel conflict ranking operator (VCRO) to divide the risk severity of near misses into three levels [38,39]. The frequency or probability of ship collision is adopted to separate the serious encounter with non-serious encounters. The degree of domain violation (DDV) and time to domain violation (TDV) are proposed to quantify risk severity to further support real-time collision avoidance decision [40]. Weng et al. proposed an ordered probit model to analyze the severity of two-ship collisions and found that ship size and visibility affect the probability of serious accidents [41]. A risk hierarchy prewarning (RHP) model based on the violation detection of a ship domain is proposed to determine risk level [42]. These methods help to enhance the understanding of the evolution of collision risk and provide a reference for a ship deciding her maneuvering strategy for collision avoidance in real time. However, most existing collision risk measures are independent of conflict resolution, so a high risk does not indicate whether a collision is inevitable or not. This could over/under-estimate the collision risk [27]. Our previous work introduced the concept of available maneuvering margins (AMM) to measure risk resolution when determining the risk level, but as the ship stability constraint is not considered, the risk resolution of a ship is overestimated [14]. In addition, these methods are mainly designed for ship pair encounters without considering traffic conditions [10]. A very limited number of methods consider multi-vessel encounters but divide them into several ship pair encounters. This simplified division ignores the interactions between ship behaviors, leading to an underestimation of the collision risk levels.

Several typical works related to CAS published in recent years are listed in Table 1. Goerlandt et al. proposed a risk-informed CAS based on fuzzy expert rules to divide the alert level into safe, caution, warning, and alarm in accordance with IMO recommendations [8]. The dynamic nature of ship actions and ship resolution are considered by adopting proximity indicators, such as reaction time and turning action, but this method is only applied in ship pair encounters. Baldauf. et al. focused specifically on the critical last phase of an encounter [16]. The last line of defense has been defined and indicates that the available maneuvers leading to the success of collision avoidance are extremely limited. The ship resolution is considered when calculating the last line of defense. Cheng et al. proposed an early warning system based on coordinated collision avoidance actions and applied it in inland waters [43]. DCPA, TCPA, and the coordination degree of collision avoidance actions of the two considered ships are the risk indicators, and this method could effectively reduce false alerts. Du et al. proposed a ship collision alert system for a stand-on ship by quantifying the action obligation of a stand-on ship as specified in the International Regulations for Preventing Collisions at Sea (COLREG) [14]. The non-linear velocity obstacle is employed to detect collision risk, and the available maneuvering margins of a ship are introduced to measure her risk resolution. When measuring the risk resolution of a ship, the impact of traffic complexity is considered, and the constraint of ship stability is ignored. Szlapczynski and Szlapczynska proposed a collision alert system based on five parameters derived from the ship domain concept [10]. The impact of late maneuvers and surrounding traffic are considered. However, one limitation is that the maneuverability of the ship is not available from the AIS data, so the risk resolution of a ship cannot be measured, which makes it difficult to directly inform the timing of the ship's evasive maneuvers to avoid a collision. Qin et al. proposed a risk hierarchy prewarning (RHP) model based on the violation detection of a ship domain [42]. Two layers of protection are constructed based on the ship domain and the possible collision domain (PCD), and the violation of each of them will activate the lower and higher alarms, respectively. The violation of inner protective layer PCD means the collision between this ship pair cannot be avoided by any maneuvers. The traffic conditions and the dynamic nature of ship actions must be considered to increase the adaptability of this model in different waters.

Table 1. Several typical works related to CAS published in recent years.

Research Work	Risk Identification	Risk Level Quantification		
	Action Dynamics	Ship Resolution	Traffic Condition	Ship Stability
Goerlandt et al., 2015	+	+	−	−
Baldauf et al., 2017	−	+	−	−
Cheng et al., 2020	+	−	−	−
Du et al., 2020	+	+	+	−
Szlapczynski and Szlapczynska, 2021	+	−	+	−
Qin et al., 2021	−	+	−	−

3. Methodology

3.1. Conceptual Framework

When two ships are approaching each other, a collision risk analysis will be conducted at the onset of the encounter. If there is no collision risk, both involved ships shall carefully check and remain vigilant until the other ships have passed safely. If collision risk occurs, the ship must prepare for evasive maneuvers based on the actual encounter. The determination of timing for taking evasive maneuvers is one crucial step. COLREGs provides guidance. A give-way ship should take early and substantial action to keep well clear if possible, see Rule 16. A stand-on ship is permitted to take evasive maneuvers if the give-way ship is evidently not maneuvering properly and effectively, as specified in Rule 17. In addition to these rules in COLREGs, the ship resolution that reflects the capability of a ship to eliminate the existing collision directly affects a navigator's decision regarding the timing to perform evasive maneuvers for collision avoidance. Our previous work has observed that the maneuvering timing is affected by a ship's COLREGs identity (a give-way ship or a stand-on ship) and her risk resolution [26].

Based on this collision avoidance process, an AMM-based CAS for ship collision avoidance is proposed to help determine the timing for a ship taking evasive maneuvers, which contains three main steps (see Figure 1), including the collision candidate detection (Step I), determination of timing for evasive maneuvers (Step II), and collision alert based on timing for evasive maneuvers (Step III). Specifically, collision candidate detection is to check whether the collision risk exists between this targeted ship pair. If a collision risk exists, Step II and Step III are activated to determine the alert level. The determination of the timing for evasive maneuvers is to quantify when the ship should take evasive maneuvers (Step II). The difference of a ship with a different COLREGs identity (a give-way ship or a stand-on ship) in determining the action timing is reflected based on the proposed risk-perception-based ship domain. Last is to determine the collision alert level based on the degree of violation of this risk perception-based ship domain (Step III). Here, we consider the ship risk resolution and ship stability limit. The methodologies for measuring these three steps are elaborated upon in Section 3.2 respectively.

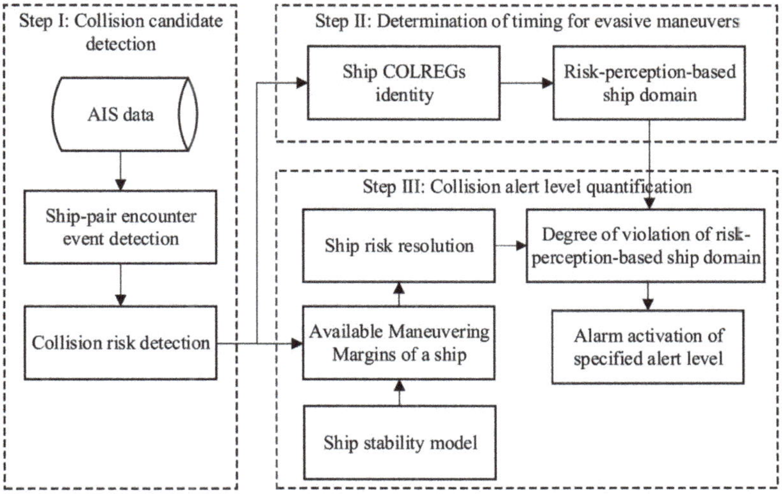

Figure 1. The conceptual framework of AMM-based ship collision alert system.

3.2. Methods Development

3.2.1. Collision Candidate Detection

To consider the dynamic nature of ship maneuvers during collision avoidance, the NLVO algorithm is adopted for collision risk detection [37], which is Step I for the construction of this AMM-based CAS, see Figure 1. By utilizing the NLVO algorithm, the collision risk for each ship pair can be detected by checking whether one ship's velocity falls into the velocity obstacle zone (S_{NL_VO}), see Figure 2.

$$IC(t) = \begin{cases} 1, \, if \; V_{TS}(t) \cap S_{NL\_VO}(t) \neq \varnothing \\ 0, \text{ else} \end{cases} \tag{1}$$

where is the index of ship conflict. V_{TS} is the velocity of the target ship (TS). S_{NL_VO} is the collection of all conflicting velocities that lead to ship collision. In Figure 2, collision risk exists for V_{TS1}, while there is no collision risk for V_{TS2}.

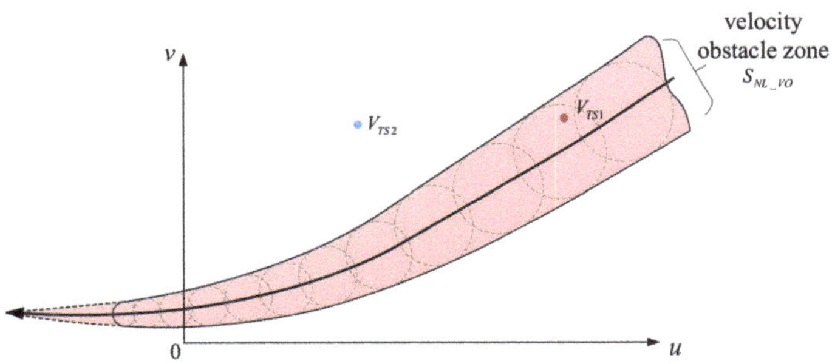

Figure 2. Ship collision risk detection based on NLVO algorithm.

3.2.2. Determination of Timing for Evasive Maneuvers

Step II is to determine the timing for evasive maneuvers, which is affected by ship COLREGs identity and the risk perceived by the navigator.

COLREGs identity is the identity of the ship during the collision avoidance process as specified in the COLREGs. It is classified as a stand-on ship (SO) or a give-way ship (GW) in terms of its action obligation for collision avoidance. As specified in Rules 16 and 17 in COLREGs, a ship with a different COLREGs identity has a diverse strategy. The ship's COLREGs identity can be determined based on their relative heading and relative bearing.

The timing of a ship taking evasive maneuvers is primarily affected by the risk perceived by the navigator [44]. AMM is selected as a proxy to reflect the risk perceived by the navigator [26]. AMM is measured based on the proportion of maneuvers of all the available maneuvers by which a ship can eliminate potential conflicts. The assumption that a ship only changes her course to avoid collision is adopted. Therefore, AMM is determined by its turning ability.

$$\begin{cases} AMM(t) = \frac{\sum \delta_s(t)}{\delta_a(t)},\ if\ \exists V(t) \in RV(\delta_s(t), t_{ob}) : V(t) \cap S_{NL\_VO}(t) = \varnothing \\ t_{ob} = \max(TCPA, 5) \end{cases} \quad (2)$$

where AMM is the value of AMM when the ship maneuvers at time t. δ_s is the adopted rudder angle that can eliminate the existing collision risk. δ_a is all the available rudder angles of a ship. RV is the OS's reachable velocity after steering with a demanded rudder angle. $\varnothing$ is an empty set. t_{ob} is the observation time window, which is determined by the time to the closest point of approach (TCPA).

Next, the risk-perception-based ship domain is defined to determine the timing for evasive maneuvers. Specifically, by statistically analyzing when ships started to take evasive maneuvers under different encounter situations over a large sample of ship encounters from AIS data, the general practice of determining the timing for a ship taking evasive maneuvers is obtained in [45], see Table 2. The lengths of small-size ships, medium-size ships, and large-size ships are 100 m or less, 100 m to 200 m, and 200 m or more, respectively. AMM_1 and AMM_2 are the upper limit and lower limit of AMM, respectively.

Table 2. The value of AMM at the boundary of risk-perception-based ship domain.

Ship Type	COLREGs Status	AMM Threshold (AMM_1/AMM_2)		
		Small-Size	Medium-Size	Large-Size
Passenger Ship	GW	0.986/0.586	0.914/0.486	0.814/0.343
	SO	0.943/0.443	0.786/0.314	0.729/0.229
Tanker	GW	0.871/0.471	0.829/0.314	0.8/0.229
	SO	0.857/0.371	0.629/0.214	0.486/0.186
Cargo Ship	GW	0.9/0.4	0.886/0.343	0.871/0.257
	SO	0.729/0.314	0.5/0.243	0.486/0.157

3.2.3. Collision Alert Level Quantification

The final step, Step III, is quantifying the collision alert level (Figure 1). The collision alert will be activated if collision risk exists and the degree of the violation of risk-perception-based ship domain determines the alert level. Violation of this risk-perception-based ship domain means that the ship's behavior is abnormal and may lead to a danger, as most ships (about 90% of ships sailing in this area) would maneuver before this moment. The degree of the violation of risk-perception-based ship domain can be measured based on ship risk resolution.

The risk resolution of a ship $AMMs$ can be measured by her available maneuvering margins with stability guarantees. A ship will heel to the opposite direction of course change when turning for collision avoidance. If the ship's heeling angle θ exceeds its threshold value θ_c, the ship will be in danger of capsizing. The ship's heeling angle can be expressed as:

$$\tan\theta = \frac{V \cdot r \cdot GB}{g \cdot GM} \quad (3)$$

where θ is the ship's heeling angle. V is the ship speed. r is the yaw rate of ship steering. GB is the distance between the center of ship gravity and the center of ship buoyancy. GM

is metacentric height, a vertical distance from the center of gravity to the metacenter. g is gravitational acceleration.

The yaw rate of ship steering r can be measured based on Nomoto model when the rudder angle is determined.

$$r = K\delta(1 - e^{-t/T}) \tag{4}$$

where δ is the rudder angle. Turning ability index K and turning lag index T vary with ship length and velocity.

Then, the ship risk resolution $AMMs$ can be modeled as follows:

$$\begin{cases} AMM_s(t) = \frac{\sum \delta_s(t)}{\delta_a(t)},\ if\ \exists V(t) \in RV(\delta_s(t), t_{ob}) : V(t) \cap S_{NL\_VO}(t) = \varnothing \\ \delta_s(t) \le \frac{\tan\theta_c \cdot g \cdot GM}{V \cdot K \cdot GB \cdot (1 - e^{-t/T})} \\ t_{ob} = \max(TCPA, 5) \end{cases} \tag{5}$$

In terms of the degree of violation of this risk-perception-based ship domain, the collision alert is divided into three levels, ranging from low risk to medium risk and high risk.

$$CAL(t) = \begin{cases} L, if\ IC(t) = 1 \& AMMs(t) \ge AMM_1 \\ M,\ if\ IC(t) = 1 \& AMM_1 > AMMs(t) \ge AMM_2 \\ H,\ if\ IC(t) = 1 \& AMMs(t) < AMM_2 \end{cases} \tag{6}$$

where CAL is collision alert level. CAL is low if collision risk exists but the $AMMs$ of a ship is higher than the upper limit AMM_1. When collision risk exists and the $AMMs$ of a ship is lower than the lower limit AMM_2, CAL is in the high level. For other situations when the risk exists, CAL is medium. AMM_1 and AMM_2 mean that 90% and 99% of the ship starts an evasive action with a higher AMM than this, respectively [19].

4. Case Study

Three typical encounter scenarios are selected from AIS data to demonstrate the feasibility of this proposed AMM-based CAS. Specifically, the first two scenarios are ship pair encounters, and the last one is a multi-vessel encounter. The ship attributes are shown in Table 3. The encounter process lasts for 30 min. A maritime mobile service identity (MMSI) uniquely identifies ship stations and is masked in Table 3 to ensure vessel can be anonymous. The encounter processes are illustrated in Figures 3–10.

Table 3. Ship attributes in two typical encounter scenarios.

Encounter Scenarios	Ship Identity	MMSI	Type	Length (m)	Width (m)
Scenario 1	OS	27335XXXX	Cargo ship	84	15
	TS	27343XXXX	Tanker	126	16
Scenario 2	OS	27333XXXX	Passenger ship	56	11
	TS	20544XXXX	Cargo ship	182	28
Scenario 3	OS	21352XXXX	Cargo ship	92	18
	TS1	21107XXXX	Cargo ship	110	14
	TS2	26125XXXX	Cargo ship	185	32

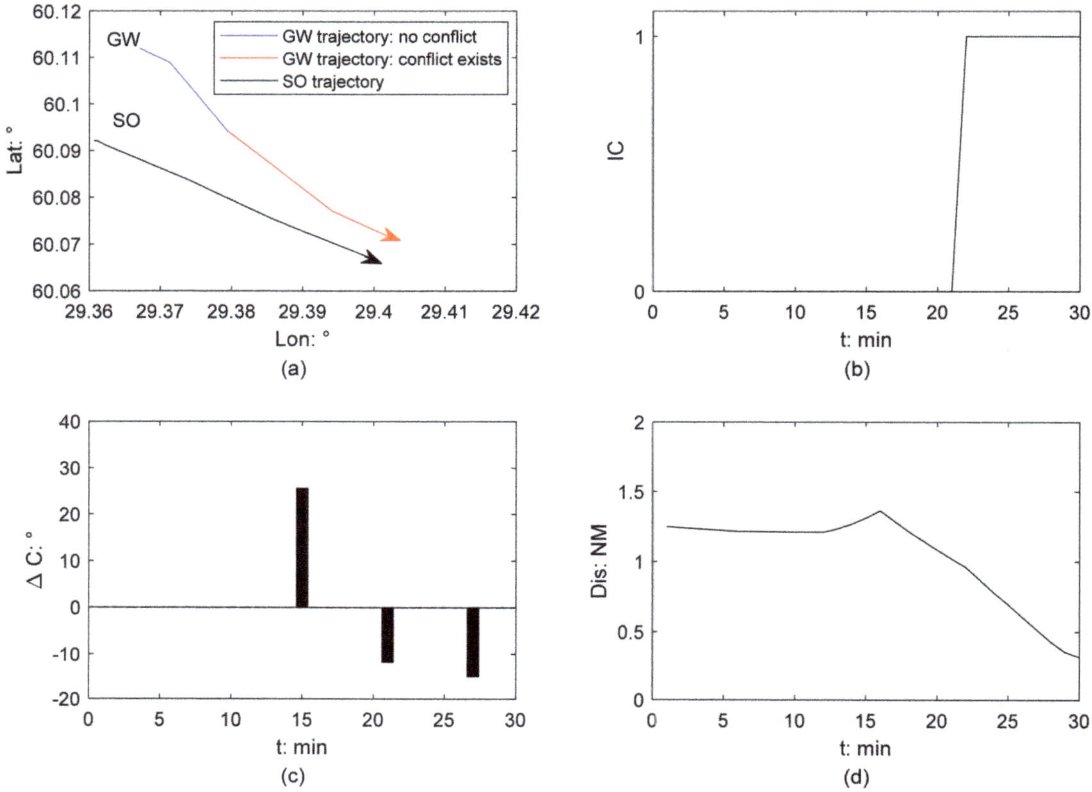

Figure 3. Ship collision detection from GW perspective in Scenario 1: (**a**) ship trajectory; (**b**) ship collision risk identification; (**c**) course change of a GW; (**d**) relative distance between this ship pair.

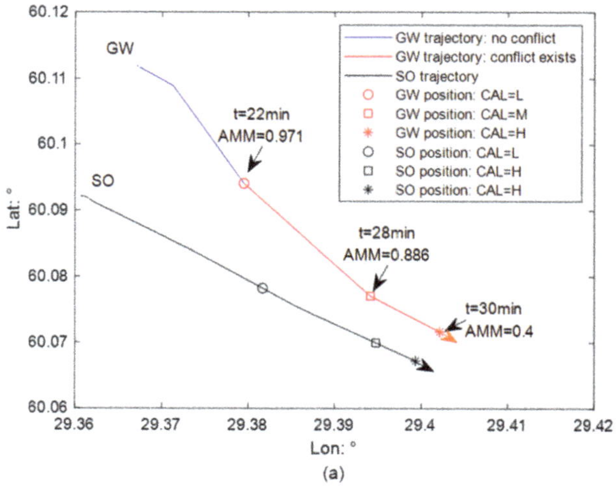

Figure 4. *Cont.*

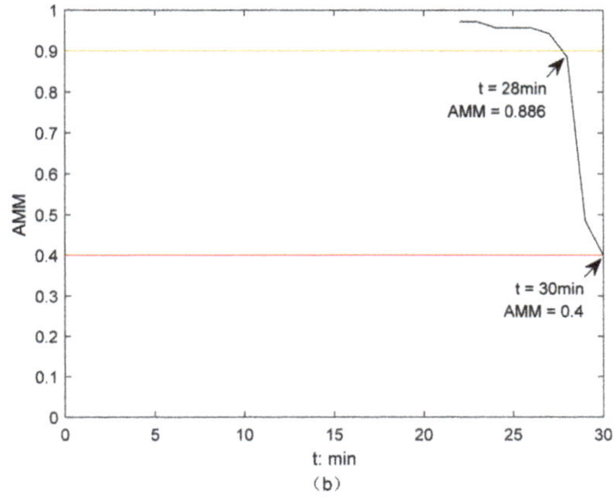

Figure 4. Illustration of the activation of AMM-based CAS from GW perspective in Scenario 1, without ship stability considered: (**a**) collision alert level determination; (**b**) the change of AMM of a GW.

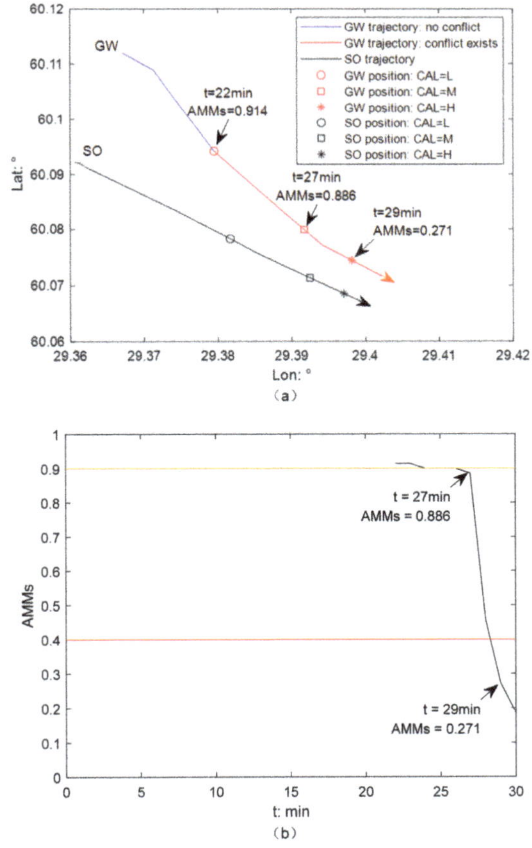

Figure 5. Illustration of the activation of AMM-based CAS from GW perspective in Scenario 1, with ship stability considered: (**a**) collision alert level determination; (**b**) the change of AMM of a GW.

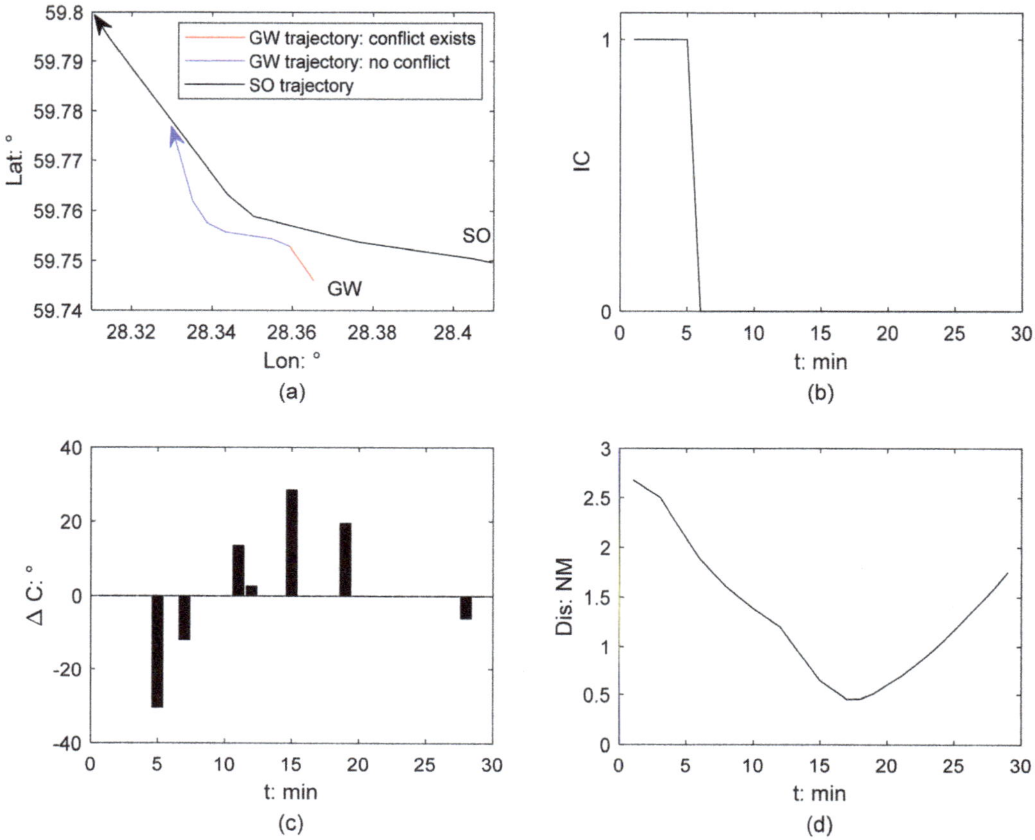

Figure 6. Ship collision detection from GW perspective in Scenario 2 (**a**) ship trajectory; (**b**) ship collision risk identification; (**c**) course change of a GW; (**d**) relative distance between this ship pair.

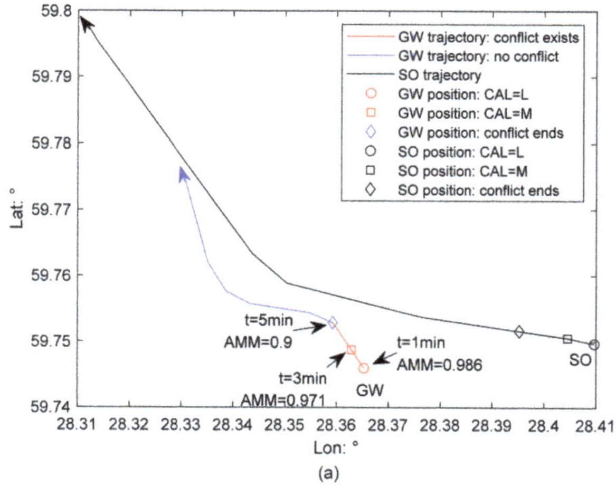

Figure 7. *Cont.*

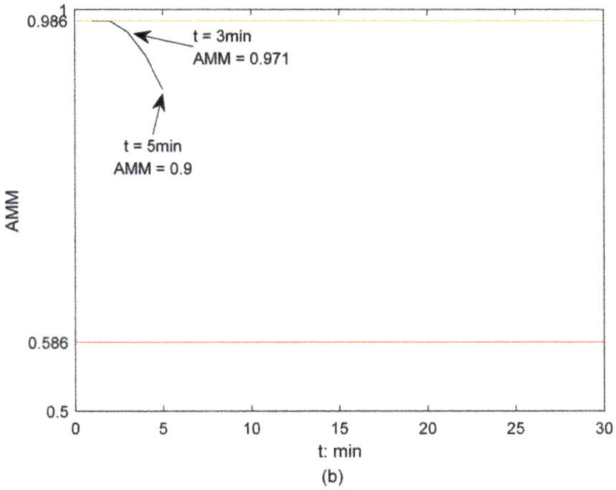

Figure 7. Illustration of the activation of AMM-based CAS from GW perspective in Scenario 2, without ship stability considered: (**a**) collision alert level determination; (**b**) the change of AMM of a GW.

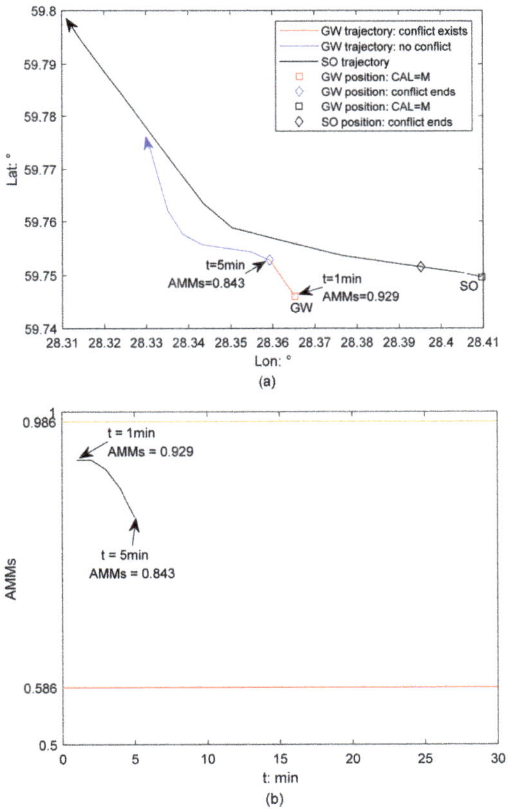

Figure 8. Illustration of the activation of AMM-based CAS from GW perspective in Scenario 2, with ship stability considered: (**a**) collision alert level determination; (**b**) the change of AMM of a GW.

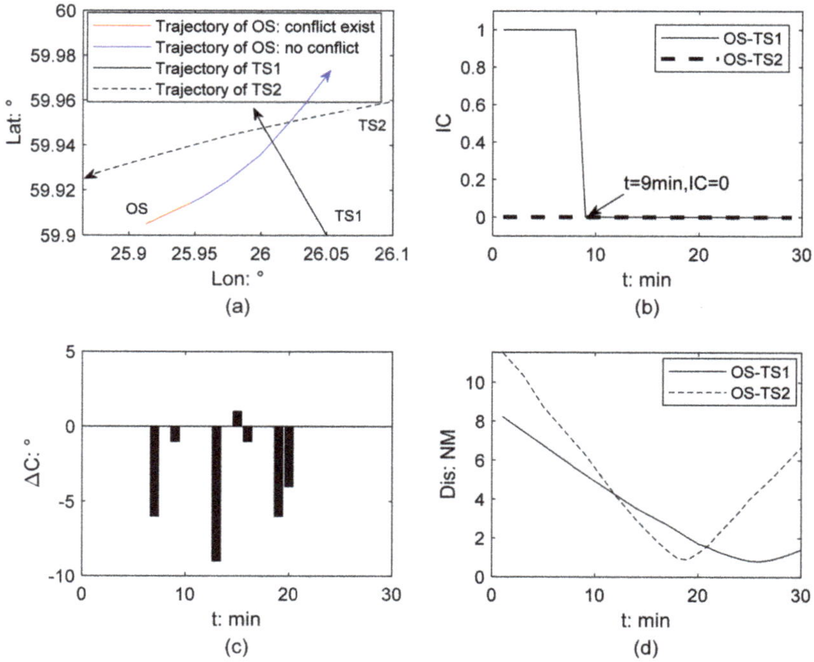

Figure 9. Ship collision detection from GW perspective in Scenario 3 (**a**) ship trajectory; (**b**) ship collision risk identification; (**c**) course change of a GW; (**d**) relative distance between each two ships.

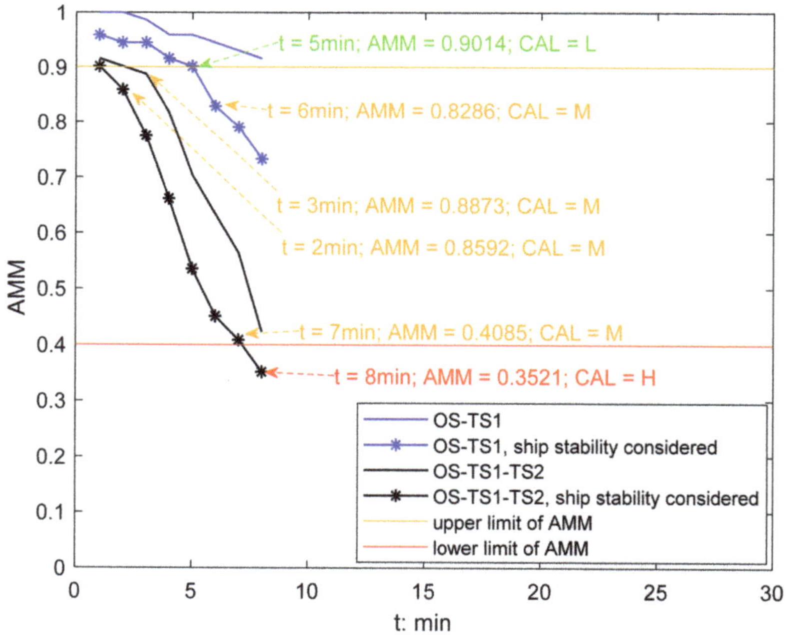

Figure 10. Illustration of the activation of AMM-based CAS from GW perspective in Scenario 3.

4.1. Ship Pair Encounter Scenarios

Figures 3 and 6 present the results of ship collision detection in Scenarios 1 and 2, respectively. According to their relative positions and relative bearings, the ship COLREGs provide the basis for determining whether a ship is a give-way (GW) or a stand-on ship (SO). The trajectory of the stand-on ship is a black line. In Figures 3a and 6a, the GW's trajectory is marked in color, in which blue and red present safe and dangerous, respectively. The arrow indicates the direction in which the ship moves forward. Figures 3b and 6b displays the change of IC, and IC = 1 means that collision risk exists, while IC = 0 means there is no collision risk. Figures 3c and 6c present the course change of a give-way ship. If ΔC is positive, it means that a ship turns to starboard, while ΔC being negative means that a ship turns to port. Figures 3d and 6d show the relative distance (Dis) between two ships.

Figures 4, 5, 7 and 8 illustrate how the AMM-based CAS from the GW perspective is activated and developed without or with the ship's stability considered, respectively.

4.1.1. Scenario 1

The cargo ship is a give-way ship because the cargo ship approaches the tanker from the tanker's port side. Figure 3 presents the result of Scenario 1 from a GW perspective. This give-way ship is regarded as own ship (OS) and the tanker is target ship (TS). The GW is a small-size cargo ship, so two limit values—AMM_1 and AMM_2—are 0.9 and 0.4, respectively, from Table 2. When measuring the stability of GW, the value of GM is set to 4.2 times the value of GB. The maximum value of ship's heeling angle θ_c is set as 35 degrees.

There is no collision risk between this ship pair at the beginning, see Figure 3b. GW's turn to portside at 21 min generates the collision risk (Figure 3b,c). From 22 min, the collision risk occurs. Afterward, the relative distance between two ships keeps decreasing, see Figure 3d,e. GW's turns to port at 28 min are not effective, as the collision hazard remains (Figure 3a,c).

Figure 4 illustrates the activation of the AMM-based CAS from the GW perspective in Scenario 1 without ship stability considered. From 22 min, collision risk occurs when the AMM of GW is 0.971 (Figure 4). Afterward, the AMM of GW continues to decrease, see Figure 4b. Before 27 min, the AMM of GW is larger than 0.9. Therefore, the CAS is activated, and CAL is low. GW's turns to port at 28 min are not effective for collision avoidance, which can be attested by the dramatic drop in her AMM from that point. From 28 min, the CAS is activated, and CAL is medium. The AMM of GW drops to 0 4 at 30 min, which reaches the lower limit AMM_2. Therefore, the CAS is activated, and CAL increases to medium from 28 min to 30 min. From 30 min, the CAS is activated, and CAL is high, as AMM of GW is less than 0.4.

Figure 5 illustrates the determination of the collision alert level from the GW perspective in Scenario 1 with ship stability considered. The CAS is activated from 22 min, when its risk resolution (AMMs) is 0.914. The CAL is low, as AMMs are larger than the upper limit value AMM_1. Afterwards, the AMMs of a GW continue to reduce (Figure 5b). From 27 min, the CAL changes to medium, as the AMMs of this GW are lower than the upper limit AMM_1 but still larger than the lower limit AMM_2. With ship stability considered, the time of the CAL mutating from low to medium was advanced by 1 min. The GW does not maneuver properly and effectively, and collision risk develops. The AMMs of the GW drops and is lower than the lower limit AMM_2 from 29 min, and therefore, the CAL increases to high. This mutation time is 1 min earlier than that when the constraint of ship stability is ignored.

4.1.2. Scenario 2

According to the relative position and relative bearing of this ship pair, GW is a small-sized passenger ship. Figure 6 presents the result of Scenario 2 from a GW perspective. This GW is regarded as OS. From Table 2, AMM_1 and AMM_2 are 0.986 and 0.586, respectively. When measuring the stability of GW, the value of GM is set to 1.5 times the value of GB. The maximum value of ship's heeling angle θ_c is set as 35 degrees.

The collision risk arises from the beginning (Figure 6a,b). GW does not take maneuvers from 3 min to 5 min (Figure 6c), so the relative distance between this ship pair reduces (Figure 6d). At 5 min, GW turns approximately 30° to port, which is sufficient to eliminate the existing collision risk. After that, there is no risk of collision; the GW turns several times to maintain a sufficient passing distance and then returns to her planned trajectory.

Figure 7 illustrates how the AMM-based CAS from the GW perspective in Scenario 2 is activated without ship stability being considered. The collision risk emerges from the beginning (Figure 6b). The AMM of the GW is 0.986 at 1 min and remains before 3 min (Figure 7b). Therefore, the CAS is activated, and CAL is low during this period. At 3 min, the AMM of the GW drops to 0.971, which is lower than the upper limit AMM_1, and the CAL changes to medium. From 3 min to 5 min, the GW sails with a constant speed and course, and her AMM continues to reduce but is still higher than the lower limit AMM_2. The CAL remains at medium level before 5 min. Due to the GW's positive and effective evasive maneuvers, there is no collision risk, and therefore, the CAS is deactivated afterwards.

Figure 8 presents the results of the activation of CAS and determination of CAL from the GW perspective in Scenario 2 with ship stability considered. At 1 min, the AMM of this GW is 0.929, which is lower than the upper limit AMM_1, and therefore the CAS is activated; the CAL is medium 2 min earlier than that with ship stability ignored. During the next five minutes, the severity of collision risk develops as the GW does not maneuver as required for collision avoidance. The AMM of the GW reduces but remains larger than the lower limit AMM_2, the CAS is activated, and CAL is medium. The CAS is deactivated since 5 min as the collision risk is eliminated due to the GW's evasive maneuvers.

4.2. Multi-Vessel Encounter Scenario

Figure 9 shows the results of the ship collision detection in Scenario 3, which is a multi-vessel encounter involving three ships, whose basic information is described in Table 3. Figure 9a displays the trajectory of three ships. Figure 9b shows that the collision risk exists only between own ship (OS) and target ship 1 (TS1) from the beginning. According to their relative positions and relative bearings, OS is a give-way ship. OS's trajectory is marked in color, in which blue and red present safe and dangerous, respectively.

Before 9 min, OS sails with constant course and speed (Figure 9c), so the collision risk remains (Figure 9b), and the relative distances between OS and other two ships continue to decrease (Figure 9d). The collision risk between OS and TS1 is eliminated from 9 min due to OS's turning to port (Figure 9c), which is proven to be positive and effective. Afterwards, there is no collision risk in this multi-vessel encounter. At 19 min, OS and TS2 reached their CPA at a distance of 0.9 nm. At 26 min, the relative distance between OS and TS1 drops to a minimum of 0.8 nm.

Figure 10 illustrates how the AMM-based CAS from the GW perspective is activated and how the CAL is developed. The GW is a small-sized cargo ship, so the upper and lower limits of AMM are 0.9 and 0.4, respectively, as shown in Table 2. The impacts of ship stability and traffic conditions on quantifying collision risk severity and collision alert level are analyzed. Four sets of experiments were conducted.

The first group neither considers the effects of ship stability nor the surrounding traffic conditions, and the experimental result is a blue line. During the period in which collision risk exists, the AMM of GW continues to reduce but is always larger than the upper limit, and therefore, the CAS is activated and the CAL is determined as low.

In the second group, only the effect of ship stability is considered, and the result is shown as a blue line with stars marked. The AMM of GW shows a steady decreasing trend and is smaller than that in Group 1 at the same moment. At 5 min, the AMM of GW is 0.9014. Prior to this, CAS is activated, and the CAL remains at a low level. At 6 min, GW's AMM violates its upper limit, dropping to 0.8286, which increased CAL from a low to a medium level. Before the collision risk is eliminated, the CAL remains at a medium level.

In the third group, only the surrounding traffic is considered, and the result is a black line. Although the surrounding ship TS2 does not directly threaten the navigation of GW, it limits the GW's risk resolution. The AMM of the GW decreases from 0.9155 at 1 min to 0.8873 at 3 min and 0.4225 at 8 min. Therefore, the CAS is activated during this period, and the CAL increases from low to medium level at 4 min.

In the fourth group, the impacts of both ship stability and surrounding traffic are considered, and the result is the black line with stars marked. At the beginning, the AMM of the GW is 0.9014, which is higher than its upper limit. The CAS is activated, and the CAL is at a low level. At 2 min, the AMM of the GW dropped to 0.8592, which lies between its upper and lower limits. The AMM of the GW continuously drops to 0.4085 at 7 min. The CAL remains at medium level from 2 min to 7 min. At 8 min, the AMM of the GW drops below its lower limit. The severity of the risk becomes severe, and at this moment, CAL escalates to high. The collision risk is eliminated since 9 min as the GW adopted effective evasive maneuvers.

4.3. Remarks

In comparison to the results of CAL determination in Scenarios 1 and 2, this proposed AMM-based CAS considering ship stability can more accurately quantify the collision risk level. Even though some extreme maneuvers, such as adopting full-rudder steering, can lead the own ship to pass safely with target ships, this full-rudder steering may lead to the capsizing of the own ship under certain environmental and loading conditions. The neglect of ship stability may lead to overestimation of the risk resolution of a ship, thus leading to an underestimation of collision risk.

This proposed AMM-based CAS is applicable in both ship pair encounters and multi-vessel encounters. Regarding a multi-vessel encounter as a linear superposition of multiple ship pair encounters could underestimate the collision risk as possible interaction effects between all ships operating nearby are ignored. In this AMM-based CAS, the impact from other ships nearby on the own ship's risk resolution is well explained. In Figure 10, the comparison of the experimental results between the first two groups and the latter two groups reveals that the AMM of the GW is relatively lower in the latter two groups. This is because the latter two groups argue that the presence of the surrounding vessel TS2 directly reduces the GW's capability of collision avoidance as some of her available operations could generate a new collision risk with TS2. The severity of the collision risk can be more accurately quantified by taking into account the complexity of traffic that caused by other vessels in the vicinity.

5. Discussion and Conclusions

This paper proposes an available maneuvering margins (AMM)-based collision alert system to alert the navigator to take evasive maneuvers timely for safe passing. This AMM-based CAS contains two main parts. The first part is the detection of collision risk. Instead of assuming the ship retains her speed and course under the threat of collision risk, the non-linear velocity obstacle (NLVO) algorithm is adopted to detect the collision risk by considering the encounter as a process. This considers the dynamic nature of ship maneuvering, so the accuracy of collision risk detection is improved. The second part is the determination of the alert level. The alert level is divided into three levels in terms of the degree of violation of risk-perception-based ship domain, which can be measured based on ship risk resolution. A ship's risk resolution is quantified by her available maneuvering margins with stability guarantees.

Three typical encounter scenarios are selected from AIS data to demonstrate the feasibility of this AMM-based CAS, and the results are promising. This proposed AMM-based CAS is applicable in both ship pair encounter and multi-vessel encounter scenarios. The collision risk can be accurately detected for these cases. It can further accurately quantify the risk level and activate the corresponding level of risk warning. Therefore, this AMM-based CAS has the potential to be applied for various purposes in complicated encounter scenarios. First, it can support the navigator to formulate a strategy for collision avoidance. Second, it could contribute to enable the safety of autonomous ships if the CAS is further developed to lay at the basis of an automatic collision avoidance system or if it is used in a shore control center. Having sufficient information on navigational safety, including the severity of collision risk and the timing for performing evasive maneuvers, is essential for both autonomous vessels and conventional ships to take the proper actions to ensure safe passage.

Nonetheless, this AMM-based CAS can be improved in the following aspects. First, this work assumes that a ship only changes course to avoid collisions. The consideration of the reduction of ship speed and course change during collision avoidance helps to improve the computational accuracy of a ship's AMM, which is one direction for our next research. Second, environmental disturbance, which directly decreases the ship's maneuverability and ship stability, must be considered in future studies. The consideration of the impact of environmental disturbance could expand the applicability of this method. Third, to the method of developing an optimal collision avoidance strategy after receiving a collision alert requires future work. Safety, economy, and comfort will be considered simultaneously. Finally, this AMM-based CAS has demonstrated its reasonableness and feasibility only in a limited test scenario, so further testing (additional scenarios of encounter cases occurring in different waters, in bridge simulators, and onboard vessels) is required before it can be used in practical contexts. Considering the difference between open waters and restricted waters, more tests in restricted waters are necessary to check whether this AMM-based CAS is still effective under such encounters.

Author Contributions: Conceptualization, L.D. and O.A.V.B.; methodology, L.D.; software, Z.S.; validation, L.D., O.A.V.B. and Z.S.; formal analysis, L.D.; investigation, Z.S.; resources, O.A.V.B.; data curation, L.D.; writing—original draft preparation, L.D.; writing—review and editing, O.A.V.B.; visualization, L.D. and Z.S.; supervision, O.A.V.B.; project administration, O.A.V.B.; funding acquisition, O.A.V.B. All authors have read and agreed to the published version of the manuscript.

Funding: This paper is supported by the Fundamental Research Funds for the Central Universities (WUT: 3120622884).

Institutional Review Board Statement: Not applicable.

Informed Consent Statement: Not applicable.

Data Availability Statement: Not applicable.

Acknowledgments: This work is supported by the Merenkulun Säätiö and Kotka Maritime Research Association. This support is gratefully acknowledged. Besides, we thank the three anonymous reviewers for their very insightful comments, which have been very instrumental to improve an earlier version of this work.

Conflicts of Interest: The authors declare no conflict of interest.

References

1. Valdez Banda, O.A.; Goerlandt, F.; Montewka, J.; Kujala, P. Winter navigation at the Baltic Sea: An analysis of accidents occurred during winters 2002–2003 & 2009–2013. In *Safety and Reliability: Methodology and Applications*; CRC Press: Leiden, The Netherlands, 2014; p. 83.
2. EMSA. Annual Overview of Marine Casualties and Incidents. European Maritime Safety Agency, 2021. Available online: https://emsa.europa.eu/accident-investigation-publications/annual-overview.html (accessed on 10 August 2022).
3. Gil, M.; Kozioł, P.; Wróbel, K.; Montewka, J. Know your safety indicator—A determination of merchant vessels Bow Crossing Range based on big data analytics. *Reliab. Eng. Syst. Saf.* **2022**, *220*, 108311. [CrossRef]
4. Zhang, W.; Zou, Z.; Wang, J.; Du, L. Multi-ship following operation in ice-covered waters with consideration of inter-ship communication. *Ocean. Eng.* **2020**, *210*, 107545. [CrossRef]
5. Zhang, M.; Conti, F.; Le Sourne, H.; Vassalos, D.; Kujala, P.; Lindroth, D.; Hirdaris, S. A method for the direct assessment of ship collision damage and flooding risk in real conditions. *Ocean. Eng.* **2021**, *237*, 109605. [CrossRef]
6. Baldauf, M.; Benedict, K.; Fischer, S.; Motz, F.; Schröder-Hinrichs, J.U. Collision avoidance systems in air and maritime traffic. *Proc. Inst. Mech. Eng. Part O J. Risk Reliab.* **2011**, *225*, 333–343. [CrossRef]

7. Simsir, U.; Amasyalı, M.F.; Bal, M.; Çelebi, U.B.; Ertugrul, S. Decision support system for collision avoidance of vessels. *Appl. Soft Comput.* **2014**, *25*, 369–378. [CrossRef]
8. Goerlandt, F.; Montewka, J.; Kuzmin, V.; Kujala, P. A risk-informed ship collision alert system: Framework and application. *Saf. Sci.* **2015**, *77*, 182–204. [CrossRef]
9. Szlapczynski, R.; Szlapczynska, J. A framework of a ship domain-based collision alert system. In *Marine Navigation*; CRC Press: Boca Raton, FL, USA, 2017; pp. 183–189.
10. Szlapczynski, R.; Szlapczynska, J. A ship domain-based model of collision risk for near-miss detection and Collision Alert Systems. *Reliab. Eng. Syst. Saf.* **2021**, *214*, 107766. [CrossRef]
11. Wang, X.; Liu, Z.; Cai, Y. The ship maneuverability based collision avoidance dynamic support system in close-quarters situation. *Ocean. Eng.* **2017**, *146*, 486–497. [CrossRef]
12. Wu, B.; Yip, T.L.; Yan, X.; Soares, C.G. Fuzzy logic based approach for ship-bridge collision alert system. *Ocean. Eng.* **2019**, *187*, 106152. [CrossRef]
13. Gil, M.; Wróbel, K.; Montewka, J.; Goerlandt, F. A bibliometric analysis and systematic review of shipboard Decision Support Systems for accident prevention. *Saf. Sci.* **2020**, *128*, 104717. [CrossRef]
14. Du, L.; Banda, O.A.V.; Goerlandt, F.; Huang, Y.; Kujala, P. A COLREG-compliant ship collision alert system for stand-on vessels. *Ocean. Eng.* **2020**, *218*, 107866. [CrossRef]
15. Yoo, Y.; Lee, J.S. Collision Risk Assessment Support System for MASS RO and VTSO Support in Multi-Ship Environment of Vessel Traffic Service Area. *J. Mar. Sci. Eng.* **2021**, *9*, 1143. [CrossRef]
16. Baldauf, M.; Mehdi, R.; Fischer, S.; Gluch, M. A perfect warning to avoid collisions at sea? *Zesz. Nauk. Akad. Mor. W Szczec.* **2017**, *49*, 53–64.
17. Mestl, T.; Tallakstad, K.T.; Castberg, R. Identifying and analyzing safety critical maneuvers from high resolution AIS data. *TransNav Int. J. Mar. Navig. Saf. Sea Transp.* **2016**, *10*, 69–77. [CrossRef]
18. Huang, Y.; van Gelder, P. Measuring ship collision risk in a dense traffic environment. *TransNav Int. J. Mar. Navig. Saf. Sea Transp.* **2019**, *13*, 737–744. [CrossRef]
19. Du, L.; Valdez Banda, O.A.; Goerlandt, F.; Kujala, P.; Zhang, W. Improving Near Miss Detection in Maritime Traffic in the Northern Baltic Sea from AIS Data. *J. Mar. Sci. Eng.* **2021**, *9*, 180. [CrossRef]
20. Sui, Z.; Wen, Y.; Huang, Y.; Zhou, C.; Du, L.; Piera, M.A. Node importance evaluation in marine traffic situation complex network for intelligent maritime supervision. *Ocean. Eng.* **2022**, *247*, 110742. [CrossRef]
21. Zhang, M.; Zhang, D.; Fu, S.; Kujala, P.; Hirdaris, S. A Predictive Analytics Method for Maritime Traffic Flow Complexity Estimation in Inland Waterways. *Reliab. Eng. Syst. Saf.* **2022**, *220*, 108317. [CrossRef]
22. Zhuo, Y.; Tang, T. An intelligent decision support system to ship anti-collision in multi-ship encounter. In Proceedings of the 2008 7th World Congress on Intelligent Control and Automation, Chongqing, China, 25–27 June 2008; pp. 1066–1071.
23. Montewka, J.; Hinz, T.; Kujala, P.; Matusiak, J. Probability modelling of vessel collisions. *Reliab. Eng. Syst. Saf.* **2010**, *95*, 573–589. [CrossRef]
24. Gil, M. A concept of critical safety area applicable for an obstacle-avoidance process for manned and autonomous ships. *Reliab. Eng. Syst. Saf.* **2021**, *214*, 107806. [CrossRef]
25. Graziano, A.; Teixeira, A.P.; Soares, C.G. Classification of human errors in grounding and collision accidents using the TRACEr taxonomy. *Saf. Sci.* **2016**, *86*, 245–257. [CrossRef]
26. Du, L.; Valdez Banda, O.A.; Huang, Y.; Goerlandt, F.; Kujala, P.; Zhang, W. An empirical ship domain based on evasive maneuver and perceived collision risk. *Reliab. Eng. Syst. Saf.* **2021**, *2013*, 107752. [CrossRef]
27. Huang, Y.; Chen, L.; Chen, P.; Negenborn, R.R.; van Gelder, P. Ship collision avoidance methods: State-of-the-art. *Saf. Sci.* **2020**, *121*, 451–473. [CrossRef]
28. Ozturk, U.; Cicek, K. Individual collision risk assessment in ship navigation: A systematic literature review. *Ocean. Eng.* **2019**, *180*, 130–143. [CrossRef]
29. Cai, M.; Zhang, J.; Zhang, D.; Yuan, X.; Soares, C.G. Collision risk analysis on ferry ships in Jiangsu Section of the Yangtze River based on AIS data. *Reliab. Eng. Syst. Saf.* **2021**, *215*, 107901. [CrossRef]
30. Pratiwi, E.; Artana, K.B.; Dinariyana, A.A.B. Fuzzy inference system for determining collision risk of ship in Madura Strait using automatic identification system. *Int. J. Mar. Environ. Sci.* **2017**, *11*, 401–405.
31. Liu, K.; Yuan, Z.; Xin, X.; Zhang, J.; Wang, W. Conflict detection method based on dynamic ship domain model for visualization of collision risk Hot-Spots. *Ocean. Eng.* **2021**, *242*, 110143. [CrossRef]
32. Szlapczynski, R.; Szlapczynska, J. Review of ship safety domains: Models and applications. *Ocean. Eng.* **2017**, *145*, 277–289. [CrossRef]
33. Park, J.; Kim, J. Predictive evaluation of ship collision risk using the concept of probability flow. *IEEE J. Ocean. Eng.* **2016**, *42*, 836–845. [CrossRef]
34. Szlapczynski, R.; Krata, P. Determining and visualizing safe motion parameters of a ship navigating in severe weather conditions. *Ocean. Eng.* **2018**, *158*, 263–274. [CrossRef]
35. Zhao, L.; Fu, X. A novel index for real-time ship collision risk assessment based on velocity obstacle considering dimension data from AIS. *Ocean. Eng.* **2021**, *240*, 109913. [CrossRef]

36. Su, C.M.; Chang, K.Y.; Cheng, C.Y. Fuzzy decision on optimal collision avoidance measures for ships in vessel traffic service. *J. Mar. Sci. Technol.* **2012**, *20*, 38–48. [CrossRef]
37. Chen, P.; Huang, Y.; Mou, J.; van Gelder, P. Ship collision candidate detection method: A velocity obstacle approach. *Ocean. Eng.* **2018**, *170*, 186–198. [CrossRef]
38. Zhang, W.; Goerlandt, F.; Montewka, J.; Kujala, P. A method for detecting possible near miss ship collisions from AIS data. *Ocean. Eng.* **2015**, *107*, 60–69. [CrossRef]
39. Zhang, W.; Goerlandt, F.; Kujala, P.; Wang, Y. An advanced method for detecting possible near miss ship collisions from AIS data. *Ocean. Eng.* **2016**, *124*, 141–156. [CrossRef]
40. Szlapczynski, R.; Szlapczynska, J. An analysis of domain-based ship collision risk parameters. *Ocean. Eng.* **2016**, *126*, 47–56. [CrossRef]
41. Weng, J.; Li, G.; Chai, T.; Yang, D. Evaluation of two-ship collision severity using ordered probit approaches. *J. Navig.* **2018**, *71*, 822–836. [CrossRef]
42. Qin, T.; Ma, G.; Li, D.; Zhou, X.; He, X.; Chen, W. Dynamic risk prewarning in ship encounter process considering domain violation. *J. Navig.* **2021**, *74*, 1416–1431. [CrossRef]
43. Cheng, Z.; Li, Y.; Wu, B. Early warning method and model of inland ship collision risk based on coordinated collision-avoidance actions. *J. Adv. Transp.* **2020**, *2020*, 5271794. [CrossRef]
44. Chauvin, C.; Lardjane, S. Decision making and strategies in an interaction situation: Collision avoidance at sea. *Transport. Res. Part F Traffic* **2008**, *11*, 259–269. [CrossRef]
45. Du, L.; Goerlandt, F.; Valdez Banda, O.A.; Huang, Y.; Wen, Y.; Kujala, P. Improving stand-on ship's situational awareness by estimating the intention of the give-way ship. *Ocean. Eng.* **2020**, *201*, 107110. [CrossRef]

Journal of
Marine Science and Engineering

MDPI

Article

Collision Avoidance Algorithm for USV Based on Rolling Obstacle Classification and Fuzzy Rules

Lifei Song [1,2], Xiaoqian Shi [1,2], Hao Sun [1,2], Kaikai Xu [1,2] and Liang Huang [3,4,*]

1 Key Laboratory of High-Performance Ship Technology, Wuhan University of Technology, Ministry of Education, Wuhan 430063, China; songlifei@whut.edu.cn (L.S.); sxq@whut.edu.cn (X.S.); sunh@whut.edu.cn (H.S.); 260381@whut.edu.cn (K.X.)
2 School of Naval Architecture, Ocean and Energy Power Engineering, Wuhan University of Technology, Wuhan 430063, China
3 Intelligent Transportation Systems Research Center, Wuhan University of Technology, Wuhan 430063, China
4 National Engineering Research Center for Water Transport Safety, Wuhan University of Technology, Wuhan 430063, China
* Correspondence: leung.huang@whut.edu.cn

Abstract: Dynamic collision avoidance between multiple vessels is a task full of challenges for unmanned surface vehicle (USV) movement, which has high requirements on real-time performance and safety. The difficulty of multi-obstacle collision avoidance is that it is hard to formulate the optimal obstacle avoidance strategy when encountering more than one obstacle threat at the same time; a good strategy to avoid one obstacle sometimes leads to threats from other obstacles. This paper presents a dynamic collision avoidance algorithm for USVs based on rolling obstacle classification and fuzzy rules. Firstly, potential collision probabilities between a USV and obstacles are calculated based on the time to the closest point of approach (*TCPA*). All obstacles are given different priorities based on potential collision probability, and the most urgent and secondary urgent ones will then be dynamically determined. Based on the velocity obstacle algorithm, four possible actions are defined to determine the basic domain in the collision avoidance strategy. After that, the Safety of Avoidance Strategy and Feasibility of Strategy Adjustment are calculated to determine the additional domain based on fuzzy rules. Fuzzy rules are used here to comprehensively consider the situation composed of multiple motion obstacles and the USV. Within the limited range of the basic domain and the additional domain, the optimal collision avoidance parameters of the USV can be calculated by the particle swarm optimization (PSO) algorithm. The PSO algorithm utilizes both the characteristic of pursuance for the population optimal and the characteristic of exploration for the individual optimal to avoid falling into the local optimal solution. Finally, numerical simulations are performed to certify the validity of the proposed method in complex traffic scenarios. The results illustrated that the proposed method could provide efficient collision avoidance actions.

Keywords: unmanned surface vehicle; velocity obstacle; collision avoidance; obstacles classification; fuzzy rules

Citation: Song, L.; Shi, X.; Sun, H.; Xu, K.; Huang, L. Collision Avoidance Algorithm for USV Based on Rolling Obstacle Classification and Fuzzy Rules. *J. Mar. Sci. Eng.* **2021**, *9*, 1321. https://doi.org/10.3390/jmse9121321

Academic Editor: Michele Viviani

Received: 27 October 2021
Accepted: 19 November 2021
Published: 23 November 2021

1. Introduction

A USV is an unmanned ship navigating on the water by autonomous or remote control. It can be widely used in maritime search and rescue, military operations, port guarding [1], oil pollution cleaning [2], and other fields in the future. Dynamic collision avoidance is the basic and vital intelligence function a USV for completing all kinds of tasks [3–6]. A unified autonomous decision-making framework is the key to multi-ship collision avoidance for USVs [7].

A variety of representative methods have been designed to implement dynamic collision avoidance for USVs, including artificial potential field, neural network, and velocity obstacle approaches. The artificial potential field with the characteristics of smoothness and security is widely used in collision avoidance planning. It may sometimes fall into a local optimal solution [8]. Therefore, artificial interference is used to avoid the local optimal solution by some studies, but this will fail on the condition of multiple obstacles [9,10]. Neural networks [11] have also been extensively applied in collision-avoidance planning problems. Neural networks have fast convergence speed and satisfactory adaptability, which improve the running efficiency of the collision avoidance model [12]. The neural network as a black-box optimization process requires sufficient sample data to avoid easily falling into an uncertain state [13]. In addition, all reasoning processes are regarded as numerical calculations, and problems of poor generalization ability and easily falling into local optimum are encountered. Particularly in the case of multi-obstacle dynamic collision avoidance, the feasible space is small and changes in real-time, requiring higher accuracy and real-time performance. Therefore, this method is of limited use in practical maritime navigation. The velocity obstacle (VO) approach was first used for robot path planning in 1998 [14], and then recreated and redeveloped continually, which led to a cooperative form of collision avoidance [15], probabilistic velocity obstacles [16], and crowd simulation [17]. The VO method forms a cone-shaped space on the obstacle, and ensures that the USV will never collide with the obstacles outside the area. The algorithm is so efficient that many studies improve it to implement a rapid collision avoidance response in the case of multiple obstacles [18]. Yoshiaki proposed combining a generated cone obstacle in the speed space of a USV with the collision avoidance rules to estimate the collision risk in different cones [19], and dynamic collision avoidance was then performed in multiple-obstacle environments. The test showed the high success rate of this method. However, the algorithm judges the collision avoidance for all obstacles at each moment and does not consider how to avoid collision when a collision-avoidance zone conflict occurs. In addition, in the simulation by Yoshiaki, the threat of multi-obstacle ships appears in different stages, and it is thus difficult to explain the effect of collision avoidance when multiple obstacles threaten at the same time.

Recently, intelligent collision avoidance systems based on fuzzy control have been developed to deal with multi-obstacle collision avoidance since it is a non-deterministic reasoning problem [20]. Fuzzy control consists of fuzzy classification and fuzzy reasoning. Regarding fuzzy classification [21], the ambiguity of collision avoidance parameters is addressed using an appropriate membership function. In terms of fuzzy reasoning [22], collision avoidance reasoning is realized by using the optimized fuzzy reasoning algorithm or combining it with other algorithms. Given that the fuzzy inference output depends on the parameters set in advance, the fuzzy control quantity currently uses empirical parameters, and its self-adaptability must be improved. Thus, scholars combine fuzzy control with a neural network algorithm to enhance real-time performance and self-learning ability. Pietrzykowski et al. proposed the concept of fuzzy ship domain and applied it to evaluate navigation safety in restricted waters [23]. Perera et al. proposed a collision-avoidance decision system based on fuzzy logic and studied the entire intelligent collision avoidance system on this basis [24,25]. The system can meet the requirements of the Convention on the International Regulations for Preventing Collisions at Sea (COLREGs), but it is only suitable for collision avoidance between two ships. Subsequently, the fuzzy-Bayesian ship intelligent collision avoidance decision/execution model was proposed [26]. The model relies on the parallel multi-decision modules of fuzzy logic and transforms the decision into continuous collision avoidance actions through a Bayesian network model. Brcko proposed a collision avoidance decision-making system based on fuzzy logic and combined it with radar [27]. It is practical in providing intelligent collision avoidance decision-making for a single ship. Si et al. designed a ship collision alarm system for vessel traffic service (VTS) using the fuzzy logic method. They provided rudder recommendations for ships [28], but the system only includes steering recommendations and is unsuitable for multi-ship encounter situations. Furthermore, the fuzzy neural network algorithm is combined with an expert system to realize intelligent collision avoidance. The problem of local convergence is reduced, but the fusion of the algorithm tends to decrease the real-time performance.

Concerning the existing limitations, this research is motivated to develop a unified and efficient strategy for multi-ship collision avoidance. A fuzzy programming method has been proposed that comprehensively considers the emergency of obstacle vessels, the safety of obstacle avoidance strategies, and the feasibility of adjusting strategies. Among them, the latter two items have not been considered in other multi-obstacle collision avoidance. Different from the existing algorithms, only two obstacles with the highest risk are considered at each time step in the process of dynamic collision avoidance. In combination with the VO algorithm and fuzzy theory, avoidance strategies of the two obstacles are optimized to reduce the collision risk of the obstacles. Through this unified strategy and rolling mechanism, all obstacles are gradually considered over time as the USV moves.

The paper is organized as follows. Section 2 introduces the calculation and classification model of obstacle collision risk based on the VO algorithm. In Section 3, the collision avoidance strategy for a single obstacle is introduced and the best avoidance scheme and secondary avoidance scheme are proposed. In Section 4, according to the urgency degree of the most urgent obstacle (MUO) and the secondary urgent obstacle (SUO), the avoidance strategies are obtained through fuzzy reasoning. Then, the strategies are further mapped into the solution space. In Section 5, simulations of trajectory prediction and obstacle avoidance are carried out, and the results are analyzed in detail to verify the algorithm's effectiveness. In Section 6, the conclusions of the study are elaborated.

2. A Collision-Avoidance Motion Model Based on VO

Figure 1 shows the collision-avoidance motion model based on the VO method. Symbol U and O represent a USV and an obstacle. The domain of the obstacle is denoted as D, which depicts the safety range of the obstacle in the form of a circle with radius d_1. D is a water area kept around the obstacle and cannot be invaded by other ships or objects. The velocity vectors of them are defined as v_R and v_O, respectively, and their relative velocity is defined as Δv, i.e., $\overrightarrow{\Delta v} = \overrightarrow{v_R} - \overrightarrow{v_O}$. UT_1 and UT_2 represent two tangents from the USV U to the obstacle O domain. Due east and due north of the USV U are referenced as the x-axis and y-axis. In this local coordinate system, several angles between UT, UO and Δv have been defined and calculated by Equation (1), where T denotes T_1 or T_2.

$$\begin{cases} \mu = \measuredangle(UT_2, UO) \\ \gamma = \sphericalangle(\Delta v, UO) \end{cases} \tag{1}$$

where the operator $\sphericalangle(a, b)$ denotes the angle where vector a rotates to vector b through the minor arc side. This angle can be positive or negative, depending on whether the rotation is in the counterclockwise or clockwise direction. a and b represent variables here. For example, UT_2 is a and UO is b in Figure 1. The operator $\measuredangle(a, b)$ denotes a positive angle between vector a and vector b.

The risk of collision between the USV and obstacle can be determined by the time to close point of approaching ($TCPA$) and the distance of close point of approaching ($DCPA$) [19]. Both two parameters are dynamically calculated to evaluate the potential collision probability at different times. Equation (2) lists the piecewise function of calculating the potential collision probability f_{em} [29].

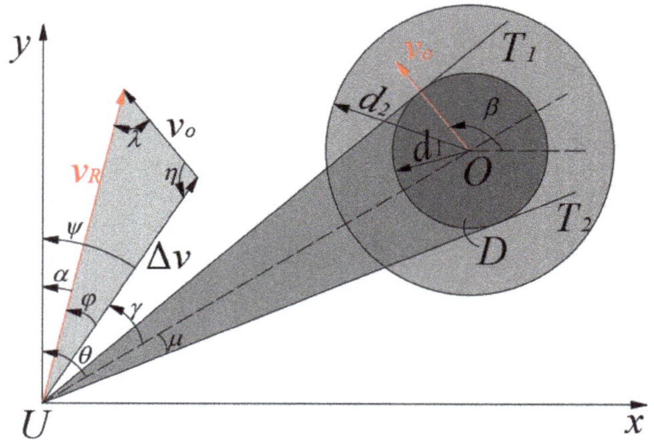

Figure 1. USV and velocity obstacle.

$$
\begin{aligned}
f_{em} &= \begin{cases} 1 & 0 < TCPA \le t_1 \\ \dfrac{t_2 - TCPA}{t_2 - t_1} & t_1 < TCPA < t_2 \\ 0 & t_2 \le TCPA \end{cases} \\
t_1 &= \begin{cases} \dfrac{\sqrt{{d_1}^2 - DCPA^2}}{\Delta v} & DCPA \le d_1 \\ 0 & DCPA > d_1 \end{cases} \\
t_2 &= \begin{cases} \dfrac{\sqrt{{d_2}^2 - DCPA^2}}{\Delta v} & DCPA \le d_2 \\ 0 & DCPA > d_2 \end{cases}
\end{aligned} \tag{2}
$$

$$
\begin{cases} TCPA = \dfrac{|UO|\cos(|\gamma|)}{|\Delta v|} \\ DCPA = |UO|\sin|\gamma| \end{cases} \tag{3}
$$

where t_1 and t_2 are parameters related to collision avoidance probability from the aspect of time. $|UO|$ is the distance between the USV and obstacle. d_1 is the secure encounter distance which equals to the domain radius of the obstacle in this article, and d_2 is the distance between the USV and obstacle when the USV must perform obstacle avoidance. If the USV starts to avoid obstacles after the distance is less than d_2, even if there is no collision, it will form an emergency situation. In this article, $d_2 = 2d_1$. d_1 and d_2 can be seen in Figure 1. The emergency situation will occur when the *DCPA* is less than d_2.

According to the collision-risk degree in Equation (2), obstacles can be classified into several categories, as shown in Table 1. For the USV, the concept of an identification zone is set to distinguish the obstacle type, which is the circular area in Figure 2. The identification zone is the collision detection area for the USV in this research, and its radius can be determined by the Automatic Identification System (AIS) [30] or radar. A ship outside the identification zone of the USV is defined as an irrelevant obstacle invisible to the USV, such as B6 and B7 in Figure 2. If a ship is inside the identification zone of the USV but poses no threat to the USV, it will be considered as a nonthreatening obstacle. The major objects of collision-avoidance for the USV are the ships inside the identification zone that satisfy Equation (4) and pose a threat to the USV. In this study, only two obstacles with the highest and second-highest collision risk are the primary concerns of collision-avoidance for the USV, denoted as the most urgent obstacle and secondary urgent obstacle.

$$abs(\gamma) \leq \mu \quad (4)$$

Table 1. The classification definition of obstacles.

Name	Abbreviation	Definition	Example
Irrelevant Obstacle	IO	Ship outside the identification zone of the USV that is invisible to the USV	B6, B7
Nonthreatening Obstacle	NO	Ship in the identification zone of the USV that does not satisfy Equation (4) and does not pose a threat to the USV	B4, B5
Threatening Obstacle	TO	Ship in the identification zone of the USV that satisfies Equation (4) and poses a threat to the USV	B1, B2, B3
Most Urgent Obstacle	MUO	Ship with the highest risk of collision f_{em} among TOs	B1
Secondary Urgent Obstacle	SUO	Ship with the second highest risk of collision f_{em} among TOs	B2

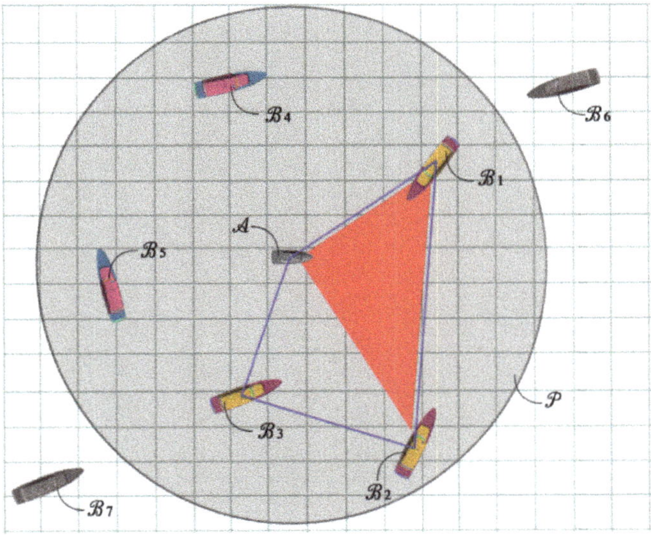

Figure 2. Classification of obstacles.

3. Collision Avoidance against Single Obstacle

When collision avoidance is performed for a single obstacle, it means the angle γ should be adjusted to satisfy $abs(\gamma) \geq \mu$. As shown in Figure 1, the value of angle γ is related to the velocity vectors of the USV and obstacle and their angles with respect to the x-axis. Geometric relationships of these parameters in the local coordinate system can be explicitly defined as in Equation (5). Then, angle γ and its derivative can be derived as an arctan function and a linear function of these parameters, as noted in Equation (6), respectively.

$$\begin{cases} v_0 \sin(\alpha-\beta) = -\Delta v \sin\varphi \\ v_R - v_0\cos(\alpha-\beta) = \Delta v\cos\varphi \\ {v_R}^2 + {v_0}^2 - 2v_R v_0 \cos(\alpha-\beta) = \Delta v^2 \\ v_R\sin(\beta-\alpha) = \Delta v\sin\eta \\ v_0 - v_R\cos(\beta-\alpha) = \Delta v\cos\eta \end{cases} \tag{5}$$

$$\begin{aligned} \gamma &= \tan^{-1}\frac{v_R\sin(\alpha-\theta) - v_0\sin(\beta-\theta)}{v_R\cos(\alpha-\theta) - v_0\cos(\beta-\theta)} \\ d\gamma &= \frac{\sin\varphi}{\Delta v}dv_R + \frac{v_R\cos\varphi}{\Delta v}d\alpha + \frac{-\sin\eta}{\Delta v}dv_o + \frac{v_o\cos\eta}{\Delta v}d\beta \end{aligned} \tag{6}$$

In view of the dynamic character of ship movement, angle γ should always meet the collision avoidance condition $|\gamma + d\gamma| \geq \mu$ at each time, as noted in Equation (7) and Equation (8).

$$\begin{cases} \Delta\gamma \geq \mu - \gamma, & if\ \gamma \geq 0 \\ \Delta\gamma \leq -\mu - \gamma, & if\ \gamma < 0 \end{cases} \tag{7}$$

or

$$\begin{cases} \Delta\gamma \leq -\mu - \gamma, & if\ \gamma \geq 0 \\ \Delta\gamma \geq \mu - \gamma, & if\ \gamma < 0 \end{cases} \tag{8}$$

Assuming that the motion state of the obstacle is unchanged, an effective measure is to adjust the velocity v_R and the course α of the USV for collision avoidance. By analyzing Equations (7) and (8), it can be obtained that the former only requires the relative velocity vector Δv to avoid the minor arc of obstacle domain D, but the latter chooses the major arc for avoiding. It is obvious that Equation (7) is better. $d\gamma$ in Equation (6) is changed to Δr and is substituted into Equations (7) and (8), respectively, in order to obtain the best collision avoidance (BCA) and secondary collision avoidance (SCA) as shown in Equations (9) and (10). For single obstacle avoiding, the USV needs to obtain the optimal solutions to parameters dv_R and $d\alpha$ at each moment.

Best Collision Avoidance:

$$\mathrm{BCA}|_A : \begin{cases} \frac{\sin\varphi}{\Delta v}x_1 + \frac{v_R\cos\varphi}{\Delta v}x_2 + \frac{-\sin\eta}{\Delta v}dv_0 + \frac{v_0\cos\eta}{\Delta v}d\beta \geq \mu - \gamma\ if\ \gamma \geq 0 \\ \frac{\sin\varphi}{\Delta v}x_1 + \frac{v_R\cos\varphi}{\Delta v}x_2 + \frac{-\sin\eta}{\Delta v}dv_0 + \frac{v_0\cos\eta}{\Delta v}d\beta \leq -\mu - \gamma\ if\ \gamma < 0 \end{cases} \tag{9}$$

Secondary Collision Avoidance:

$$\mathrm{SCA}|_A : \begin{cases} \frac{\sin\varphi}{\Delta v}x_1 + \frac{v_R\cos\varphi}{\Delta v}x_2 + \frac{-\sin\eta}{\Delta v}dv_0 + \frac{v_0\cos\eta}{\Delta v}d\beta \leq -\mu - \gamma\ if\ \gamma \geq 0 \\ \frac{\sin\varphi}{\Delta v}x_1 + \frac{v_R\cos\varphi}{\Delta v}x_2 + \frac{-\sin\eta}{\Delta v}dv_0 + \frac{v_0\cos\eta}{\Delta v}d\beta \geq \mu - \gamma\ if\ \gamma < 0 \end{cases} \tag{10}$$

where two variables x_1 and x_2 denote collision avoidance parameter dv_R and $d\alpha$, and $|_A$ represents a collision avoidance scheme of USV for the obstacle.

$$\begin{aligned} \text{Goal Function} : \{x_1, x_2 \,|\, min\ f(x_1, x_2) = w_1x_1 + w_2x_2\} \\ s.t.\ : (x_1, x_2) \in \mathrm{BCA}|_A \end{aligned} \tag{11}$$

In this study, the particle swarm optimization (PSO) algorithm is adopted to train and obtain the optimal values of collision avoidance parameters. The goal function is defined as a weighted linear function, as shown in Equation (11), where w_1 and w_2 are weights of dv_R and $d\alpha$. The goal function aims to minimize the change in the velocity and course of the USV by adjusting the weights and the priority of changing speed and course. The solution (x_1, x_2), which corresponds to the optimal Δv_R and $d\alpha$ (dv_R and $d\alpha$ in discrete version), could enable the USV to successfully avoid the circular domain of obstacle (d_1). It is worth noting that the course should not be restricted and can be selected randomly in the range of $[-\pi, \pi]$. The reason is that collision can only be avoided by drastically altering the course of the USV while the USV is dangerously close to the obstacle. If $d\alpha$ is defined strictly, Equation (11) may not have a solution.

In some cases, the USV may be incapable of obtaining the optimal solutions to avoid the circular obstacle domain due to the limitations of its manipulating capability. Therefore, the collision-avoidance motion of the USV is also affected by the limit values of its speed, acceleration, and angular acceleration that are denoted as $\overline{v}_R$, $\overline{\alpha}$ and $\overline{\omega}$, respectively. In the optimization process, if the expected solutions of collision avoidance parameters exceed the upper limit of operational capabilities of the USV, they should be revised based on the limit values of its speed, acceleration, and angular acceleration. Equations (12) and (13) have listed different calibration methods on the condition that the expected collision avoidance parameters exceed their extreme values.

$$\Delta v_R = \frac{v_R}{|v_R|} min(\overline{v}_R - v_R, \overline{\alpha})$$
$$\Delta\alpha = \begin{cases} min\left(\overline{\omega}, max\left(0, \frac{(\mu-\gamma)\Delta v + \Delta v_0 \sin\eta - v_0 \Delta\beta \cos\eta - \Delta v_R \sin\varphi}{v_R \cos\varphi}\right)\right) \; if\ \gamma \geq 0 \\ max\left(-\overline{\omega}, min\left(0, \frac{(-\mu-\gamma)\Delta v + \Delta v_0 \sin\eta - v_0 \Delta\beta \cos\eta - \Delta v_R \sin\varphi}{v_R \cos\varphi}\right)\right) \; if\ \gamma < 0 \end{cases} \quad , when\ \Delta v_R > \overline{\alpha}\ or\ v_R + \Delta v_R > \overline{v}_R \tag{12}$$

$$\Delta\alpha = \frac{\Delta\alpha}{|\Delta\alpha|}\overline{\omega}$$
$$\Delta v_R = \begin{cases} min\left(\overline{\alpha}, max\left(0, \frac{(\mu-\gamma)\Delta v + \Delta v_0 \sin\eta - v_0 \Delta\beta \cos\eta - \Delta v_R \cos\varphi}{\sin\varphi}\right)\right) \; if\ \gamma \geq 0 \\ min\left(\overline{\alpha}, max\left(0, \frac{(-\mu-\gamma)\Delta v + \Delta v_0 \sin\eta - v_0 \Delta\beta \cos\eta - \Delta v_R \cos\varphi}{\sin\varphi}\right)\right) \; if\ \gamma < 0 \end{cases} \quad , when\ \Delta\alpha > \overline{\omega} \tag{13}$$

The process of using the PSO algorithm to solve the collision avoidance against a single obstacle is as follows:

Step 1. Initialize the particles swarm.

Step 2. Judge if the particle satisfies the avoidance condition in Equation (9). If the condition is satisfied, calculate the fitness value by Equation (11). If it is not satisfied, the fitness value of this particle is set to infinity.

Step 3. All particles update the position and velocity of their two dimensions toward the best particle in the swarm.

Step 4. Check if the terminal condition is fulfilled. If it is fulfilled, output the two dimensions, Δv_R and $\Delta\alpha$, of the best particle. Then proceed to Step5. If not, then go back to Step 2.

Step 5. Check if Δv_R and $\Delta\alpha$ exceed the $\overline{\alpha}$ and $\overline{\omega}$. If they do, follow Equations (12) and (13), and achieve the avoidance strategy Δv_R and $\Delta\alpha$ at the local cycle by compromising to the motion capacity of the USV. If no, Δv_R and $\Delta\alpha$ are the avoidance parameters.

4. Collision Avoidance against Multiple Obstacles

4.1. Four Obstacle Avoidance Schemes

In the context of maritime traffic analysis, the concept of "ship domain" is proposed to evaluate near ship collision scenarios, which characterizes the safety range of a ship. Combined with the identification zone defined in Section 2, only the most urgent obstacle (MUO) and secondary urgent obstacle (SUO) are viewed as the collision avoidance objects when multiple threatening obstacles exit in the identification zone of the USV.

Based on the definitions of BCA and SCA above, there are four obstacle avoidance schemes, namely $BCA|_{MUO}$, $BCA|_{SUO}$, $SCA|_{MUO}$, and $SCA|_{SUO}$. Each obstacle avoidance scheme will determine a domain of the optimization target in Equation (11). Then, four domains can be obtained and combined to form three safe navigable areas for the USV to avoid two obstacles, as shown in Figure 3. In the figure, the BCA for obstacle I is to enter collision avoidance area ① and the SCA is to enter area ②. The BCA for obstacle II is to enter area ③ and the SCA is to enter area ②. The goal of multi-obstacle collision avoidance therefore refers to determining the optimal solution $(\Delta v_R, \Delta\alpha)$ in the definition domain called Basic Domain, as expressed in Equation (14).

$$\text{S.t. } (x_1, x_2) \in BCA|_{\text{MUO}} \cup BCA|_{\text{SUO}} \cup SCA|_{\text{MUO}} \cup SCA|_{\text{SUO}} \quad (14)$$

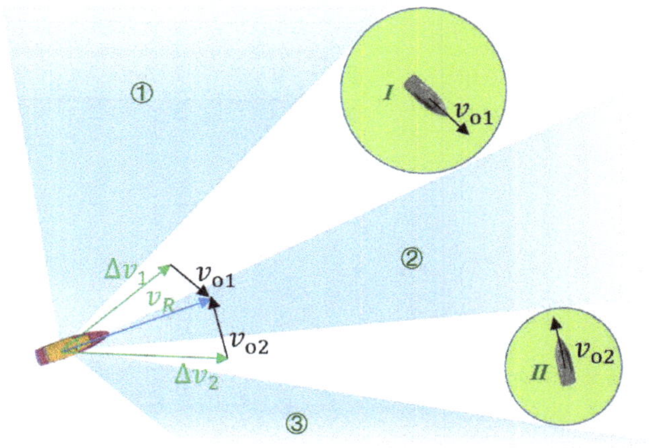

Figure 3. Three collision avoidance areas for MUO and SUO.

4.2. Multi-Obstacle Collision Avoidance Strategy

The determination of a multi-obstacle collision avoidance strategy should take account of two basic demands, which are safety of avoidance strategy (SAS) and feasibility of strategy adjustment (FSA).

SAS is an indicator of afterward evaluation, which indicates the safe state of the USV after the collision avoidance parameter $(\Delta v_R, \Delta\alpha)$ is performed. In the case of multi-ship interference where the ship does not necessarily comply with COLREGs, the avoidance action used by the USV may not be suitable for all obstacles. It is necessary to evaluate the safety of the USV after performing a collision avoidance operation. As shown in Figure 4, USV speed (v_R) is firstly adjusted to $v_R{}'$ for obstacle I avoidance and the resultant velocity correspondingly changes from Δv_1 to $\Delta v_1{}'$, which satisfies the BCA for obstacle I. However, the resultant velocity of the USV and obstacle II $\Delta v_2{}'$ is in the collision range, which means the BCA for obstacle I poses a threat to obstacle II. The safe state of this obstacle avoidance strategy is therefore unsatisfactory. As a result, this research proposes two adjustment ways to ensure the safety of the avoidance strategy.

Definition 1. *AS1 (adjustment strategy 1):* $(\Delta v_R, \Delta\alpha)$ *satisfies* $abs(\gamma|_{\text{MUO}} > \mu)$ *where* $\gamma|_{\text{MUO}} = R(\text{RO}, \Delta v|_{\text{MUO}})$, *and* $\Delta v|_{\text{MUO}} = v_{\text{R}} + \Delta v_{\text{R}} - v_{\text{o}}|_{\text{MUO}}$, *which means USV avoids the MUO after its movement adjustment.*

Definition 2. *AS2 (adjustment strategy 2):* $(\Delta\widetilde{v_R}, \Delta\widetilde{\alpha})$ *satisfies* $abs(\overline{\gamma}|_{\text{SUO}} > \mu)$ *where* $\gamma|_{\text{SUO}} = R(\text{RO}, \Delta v|_{\text{MUO}})$, *and* $\Delta v|_{\text{SUO}} = v_{\text{R}} + \Delta v_{\text{R}} - v_{\text{o}}|_{\text{SUO}}$, *which means USV avoids the SUO after its movement adjustment.*

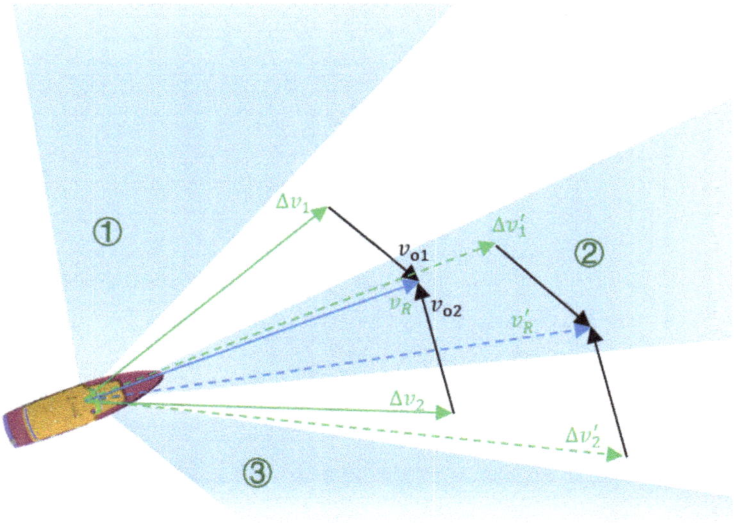

Figure 4. Unreasonable situation after collision avoidance parameter adjustment.

FSA acts as an indicator of pre-evaluation, which indicates whether a USV has an adequate response time to adjust collision avoidance parameters $(\Delta v_R, \Delta\alpha)$ before the collision, on condition of its acceleration and steering limitations. There are two conditions defined to ensure the appropriate adjustment time by changing speed and course.

Definition 3. *ST1 (strategy time 1): $(\Delta v_R, \Delta\alpha)$ must satisfy $TCPA\big|_{\text{MUO}} > max\left(\frac{\Delta v_R}{\Delta \tilde{v}_R}, \frac{\Delta\alpha}{\Delta\tilde{\alpha}}\right)\Big|_{\text{MUO}}$, which means USV will not collide with MUO before adjusting to the required velocity vector.*

Definition 4. *ST2 (strategy time 2): $(\Delta v_R, \Delta\alpha)$ must satisfy $TCPA\big|_{\text{SUO}} > max\left(\frac{\Delta v_R}{\Delta \tilde{v}_R}, \frac{\Delta\alpha}{\Delta\tilde{\alpha}}\right)\Big|_{\text{SUO}}$, which means USV will not collide with SUO before adjusting to the required velocity vector.*

In this research, fuzzy rules based on multi-obstacle collision avoidance strategy is proposed to avoid MUO and SUO for the USV. The *TCPA* of MUO and SUO, denoted as $TCPA\,|_{\text{MUO}}$ and $TCPA\,|_{\text{SUO}}$, respectively, can be represented by three fuzzy linguistic variables: "Emergency (EG)", "Easy (EZ)", and "Far (FA)". The membership function of fuzzy linguistic variables for urgency degree of collision avoidance is defined as trapezoidal functions shown in Figure 5. Four important time parameters, including T_{near}, T_{far}, T_{mid}, and T_{saf}, are defined for the selection of three membership functions, which can be calculated by Equation (15). The collision avoidance priority level of MUO and SUO can be then determined by fuzzy reasoning rules in Table 2. The inputs of fuzzy reasoning rules are the urgency degree of fuzzy linguistic variables for MUO and the SUO, and the output is the fuzzy subset of the avoidance priority level. The fuzzy subset indicates the following collision avoidance intention.

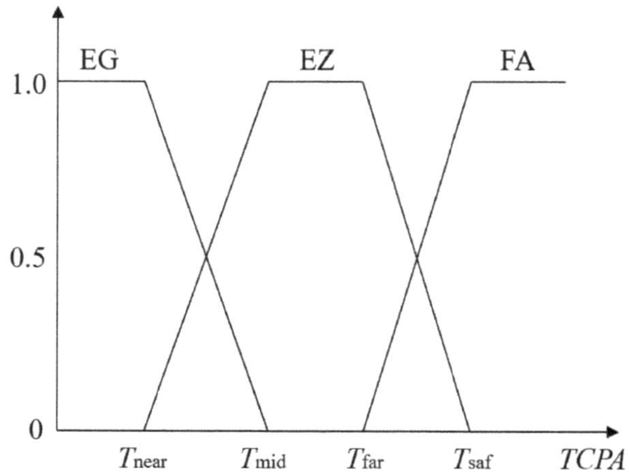

Figure 5. Membership function of the fuzzy linguistic variable for urgency degree.

Table 2. Fuzzy reasoning rules.

		TCPA I MUO		
		EG	**EZ**	**FA**
TCPA I SUO	EG	SO	——	——
	EZ	SO	SN	——
	FA	OT	OT	SN

OT: Equal avoiding priority to both MUO and SUO.
SO: Avoiding both MUO and SUO but giving priority to MUO.
SN: Avoiding MUO only.

$$T_{near} = \frac{\sqrt{{d_1}^2 - DCPA^2}}{\Delta v}, \; T_{far} = \frac{\sqrt{{d_2}^2 - DCPA^2}}{\Delta v}, \; T_{mid} = \frac{T_{near} + T_{far}}{2}, \; T_{saf} = \frac{3T_{far} - T_{near}}{2} \tag{15}$$

Table 3 lists the relationships of collision avoidance demands (SAS and FAS) and avoidance priority levels. SAS and FAS can be jointly viewed as the Additional Domain for collision avoidance. Combined with the Basic Domain in Equation (14), the solution domain can be determined, and the optimization goal can be then represented as Equation (16). The equation is used to obtain the optimal Δv_R and $\Delta \alpha$.

$$\begin{gathered} \{x_1, x_2 \mid minf(x_1, x_2) = w_1 x_1 + w_2 x_2\} \\ \text{S.t. } (x_1, x_2) \in BAP|_{\text{MUO}} \cup BAP|_{\text{SUO}} \cup SAP|_{\text{MUO}} \cup SAP|_{\text{SUO}} \\ (x_1, x_2) \in AS1 \cap AS2 \text{ or } AS1 \text{ or } ST1 \cap ST2 \text{ or } ST1 \end{gathered} \tag{16}$$

Table 3. Relationships of collision avoidance demands and avoidance priority level.

Priority Level	OT	SO	SN
SAS	$AS1 \cap AS2$	$AS1$	$AS1$
FAS	-	$ST1 \cap ST2$	$ST1$

Proof. The convergence proof of the above algorithm is as follows:

(1) The two-dimensional particles in the PSO, x_1 and x_2, are optimization objectives, respectively Δv_R and $\Delta \alpha$ which to be solved in Equation (16). According to Equations (14) and (16),

the definition domain is three parts of the continuous region marked with ①, ② and ③, as shown in Figure 3. As long as $UO > d_2$ (in Figure 1) is satisfied, the continuous region certainly exists, and the definition domain inevitably is a nonempty set.

(2) The initial global optimal value in the PSO algorithm is $+infinite$. According to Equation (16), as long as any element is found in the nonempty set of the definition domain in the iterative process of PSO, the global optimization will replace the original $+infinite$ and obtain the optimized x_1 and x_2.

(3) In each cycle, any optimal solution obtained by the PSO algorithm may be changed to the limit values of the USV according to Equations (12) and (13) due to the limitations of its manipulating capability. This means a pair of executable parameters for the USV is definitely obtained in each cycle, regardless of the astringency of PSO.

(4) According to the Table 1, MUO and SUO are not fixed on a specific obstacle. They will roll and transfer according to the situation, and the definition domain will change, but they are always nonempty sets. □

In conclusion, the algorithm has convergence.

The process of using fuzzy rules to the formulate collision avoidance strategy is as follows:

Step 1. Confirm the MUO and SUO based on Equations (1)–(4).

Step 2. Calculate fuzzy variables based on Figure 5.

Step 3. Determine the fuzzy reasoning according to Table 2.

Step 4. Obtain the solution domain according to fuzzy reasoning, as shown in Equation (16).

Step 5. Calculate the optimal solution in the domain with the PSO algorithm.

4.3. The Calculation Process of Collision Avoidance

Based on the strategy above, the calculation process of multi-obstacle collision avoidance can be divided into three steps, including obstacle emergency degree calculation, obstacle classification, and optimal collision avoidance parameters' calculation.

Figure 6 lists the workflow of collision avoidance calculation. In a calculation period, the emergency degree for all ships around the USV will be firstly estimated. When a ship is located in the identification zone of the USV and poses a threat to the USV, it will be viewed as the TO for the USV. All TOs are ordered and classified based on their emergency degrees. Only two of the most urgent obstacles, MUO and SUO, will be determined and selected for collision avoidance in this period.

After that, fuzzy linguistic variables based on the *TCPA* are firstly calculated for these two obstacles with the membership function in Figure 5. Combining the variables of two obstacles; then, the avoidance priority level can be determined based on Table 2. Meanwhile, two collision avoidance demands (SAS and FAS) are calculated based on the dynamic performance of the USV in the condition of avoidance priority level. With these conditions, both the basic domain and the additional domain of collision avoidance parameter can be determined using Equations (1)–(10). The PSO algorithm is further applied to estimate the optimal solution in the domain.

In general, the calculation process of collision avoidance is dynamic. The MUO and SUO will be continuously reselected in each calculation period as the USV moves. The emergency degree of the MUO and SUO in a previous period may gradually decrease while their avoidance priority in the collision avoidance operation becomes lower in the next period. The new MUO and SUO may be other obstacles that were not considered in the previous calculation period, while the threats they pose to the USV continue to increase in the following period. As a result, all obstacles will be logically considered for collision avoidance step by step in the scenario of multi-obstacle collision avoidance. In this way, computational efficiency and performance are greatly increased.

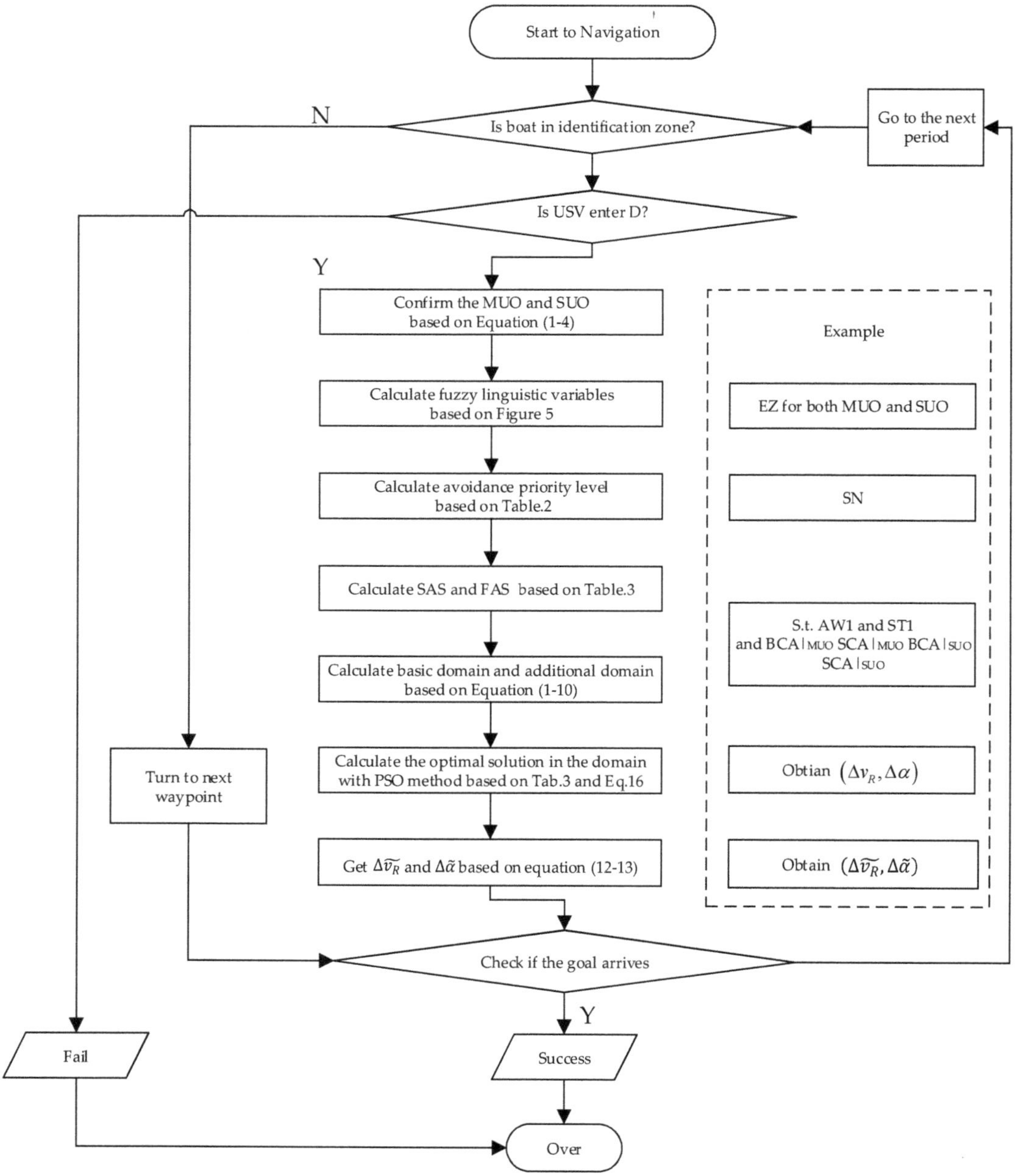

Figure 6. Flowchart of the multi-obstacle collision avoidance algorithm.

Rules: The termination rules in the above calculation process are as follows:

Rule1. The termination rules of the whole algorithm simulation or test process are:

(1) The USV arrives within 50 m of the goal, and there is no risk of collision with the obstacles.

(2) The USV fails to avoid collision and collides with any obstacle, i.e., the USV enters the collision circle of the obstacle (U in D).

Rule2. The termination rules in the PSO algorithm are:

(1) The global optimal solution of the particle swarm is continuously stable for five cycles.

(2) If (1) is not satisfied, the maximum number of iterations is 100.

5. Simulation Experiment and Analysis

To verify the approach, a 3D simulation environment based on Unity for USV collision avoidance is built by integration of Visual Studio and MATLAB. In all simulation experiments, the calculation period for the USV is set to 5 s. The dynamic performances of USV are, $\overline{v}_R = 6$ m/s, $\overline{\alpha} = 0.6$ m/s^2, $\overline{\omega} = 3$ deg/s^2. These parameters come from the maneuvering simulation and experiment, which are in still water without considering the influence of environmental factors such as waves. The USV can perform collision avoidance only based on the pre-existing motion of obstacles and can be unaware of continuous movement of obstacles.

In the simulation environment, a local coordinate system centered at the USV is firstly set up. In the system, the USV is represented by a black circle with an initial position (0, 0) and an eastward velocity. Four ships around the USV are set to potential obstacles for collision avoidance. SHIP1 is represented by a green pentacle and located at (−20, −100) with velocity to the east by south. SHIP2 is represented by a blue-green hexagon and is located at (1350, 380), whose velocity points to the west. SHIP3 is represented by a magenta triangle with the initial position (1200, −250) and velocity to the west by north. SHIP4 is enclosed by blue square locates at the position (730, −700) with velocity to the east by north.

Assuming that the USV has finished global path planning in advance, the sub-global goal of the USV is to move to the position (1300, 0). Figure 7 shows the movement track of the USV and four ships within 140 s. During this time, multiple ships have interacted with the USV simultaneously and have been viewed as normal TO, MUO, or SUO, alternately. The USV has adopted continuous collision avoidance strategies to reach its target position successfully.

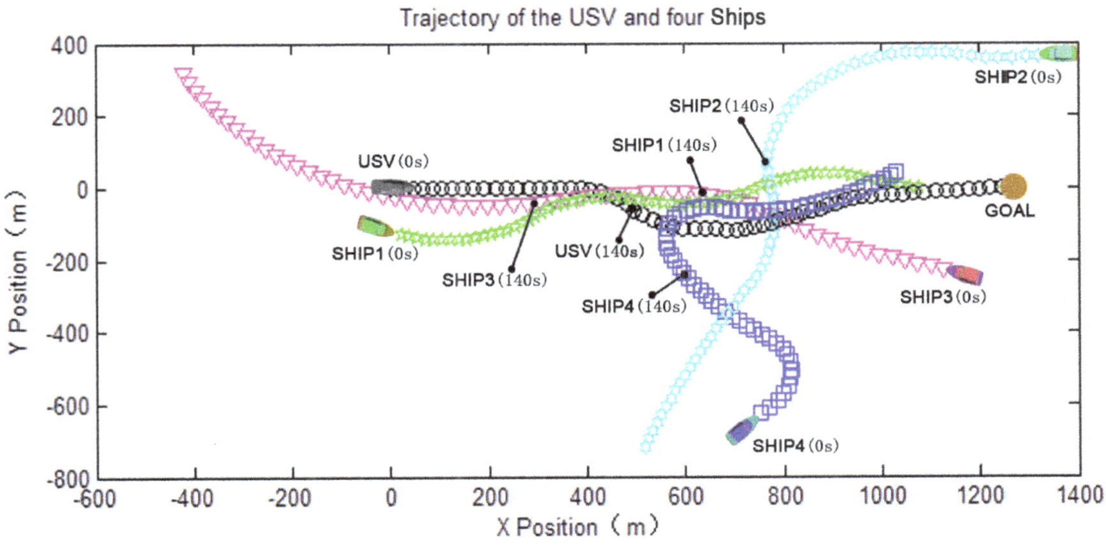

Figure 7. The movement track of USV and four obstacles.

Figure 8 reflects the movement relationship of the USV and ships in the 140th second. The light-colored areas enclosing the USV and traffic ships indicate their domains, i.e., a cyan area for the USV's domain. At this moment, all four ships are located within the identification zone of the USV and no IOs exist. SHIP3 moving away from the USV poses no threat to the USV and can be viewed as an IO. The other three ships are in danger of collision. According to the order of their emergency degree, SHIP1, SHIP4, and SHIP2 are MUO, SUO, and NO respectively. Based on the proposed method, only SHIP1 and SHIP4 should be considered for collision avoidance in this calculation period.

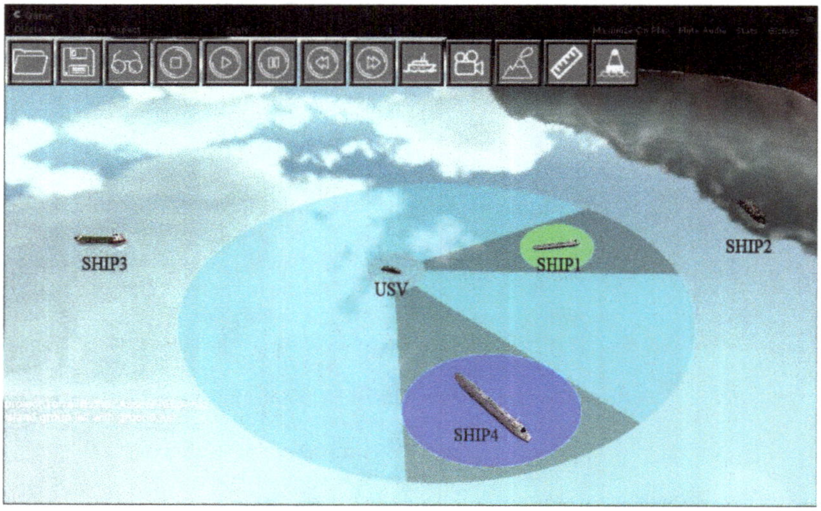

Figure 8. Motion relationship between the USV and traffic ships in the 140th second.

Figure 9 shows the variation curves of obstacle classification of four ships in all calculation periods. It can be seen that the collision risk degree of each ship varies with the course of the USV. The obstacle type of each ship alternately changes among the five types. A general trend is that the collision risk of the ships around the USV may increase first and then decrease while the USV constantly changes its parameters. The obstacle type for each ship will be temporally classified and redetermined in the next period. Obstacles with higher collision risk may gradually decrease from MUO to SUO, and then to NO or IO, i.e., SHIP1 and SHIP3. The irrelevant or nonthreatening ship may change to TO and then into MUO. In a word, all ships are considered for collision avoidance step by step, and the global emergency degree constantly decreases even if there is still a MUO or SUO for the USV.

Figure 10 shows the variation curves of course and velocity of the USV in all calculation periods. It can be seen that both speed and course are constantly adjusted for collision avoidance. However, the range of parameter adjustment is limited by the performance of the USV. The slopes of both two curves have upper limit values, indicating $\overline{a}$ and $\overline{\omega}$ play a leading role in variation rate limitation. The speed curve also has an upper limit value, indicating $\overline{v}_R$ limits the variation range. Such continuous and constrained motion curves are not only coincided with the actual movement of the USV, but are also convenient to control and track.

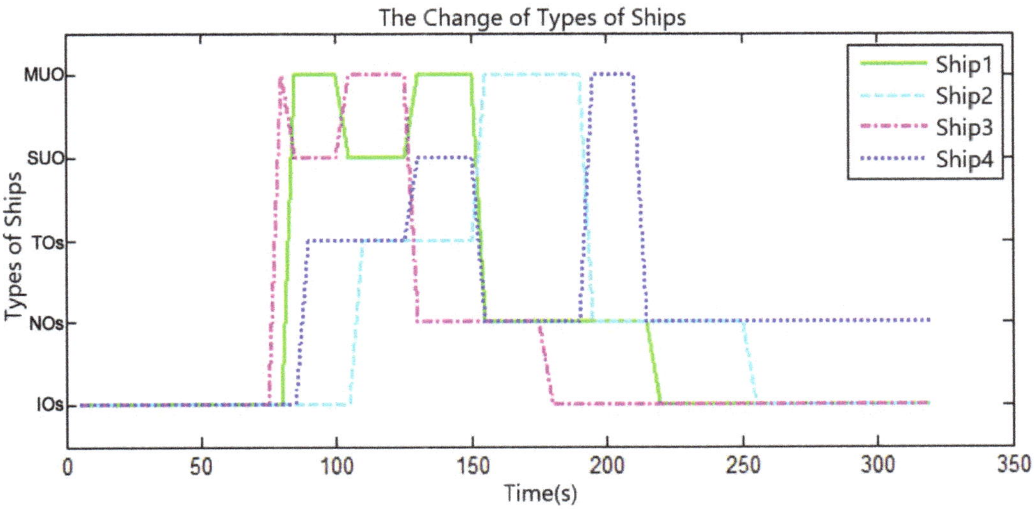

Figure 9. Alternation of obstacle types of ships as USV moves.

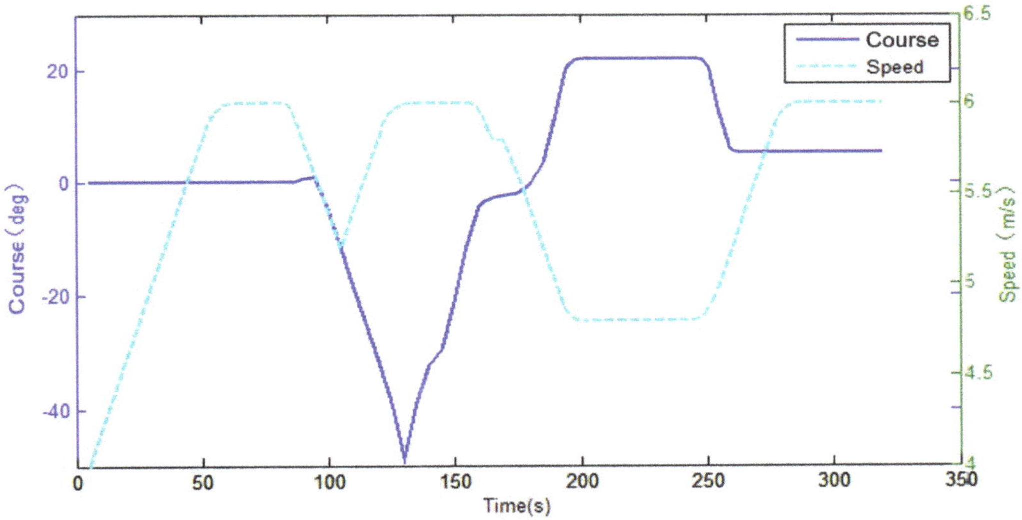

Figure 10. Variation curves of course and speed of the USV.

6. Conclusions

This study proposes a dynamic collision avoidance algorithm based on rolling obstacle classification and fuzzy rules to solve the two main problems, i.e., the low efficiency and easy failure of collision avoidance. Different from the existing multi-obstacle collision avoidance algorithms, the proposed algorithm comprehensively considers the emergency of obstacle ships, the safety of obstacle avoidance strategy, and the feasibility of adjusting strategy. In the process of dynamic collision avoidance, only two obstacles with the highest risk are considered in each time step, and the avoidance strategies of the two obstacles are optimized to reduce the collision risk of obstacles. Through the rolling mechanism of the algorithm, all obstacles will be gradually considered along with the USV's moves. Simulation results indicate that the proposed algorithm is valid and efficient for the USV.

In further research, avoidance strategies for more complex scenarios will be discussed, i.e., the USV is encircled and crashed by multiple obstacles purposefully.

Author Contributions: Data curation, X.S.; funding acquisition, L.S.; methodology, L.S.; software, H.S.; visualization, K.X.; writing—original draft, L.S. and L.H.; writing—review & editing, L.H. All authors have read and agreed to the published version of the manuscript.

Funding: This research was funded by the National Natural Science Foundation of China, Grant No. 51809203, Grant No. 41801375.

Acknowledgments: The work was supported by the grant from the National Natural Science Foundation of China (Grant No. 51809203, Grant No. 41801375).

Conflicts of Interest: The authors declare no conflict of interest.

References

1. Chen, J.H.; Zheng, T.X.; Garg, A.; Xu, L.; Li, S.F.; Fei, Y.J. Alternative maritime power application as a green port strategy: Barriers in China. *J. Clean. Prod.* **2019**, *213*, 825–837. [CrossRef]
2. Chen, J.H.; Zhang, W.P.; Wan, Z.; Li, S.F.; Huang, T.C.; Fei, Y.J. Oil spills from global tankers: Status review and future governance. *J. Clean. Prod.* **2019**, *227*, 20–32. [CrossRef]
3. Li, W.F.; Ma, W.Y. Simulation on Vessel Intelligent Collision Avoidance Based on Artificial Fish Swarm Algorithm. *Pol. Marit. Res.* **2016**, *23*, 138–143. [CrossRef]
4. Qu, H.; Xing, K.; Alexander, T. An improved genetic algorithm with coevolutionary strategy for global path planning of multiple mobile robots. *Neurocomputing* **2013**, *120*, 509–517. [CrossRef]
5. Borkowski, P.; Pietrzykowski, Z.; Magaj, J. The Algorithm of Determining an Anti-Collision Manoeuvre Trajectory Based on the Interpolation of Ship's State Vector. *Sensors* **2021**, *21*, 5332. [CrossRef] [PubMed]
6. Huang, Y.; Chen, L.; Chen, P.; Negenborn, R.R.; van Gelder, P.H.A.J.M. Ship Collision Avoidance Methods: State-of-the-art. *Saf. Sci.* **2020**, *121*, 451–473. [CrossRef]
7. Asl, A.N.; Menhaj, M.B.; Sajedin, A. Control of leader follower formation and path planning of mobile robots using asexual reproduction optimization (ARO). *Appl. Soft Comput.* **2014**, *14 Pt C*, 563–576.
8. Khatib, O. Real-time obstacle avoidance for manipulators and mobile robots. *Int. J. Robot. Res.* **1986**, *5*, 90–98. [CrossRef]
9. Wang, J.; Wu, X.; Xu, Z. Potential-based obstacle avoidance in formation control. *J. Control. Theory Appl.* **2008**, *6*, 311–316. [CrossRef]
10. Yang, Y.; Wang, S.X.; Wu, Z.L.; Wang, Y.H. Motion planning for multi-HUG formation in an environment with obstacles. *Ocean Eng.* **2011**, *38*, 2262–2269. [CrossRef]
11. Liu, Y.H.; Shi, C.J. A fuzzy-neural inference network for ship collision avoidance. In Proceedings of the IEEE Third International Conference on Machine Learning and Cybernetics, Guangzhou, China, 18–21 August 2005; pp. 4754–4759.
12. Gao, M.; Shi, G.Y. Ship-Collision Avoidance Decision-Making Learning of Unmanned Surface Vehicles with Automatic Identification System Data Based on Encoder—Decoder Automatic-Response Neural Networks. *J. Mar. Sci. Eng.* **2020**, *8*, 754. [CrossRef]
13. Ayala, H.V.H.; Habineza, D.; Rakotondrabe, M.; dos Santos Coelho, L. Nonlinear black-box system identification through coevolutionary algorithms and radial basis function artificial neural networks. *Appl. Soft Comput.* **2019**, *87*, 105990. [CrossRef]
14. Fiorini, P.; Shiller, Z. Motion planning in dynamic environments using velocity obstacles. *Int. J. Robot. Res.* **1998**, *17*, 760–772. [CrossRef]
15. Berg, V.D.J.; Lin, M.; Manocha, D. Reciprocal velocity obstacles for real-time multi-agent navigation. In Proceedings of the 2008 International Conference on Robotics and Automation, Pasadena, CA, USA, 19–23 May 2008; 1928–1935.
16. Kluge, B.; Parssler, E. Reflective navigation: Individual behaviors and group behaviors. In Proceedings of the IEEE International Conference on Robotics & Automation, New Orleans, LA, USA, 26 April–1 May 2004; Volume 4, pp. 4172–4177.
17. Berg, V.D.J.; Patil, S.; Sewall, J.; Manocha, D.; Lin, M. Interactive navigation of multiple agents in crowded environments. In Proceedings of the 2008 symposium on Interactive 3D Graphics and Games, Redwood City, California, USA, 15–17 February 2008; pp. 139–147.
18. Song, A.L.; Su, B.Y.; Dong, C.Z.; Shen, D.W.; Xiang, E.Z.; Mao, F.P. A two-level dynamic obstacle avoidance algorithm for unmanned surface vehicles. *Ocean Eng.* **2018**, *170*, 351–360. [CrossRef]
19. Kuwata, Y.; Wolf, M.T.; Zarzhitsky, D.; Huntsberger, T.L. Safe maritime autonomous navigation with COLREGS, using velocity obstacles. *IEEE J. Ocean. Eng.* **2014**, *39*, 110–119. [CrossRef]
20. Mou, J.M.; Li, M.X.; Hu, W.X.; Zhang, X.H.; Gong, S.; Chen, P.F.; He, Y.X. Mechanism of dynamic automatic collision avoidance and the optimal route in multi-ship encounter situations. *J. Mar. Sci. Technol.* **2021**, *26*, 141–158.
21. Valdes-Vela, M.; Toledo-Moreo, R.; Terroso-Saenz, F.; Zamora-Izquierdo, M.A. An application of a fuzzy classifier extracted from data for collision avoidance support in road vehicles. *Eng. Appl. Artif. Intell.* **2013**, *26*, 173–183. [CrossRef]

22. Qin, L.; Shen, Y. Research on Safety Classification for Vehicle Anti-collision Data by Improved Interval Fuzzy Reasoning. In Proceedings of the 2021 3rd International Conference on Robotics and Computer Vision (ICRCV), Beijing, China, 6–8 August 2021; pp. 82–85.
23. Pietrzykowski, Z.; Uriasz, J. The Ship Domain-A Criterion of Navigational Safety Assessment in an Open Sea Area. *J. Navig.* **2009**, *62*, 93–108. [CrossRef]
24. Perera, L.P.; Carvalho, J.P.; Soares, C.G. Autonomous Guidance and Navigation based on COLREGs Rules and Regulations of Collision Avoidance. In Proceedings of the International Workshop "Advanced Ship Design for Pollution Prevention", Split, Croatia, 23–24 November 2009; pp. 205–216.
25. Perera, L.P.; Carvalho, J.P.; Guedes, S.C. Fuzzy Logic Based Derision Making System for Collision Avoidance of Ocean Navigation under Critical Collision Conditions. *J. Mar. Sci. Technol.* **2011**, *16*, 84–99. [CrossRef]
26. Perera, L.P.; Carvalho, J.P.; Guedes, S.C. Intelligent Ocean Navigation and Fuzzy-Bayesian Decision/Action Formulation. *IEEE J. Ocean. Eng.* **2012**, *37*, 204–219. [CrossRef]
27. Brcko, T.; Svetak, J. Fuzzy Reasoning as a Base for Collision Avoidance Decision Support System. *Promet Traffic Transp.* **2013**, *25*, 555–564. [CrossRef]
28. Kao, S.L.; Lee, K.T.; Chang, K.Y.; Ko, M.D. A fuzzy logic method for collision avoidance in vessel traffic service. *J. Navig.* **2007**, *60*, 17–31. [CrossRef]
29. Tang, C.; Xu, Z.; Liu, Z.; Liu, J. Research on Collision Probability Model Based on Ship Domain. In Proceedings of the Second International Conference on Transportation Information & Safety, Wuhan University of Technology, Wuhan, China, 28–30 June 2013; pp. 2307–2313.
30. Huang, L.; Wen, Y.; Zhang, Y.; Zhou, C.; Zhang, F.; Yang, T. Dynamic calculation of ship exhaust emissions based on real-time AIS data. *Transp. Res. Part D Transp. Environ.* **2020**, *80*, 102277. [CrossRef]

Journal of
Marine Science
and Engineering

MDPI

Article

A Novel Decision Support Methodology for Autonomous Collision Avoidance Based on Deduction of Manoeuvring Process

Ke Zhang [1,2], Liwen Huang [1,2], Xiao Liu [1,2,*], Jiahao Chen [1,2], Xingya Zhao [1,2], Weiguo Huang [1,2] and Yixiong He [1,2,*]

[1] School of Navigation, Wuhan University of Technology, Wuhan 430063, China; kezhang_rq@whut.edu.cn (K.Z.); lwhuang@whut.edu.cn (L.H.); jh.chen666@foxmail.com (J.C.); 128794@whut.edu.cn (X.Z.); huangweiguo@whut.edu.cn (W.H.)

[2] Hubei Key Laboratory of Inland Shipping Technology, Wuhan 430063, China

* Correspondence: lxiao@whut.edu.cn (X.L.); heyixiong7@whut.edu.cn (Y.H.)

Abstract: In the last few years, autonomous ships have attracted increasing attention in the maritime industry. Autonomous ships with an autonomous collision avoidance capability are the development trend for future ships. In this study, a ship manoeuvring process deduction-based dynamic adaptive autonomous collision avoidance decision support method for autonomous ships is presented. Firstly, the dynamic motion parameters of the own ship relative to the target ship are calculated by using the dynamic mathematical model. Then the fuzzy set theory is adopted to construct collision risk models, which combine the spatial collision risk index (SCRI) and time collision risk index (TCRI) in different encountered situations. After that, the ship movement model and fuzzy adaptive PID method are used to derive the ships' manoeuvre motion process. On this basis, the feasible avoidance range and the optimal steering angle for ship collision avoidance are calculated by deducting the manoeuvring process and the modified velocity obstacle (VO) method. Moreover, to address the issue of resuming sailing after the ship collision avoidance is completed, the Line of Sight (LOS) guidance system is adopted to resume normal navigation for the own ship in this study. Finally, the dynamic adaptive autonomous collision avoidance model is developed by combining the ship movement model, the fuzzy adaptive PID control model, the modified VO method and the resume-sailing model. The results of the simulation show that the proposed methodology can effectively avoid collisions between the own ship and the moving TSs for situations involving two or multiple ships, and the own ship can resume its original route after collision avoidance is completed. Additionally, it is also proved that this method can be applied to complex situations with various encountered ships, and it exhibits excellent adaptability and effectiveness when encountering multiple objects and complex situations.

Keywords: autonomous ship; collision avoidance; ship manoeuvrability; velocity obstacle; deduction of the manoeuvring process

Citation: Zhang, K.; Huang, L.; Liu, X.; Chen, J.; Zhao, X.; Huang, W.; He, Y. A Novel Decision Support Methodology for Autonomous Collision Avoidance Based on Deduction of Manoeuvring Process. *J. Mar. Sci. Eng.* **2022**, *10*, 765. https://doi.org/10.3390/jmse10060765

Academic Editors: Myung-Il Roh and Marco Cococcioni

Received: 25 April 2022
Accepted: 26 May 2022
Published: 1 June 2022

1. Introduction

1.1. Background

In recent years, autonomous ships have received a lot of attention and development in the maritime industry. The International Maritime Organization has been committed to the research on the relevant technologies and regulations of Maritime Autonomous Surface Ships (MASS) [1]. Improving the intelligence level of ships is crucial to the safety of ship navigation. Although some advanced technologies and methods have been developed and applied to ships, collision accidents still happen from time to time. Actually, a report published by the European Maritime Safety Agency (EMSA, 2020) showed that contact and collision incidents of ship accounted for 32% of all navigational casualties between 2014 and 2019 [2]. Ship collision accidents are a major threat to the safety of maritime navigation, and may cause serious casualties, economic losses and marine environmental pollution, etc. Therefore, to reduce the navigational risk and casualties caused by human factors, it is of great significance to develop a novel autonomous collision avoidance decision support methodology or system to help ships to make secure and expeditious decisions to avoid collisions.

Making a ship collision avoidance decision is a complex process, especially in the multi-object environment and restricted waters. At present, many technologies and algorithms have been developed for the autonomous collision avoidance problem of MASS, such as the Artificial Potential Field (APF) [3,4], Particle Swarm Optimization (PSO) [5], Rapidly exploring Random Tree (RRT) [6], velocity obstacle (VO) [7], genetic algorithm (GA) [8], Artificial Neural Network (ANN) [9], Deep Reinforcement Learning (DRL) [10], and so on. In the future, all kinds of vessels, including autonomous vessels and MASS, will be expected to follow the existing guidelines based on good seamanship and the International Regulations for Preventing Collisions at Sea (COLREGs). Considering that autonomous collision avoidance is a highly complex problem, our primary aim in this study is to design a ship manoeuvring process deduction-based dynamic adaptive autonomous collision avoidance decision support methodology for autonomous ships and MASS, which take into account the COLREGs and good seamanship mentioned above as well as ship manoeuvrability.

1.2. Related Studies

As a key technology to realize ship automation, intelligent collision avoidance technology has attracted more and more attention from researchers in recent years [11]. Diverse solutions can be found in the literature related to ship collision avoidance [12,13]. In general, these collision avoidance methods can be divided into path generation methods and intelligent optimization methods. The path generation algorithms mainly include the A* algorithm, the APF algorithm, the RRT algorithm and the VO algorithm. The A* algorithm is an intelligent search algorithm that mainly considers the start position and the destination, which has better performance and accuracy. The APF model has been extensively used in the field of the autonomous collision avoidance of ships [14]. Lv et al. [4] and Lazarowska et al. [15] proposed a method for safe vessel trajectory planning based on the APF model, respectively. At present, the RRT algorithm and its modified algorithm are widely adopted for ship optimal path planning [6]. Based on the problem of optimal path planning for ships in the perspective of real-time applications, Zaccone et al. [16] proposed an optimal path planning algorithm for autonomous ships based on modified RRT. Chiang et al. [17] proposed a COLREGs-compliant RRT-based motion planner for Autonomous Surface Vehicles' navigation. This algorithm has a higher navigation success rate and COLREGs compliance compared to other methods. The VO algorithm is a classic collision avoidance algorithm in the field of mobile robots. At present, many scholars have applied it to the study of ship collision avoidance. Huang et al. [18] built a collision avoidance decision system based on a non-linear VO model. It can assist the marine navigator to make collision avoidance decisions. Chen et al. [19] presented an improved time discretized non-linear velocity obstacle algorithm to detect multi-ship encounter situations using historical automatic identification system (AIS) data.

In the last two decades, researchers have put forward many new intelligent optimization methods and achieved fine results. Ni et al. [8] generated a collision-free optimal path for autonomous ships based on multiple genetic algorithms. Xie et al. [20] presented a collision avoidance method based on an improved Q-learning beetle swarm antenna search algorithm and ANN for USV. An autonomous collision avoidance decision system based on the ANN and fuzzy logic methods was designed by Ahn et al. [21]. This system can calculate ship collision risk in real time. As an important learning method of machine learning, DRL has been widely used for intelligent autonomous systems due to its excellent adaptive and self-learning capabilities for complex systems. Based on the DRL model,

Zhao et al. [22,23] established a novel collision avoidance decision system for autonomous ships. However, this system is only suitable for two-ship encounters, not for multi-ship scenarios, and restricted waters. In order to solve the above problem, Sawada et al. [1] proposed an automatic collision avoidance algorithm for ships based on the DRL algorithm. Shen et al. [24] designed a new method based on deep Q-learning to realise the automatic collision avoidance of ships. This model is also suitable for restricted waters. However, most studies in this field only focus on the computation of collision-free paths without obeying the rules of the COLREGs.

As a significant component of realizing autonomous ships, collision avoidance decision-making systems have attracted more attention from researchers in recent years. At present, many scholars are carrying out research work related to the development of collision avoidance decisions systems, such as a collision avoidance decision-making system, autonomous collision avoidance system, etc. [9,25–27]. Zhang et al. [28] and Mizythras et al. [29] introduced a distributed anti-collision decision support system. Among these, Mizythras's system takes into account the ship's manoeuvrability and propulsion system performance. Wang et al. [25] proposed a collision avoidance decision-making system designed for autonomous ships. However, the states, actions and trigger conditions defined in the FSM model are easily affected by subjectivity and have great uncertainty. Pietrzykowski et al. [30] presented a summary of the research on navigational decision support systems. This also pointed out that the usability of navigational decision support systems on vessels has been confirmed by the actual users of the navigational decision support system.

Ship manoeuvrability has a very important impact on the safety of ship navigation and the process of ship collision avoidance. However, at present, most of the above researches have only focused on the intelligent algorithms of being collision free and ignored the kinematical constraints of ships. As a matter of fact, research results will be more reliable if ship manoeuvrability is fully considered in the process of collision avoidance decision making. Li et al. [31] presented a dynamic path planning model based on the Morphin algorithm, which considers a ship's manoeuvrability. However, the proposed model is greatly affected by human factors, and it is difficult to balance the relationship between the time cost of the model's construction and the number of layers of the search tree. Wang et al. [32] developed a dynamic support system for ship collision avoidance by combining the ship manoeuvring motion model and the control mechanism of ships' manoeuvring motion. However, this system is only suitable for a two ship encounter scenario and not applicable to the collision avoidance of the ships at different speeds. Generally, the complete ship collision avoidance manoeuvring process includes three stages: ships' manoeuvring process, course keeping and returning to the original route. After the collision avoidance is completed, the ship should return to the original route. In other words, the own ship (OS) is considered to be back on the original route when the target ship (TS) is finally past and clear. Regarding the problem of resuming sailing, many scholars [1,25,33,34] have transformed it into the constraint conditions of the given decision variables, thereby obtaining a course manoeuvring degree to resume the original route.

1.3. Contributions and Outlines

Despite a lot of research work and achievements being completed on ship collision avoidance, there are still some unignorable shortcomings in the available studies: most of the researches only focus on the intelligent algorithms of being collision free without obeying the rules of the COLREGs, and they seldom consider the impact of the TS's action uncertainty on collision avoidance decisions. Moreover, although some studies consider the constraints of the COLREGs rules, most of them rarely consider the ship's manoeuvrability, COLREGs, good seamanship and uncertainty of the TS's movement for collision avoidance at the same time.

The main motivation and contribution of this work is to present a decision support methodology of dynamic adaptive autonomous collision avoidance based on the ship manoeuvring process deduction method for autonomous ships. This methodology can solve the problem of the autonomous collision avoidance conundrum when encountering multiple objects and complex situations. Furthermore, it takes full account of various factors, including the COLREGs, ship manoeuvrability, good seamanship, multiple objects and complex encounter situations, and the TS's uncoordinated or temporary actions.

Given the above, some highlights of our paper are:

(1) A ship manoeuvring process deduction-based dynamic adaptive autonomous collision avoidance decision support methodology is established.
(2) The COLREGs rules, good seamanship and ship manoeuvrability are comprehensively considered in the decision support methodology for autonomous collision avoidance.
(3) A new collision risk evaluation model is constructed on the basis of the fuzzy set method to synthesize the SCRI and TCRI. This model considers the ship's movement process and calculates the collision risk among ships in a dynamic way.
(4) The impact of the TS's action uncertainty on the collision avoidance decision is considered by the ship collision avoidance decision systems constructed in this paper.
(5) The method constructed in this research is suitable for two-ship and multi-ship encounter situations, which is indispensable for collision avoidance systems for MASS.

The remaining sections of the paper are arranged as follows: In Section 2, the structure of the proposed method is clarified. In Section 3, the functionality of the collision avoidance decision support methodology is presented, describing the formulation for the collision risk calculation, autonomous collision avoidance method, resume-sailing model and collision avoidance algorithm implementation process. In Section 4, the proposed decision support methodology is validated through five cases. Subsequently, the experimental results analysis and discussions are made in Section 5, and in Section 6, the conclusion of this study is drawn.

2. Proposed Framework

In order to cope with the unpredictable manoeuvres of the TS adaptively, based on the ship manoeuvring mathematical model group (MMG) model and fuzzy adaptive proportion integration differentiation (PID) method, this paper proposes a ship manoeuvring process deduction-based dynamic adaptive autonomous collision avoidance decision support method for autonomous ships. Among them, the information of the OS and the TS obtained from the automatic identification system (AIS) and automatic radar plotting aids (ARPA) is the input, and the output is the collision avoidance decision scheme. The specific implementation steps are as follows:

Firstly, the system obtains the basic collision avoidance parameter information by AIS and ARPA. This information includes the ship velocity, ship position, course, true bearing and so on. Then, the collision risk judgment is made. If collision risks exist, the collision avoidance decision-making scheme and the resume-sailing angle are calculated based on the constructed collision avoidance decision-making model; if there are no collision risks, the OS keeps its current course and speed. Finally, a feasible collision avoidance decision scheme is given based on the constructed collision avoidance decision model. The process includes the execution of a control system consisting of the MMG model and PID. Based on the real-time updated ship information, the collision avoidance manoeuvring decision is executed cyclically.

The errors of the collision avoidance decision scheme in this method mainly include three aspects: the information error of the ship collision avoidance parameters, the error of the MMG model parameters and the error of the PID parameters. The effects of these errors can be eliminated or reduced by: improving equipment accuracy, selecting an appropriate MMG model to improve its adaptability and having prior knowledge to optimize controller parameters. For other errors, they can be compensated by designing an adaptive system. Based on the ship collision avoidance information at the current time T_0, the adaptive system will calculate and execute the optimal decisions scheme during a fixed time interval Δt. The error from T_0 to $T_0 + \Delta t$ will be compensated if the system circularly recalculates and executes the new optimal decisions scheme based on the information after time step Δt.

The time series rolling calculation method by quickly updating input information is taken in the adaptive method. The influence of the residual error on the manoeuvring scheme can therefore be compensated.

The framework of this research mainly consists of three parts. The first part is calculating the dynamic movement parameters of the OS and the TSs in real time and determining the collision risk between ships in different encounter situations. In the second part, the dynamic adaptive autonomous collision avoidance model is developed by combining the ship motion mathematical model, the fuzzy adaptive PID control model and the modified VO method. In the third part, in order to solve the resume-sailing problem, the LOS guidance system is adopted to resume the original route of the OS. The ship collision avoidance decision-making system considers various factors including the ship's manoeuvrability, ship encounter situation, good seamanship and COLREGs. The framework of the decision support methodology for autonomous collision avoidance is drawn in Figure 1.

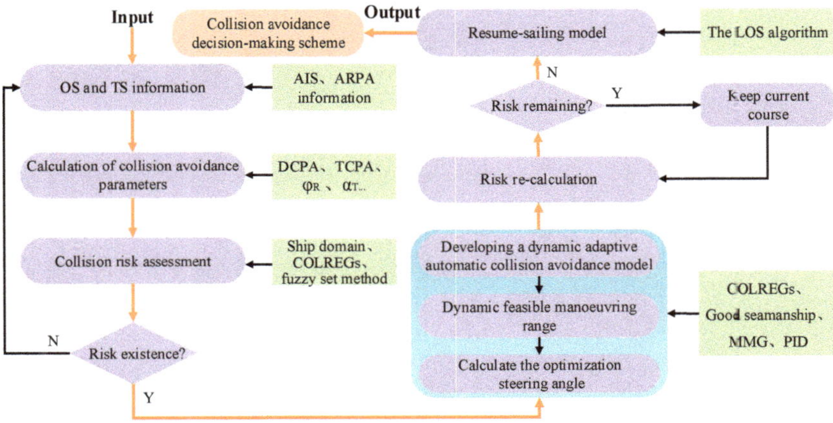

Figure 1. The framework of decision support methodology for autonomous collision avoidance.

3. Decision Support Methodology

3.1. The Dynamic Calculation Model of Collision Avoidance Parameters

The determination of ship collision avoidance parameters is the basis of collision risk calculation and ship collision avoidance decisions. In this section, collision avoidance parameters between ships are calculated dynamically based on ship manoeuvring motion model. The ship collision avoidance parameter is shown below.

Figure 2 shows the relative position of two ships in the applied coordinate systems for a typical crossing situation. The speed and course of each ship are represented as V_{OS}, V_{TS}, φ_0 and φ_T, respectively. Supposing that the initial position and relative distance of the OS is $OS(0,0)$ and R_0, respectively. Relative bearing angle between OS and TS is α_{OT}.

The relative speed of the OS and TS on the X-axis and Y-axis can be calculated by

$$\begin{cases} v_{x_R} = v_{TS}\sin\varphi_T - v_{OS}\sin\varphi_0 \\ v_{y_R} = v_{TS}\cos\varphi_T - v_{OS}\cos\varphi_0 \end{cases} \tag{1}$$

$$V_{OT} = \sqrt{v_{x_R}^2 + v_{y_R}^2} \tag{2}$$

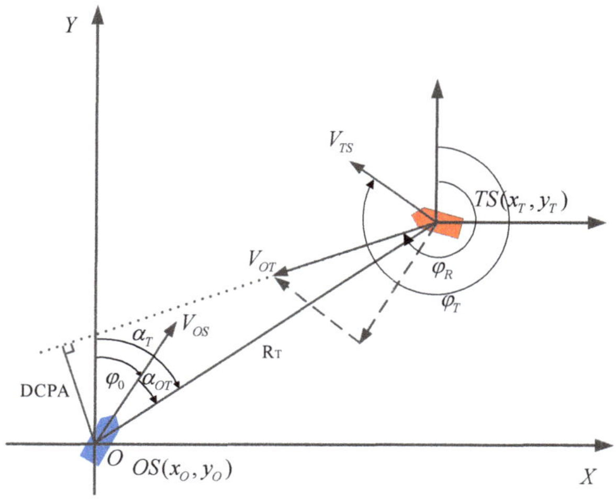

Figure 2. Motion parameters of two ships in a typical encounter situation.

At time t after the OS takes evasive action, the position $TS(x_T, y_T)$ of the OS is

$$\begin{cases} x_0(t) = \int_0^t V_{OS} \sin \varphi_0 dt \\ y_0(t) = \int_0^t V_{OS} \cos \varphi_0 dt \end{cases} \tag{3}$$

Then the relative displacement from the TS to the OS is expressed as follows

$$\begin{cases} \Delta x = R_0 \sin(\varphi_0 + \alpha_{OT}) + \int_0^t (V_{OS} \sin \varphi_0 - V_{TS} \sin \varphi_T) dt \\ \Delta y = R_0 \cos(\varphi_0 + \alpha_{OT}) + \int_0^t (V_{OS} \cos \varphi_0 - V_{TS} \cos \varphi_T) dt \end{cases} \tag{4}$$

At time t, the relative distance R_T between the OS and TS can be obtained as follows

$$R_T = \sqrt{\Delta x^2 + \Delta y^2} \tag{5}$$

Then, as indicated in Figure 2, the distance to closest point of approach (*DCPA*) between OS and TS can be calculated as follows

$$DCPA = R_T \sin(\varphi_R - \alpha_T - \pi) \tag{6}$$

and the time to closest point of approach (*TCPA*) is expressed as

$$TCPA = R_T \cos(\varphi_R - \alpha_T - \pi)/V_{OT} \tag{7}$$

where φ_R is relative course, α_T is the true relative bearing to TS, which can be given as follows

$$\varphi_R = \begin{cases} \arctan(v_{x_R}/v_{y_R}), & v_{xR} \geq 0 \cap v_{yR} \geq 0 \\ \arctan(v_{x_R}/v_{y_R}) + \pi, & (v_{xR} < 0 \cap v_{yR} < 0) \cup (v_{xR} \geq 0 \cap v_{yR} < 0) \\ \arctan(v_{x_R}/v_{y_R}) + 2\pi, & v_{xR} < 0 \cap v_{yR} \geq 0 \end{cases} \tag{8}$$

$$\alpha_T = \begin{cases} \arctan(\Delta x/\Delta y), & v_{xR} \geq 0 \cap v_{yR} \geq 0 \\ \arctan(\Delta x/\Delta y) + \pi, & (v_{xR} < 0 \cap v_{yR} < 0) \cup (v_{xR} \geq 0 \cap v_{yR} < 0) \\ \arctan(\Delta x/\Delta y) + 2\pi, & v_{xR} < 0 \cap v_{yR} \geq 0 \end{cases} \tag{9}$$

3.2. Collision Risk Method

The collision risk index (CRI) is used to assess the probability and severity of a ship collision with other ships in the vicinity, and its value ranges from 0 to 1 [25]. The value of CRI can be affected by various kinds of factors. In this section, considering the ship domain, relative position, *DCPA*, *TCPA*, manoeuvrability and vessel velocity, the fuzzy set method is adopted to construct a new collision risk model, which combines SCRI and TCRI in different encounter situations. The collision risk model for different situations is constructed below according to the types of encounter situations.

(a) Head-on situation

The SCRI is a measure of the collision probability, which can be determined by taking the minimum safe distance between two ships in danger of collision as the main indicator. Accordingly, it can be measured by whether TS will eventually enter the OS's domain. According to the COLREGs and good seamanship, the value of the SCRI is either 1 or 0. The SCRI is expressed as follows

$$u_{sH} = \begin{cases} 1, \exists (x,y)^t \in SDom^t, t \in [0, TCPA] \\ 0, \forall (x,y)^t \notin SDom^t, t \in [0, TCPA] \end{cases} \tag{10}$$

where, u_{sH} is the membership function of the fuzzy set U_{sH}, $(x,y)^t$ represents the position coordinates of the TS at time t, $SDom^t$ represents the sets of location point elements at time t in the OS's domain. For this encounter situation, this paper uses the elliptical ship domain, where semi-major axis is 8 L (8 times ship length) and semi-minor axis is 4 L.

The TCRI is affected by factors such as time to close situation (*TCS*), *DCPA*, etc. On the basis of referring to existing research [35], the formula for determining the TCRI is

$$u_{tH} = \begin{cases} 1, & TCS \le 0 \\ \left(\frac{k-1}{k-1+\frac{TCS}{TCPA}} \right)^{3.03}, & TCS > 0 \; and \; D < D_s \\ 0, & D \ge D_s \end{cases} \tag{11}$$

where, $k = \sqrt{R_T^2 - DCPA^2} / \sqrt{D^2 - DCPA^2}$, u_{tH} is the membership function of the fuzzy set U_{tH}, *TCS* represents the time from the current moment to the first time point of a close-quarter situation. R_T is the distance between OS and TS. D_s is a constant with a value of 5 nm.

Thus, the collision risk model in the head-on situation can be written in Equation (12)

$$u_{CRI} = u_{sH} \otimes u_{tH} \tag{12}$$

where $\otimes$ is the risk synthesis operator.

(b) Overtaking situation

According to the requirements of the rules on the formation of overtaking situation, when the rear ship catches up with the preceding ship and the distance is less than 3 nm, the distance condition for the formation of overtake is satisfied [35]. Therefore, in overtaking situation, the value of R_T in the TCRI is 3 nm. Other parameters are the same as the collision risk model for head-on situation, and for elliptical ship domain, where semi-major axis is 5 L and semi-minor axis is 4 L.

(c) Crossing situation

According to the definition of SCRI, the SCRI in the crossing encounter situation is the same as the model of the other two encounter situations. The TCRI of crossing situation should satisfy both conditions of $D < 5$ nm and $TCS \le 20$ min. Therefore, the TCRI in the crossing encounter situation takes the smaller of these two values.

For the first case, when the potential collision risk exists, $D < D_s$ and $TCPA > 0$, the TCRI under the crossing encounter situation is the same as Equation (11).

$$u_{tC}^1 = u_{tH} \tag{13}$$

For the second case, when the potential collision risk exists, $TCS \leq 20$ min and $TCPA > 0$, the TCRI under the crossing encounter situation is as follows

$$u_{tC}^2 = \begin{cases} 1, & TCS \leq 0 \\ (1 - \frac{TCS}{1200})^{3.03}, & 0 < TCS < 20 \\ 0, & TCS \geq 20 \end{cases} \tag{14}$$

The TCRI of crossing situation is

$$u_{tC} = \min(u_{tC}^1, u_{tC}^2) \tag{15}$$

Thus, the collision risk model in the crossing situation can be written in Equation (16)

$$u_{CRI} = u_{sC} \otimes u_{tC} \tag{16}$$

where $\otimes$ is the risk synthesis operator.

3.3. Autonomous Collision Avoidance Method

VO algorithm was first proposed by Fiorini and Shiller [36], and is an effective and simple method collision avoidance method, such as robots' collision avoidance and ships' obstacle avoidance [7,18,37]. According to its principle, VO can calculate the speed sets of all ships that may cause collision risk. Thus, in this paper, the MMG, fuzzy adaptive PID control model and modified VO algorithm are used to derive the ships' manoeuvre motion process. On this basis, the feasible avoidance range and optimal steering angle of ships' collision avoidance can be calculated.

Assume that the position and velocity of OS and TS are denoted as P_{OS}, P_{TS}, V_{OS} and V_{TS} , respectively, D represents the safe distance between OS and TS. Hence, the possible position of OS when a collision happens is termed as "Conflict Position $ConfP$".

$$ConfP(O, D) = \{P \| P_{TS} - P_{OS}\| \leq D\} \tag{17}$$

where $\|\cdot\|$ is the geographic distance between two vessels. P is denoted as a position. If the distance between the OS and TS is less than the threshold D, a collision will definitely occur. In other words, the two ships will collide at time t_n, with the following conditions fulfilled

$$P_i(t_n) \in P_j(t_n) \otimes ConfP(O, D) \tag{18}$$

where operation $\otimes$ is the Minkowski addition, which means adding P_j to each element in $ConfP$. Assuming that kinematic information of both ships is known, Equation (18) can be substituted with Equation (19)

$$VO_{OS|TS} = U_t^N \left(\frac{P_{TS}(t) - P_{OS}(t_0)}{t - t_0} \right) \otimes \frac{ConfP(O, D)}{t - t_0} \tag{19}$$

where N is an infinite number. If the OS keeps up this vector velocity all the time, there will definitely be collisions in the future ($t_0 \to N$).

In Figure 3, the TS and OS form a starboard crossing situation. Assuming that the steering angles of OS to avoid the TS to the right and left are θ_1 and θ_2, respectively, when OS alters to the starboard side or port side, the target course is $C + \theta_1$ and $C + \theta_2$. If the ship's manoeuvrability is not considered, the critical trajectories of the OS are straight lines L_1' and L_2'. From the moment the ship steers to the starboard to time T_1, the OS and TS are located at point A_1 and B_1, respectively, and at this time the OS just passes through the fore of the TS. From the moment when OS steers to the port to time T_2, the OS and TS are located at point A_2 and B_2, respectively, and at this time the OS just passes through the aft of the TS. Due to the non-linear characteristics of the ship's manoeuvring model, it takes a period of time for the ship to maintain heading stability. Therefore, if the ship's manoeuvring is considered and the fuzzy adaptive PID method is used to control the steering of the ship, the ship trajectory is the curve line L_1 and L_2.

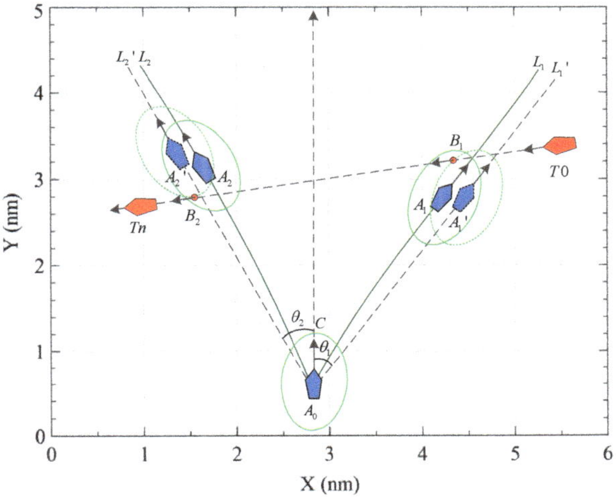

Figure 3. Dynamic feasible manoeuvring range.

In this case, the TS will enter the OS's ship domain if $C + \theta_1$ and $C + \theta_2$ are used as the target heading angle for collision avoidance. In summary, if the ship manoeuvrability constraints of the OS are not considered, and the collision avoidance scheme is performed according to the original collision avoidance angle, the two ships will collide.

If the steering angles of OS to the left or right are less than θ_1 and θ_2, respectively, the TS will enter the OS's ship domain. Then the steering angle interval (θ_1, θ_2) is the obstacle range of the speed vector from the TS to OS. For multi-ship encounter situation, the feasible manoeuvring range of OS is the complement of the union of the speed vector obstacle range for each TS

$$C_{range} = \overline{U_{i=1}^{n}\theta_i}\ ,(i = 1, 2) \tag{20}$$

For any course altering angle θ_i, as long as the course of OS is not in the speed obstacle range of all TSs when the redirection is completed, then this altering angle belongs to the feasible manoeuvring range. The algorithm proposed to obtain the dynamic feasible manoeuvring range is denoted in Algorithm 1.

Theoretically, any course altering angle within the feasible manoeuvring range can make the giving-way ship safely avoid all TSs. However, considering the safety and economy in navigation, the course altering angle should not be too large or too small, so it is very necessary to choose an appropriate altering angle. Based on this, this study introduces the surplus amount θ_k, the optimization steering angle θ of ship collision avoidance can be expressed by Equation (21).

$$\theta = \begin{cases} \theta_{min} + \theta_k, & \theta \in C_{range} \\ \theta_{min}, & \theta \notin C_{range} \end{cases} \tag{21}$$

where θ_{min} is the minimum altering angle in the feasible manoeuvring range, and θ_k is the surplus amount, which can be adjusted according to specific circumstances.

Algorithm 1. Algorithm for calculating dynamic feasible manoeuvring range.

Input: the position $P_{OS}^{(0)}$, $P_{TS}^{(0)}$; the speed $v_{OS}^{(0)}$, $v_{TS}^{(0)}$; the course $C_{OS}^{(0)}$, $C_{TS}^{(0)}$
Output: dynamic feasible manoeuvring range $[\theta_{i1}, \theta_{i2}]$

Initialize the i = 1, $\Delta t = 1$
Initialize the θ = -90
for $u_{CRI} > 0$ **do**
  for $i \leq 361$ **do**
    Calculate the target course $C_t = C_0^{(0)} + \theta$
    Put target course C_t into course control system
    Update: $P_{OS}^{(i*\Delta t)}$, $P_{TS}^{(i*\Delta t)}$ $v_{OS}^{(i*\Delta t)}$, $v_{TS}^{(i*\Delta t)}$, $C_{OS}^{(i*\Delta t)}$, $C_{TS}^{(i*\Delta t)}$
    for each TS **do**
      Calculate whether the TS enters OS's ship domain
      If TS enter OS's ship domain **then**
        $\theta = \theta + 0.5$
        i = i + Δt
      else
        break
    end for
  end for
end for
return: feasible manoeuvring range $(\theta_{i1}, \theta_{i2})$

3.4. Resume-Sailing Model

The LOS algorithm is a classic trajectory control algorithm, which is not model dependent. The target heading is only related to the unmanned ship's real-time position and target course [22]. In this study, the LOS guidance strategy, which is widely used in path tracking, is adopted for the problem of resuming route and heading keeping.

The sketch map of the resume-sailing model is shown in Figure 4, where the LOS position P_{LOS} is the point along the path that the ship should point to. OS is sailing from the start position $P_s(x_s, y_s)$ to destination $P_n(x_n, y_n)$. After completing the avoidance process, the ship applies the LOS strategy to resume route. Make the OS's current position be located at the centre of a circle with a radius of n times its length. The circle intersects the line between $P_s(x_s, y_s)$ and $P_m(x_m, y_m)$, and P_{LOS}, the closest point to $P_m(x_m, y_m)$ is selected as the turning point.

In LOS guidance system, OS is guided to resume route by the minimum error ψ_P between the actual heading angle ψ and the LOS angle ψ_{LOS}. The LOS angle ψ_{LOS} can be calculated by the following equation

$$\psi_{LOS} = arcsin(\frac{x_{LOS} - x}{R_{LOS}}) \tag{22}$$

where R_{LOS} is the radius of the circle, which satisfies the following Equation

$$R_{LOS}^2 = (x_{LOS} - x)^2 + (y_{LOS} - y)^2 \tag{23}$$

Then the resume-sailing angle is tracking error; it can be calculated by

$$\psi_P = \psi - \psi_{LOS} \tag{24}$$

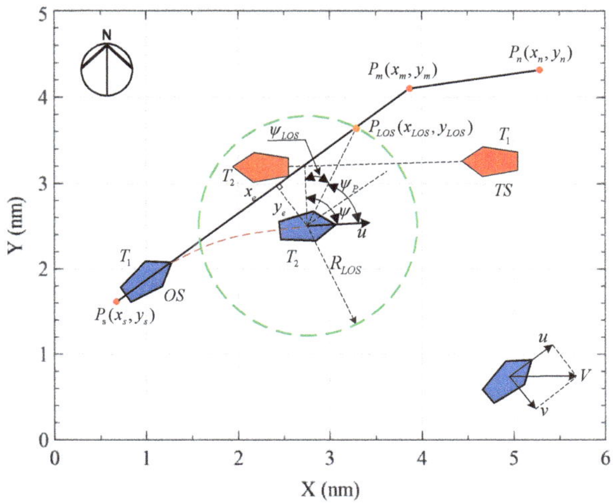

Figure 4. The LOS guidance strategy.

3.5. Design of Adaptive Autonomous Collision Avoidance Algorithm

In order to realize the ship's adaptive autonomous collision avoidance in complex encounter situations, a ship manoeuvring process deduction-based dynamic adaptive autonomous collision avoidance decision support method (Figure 5) is constructed in this study. This system can acquire the dynamic and static information of the TSs in real time and calculate the course altering angle required at the current time and input it into the course control system with the interval of fixed calculation step $\Delta t = 1$ s. The ship motion model is used to deduce the movement trend of the OS and the TSs within a certain period of time, and the information of the OS and the TSs is updated in real time through rolling calculations to realize autonomous collision avoidance.

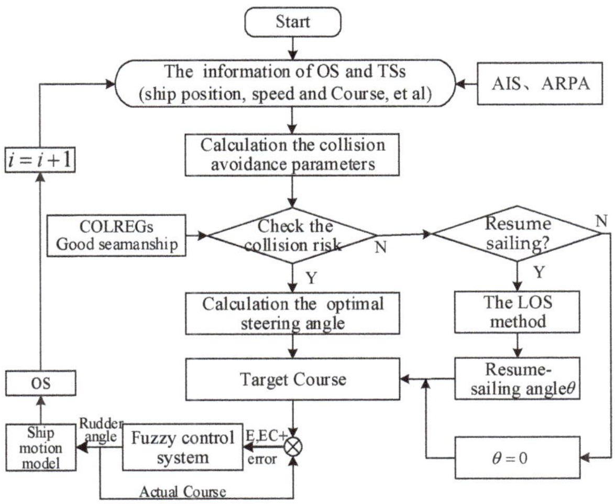

Figure 5. Flowchart of the adaptive autonomous collision avoidance algorithm.

The specific implementation process is as follows:

Step 1: Obtain the static and dynamic information of ships, and calculate the collision avoidance parameters considering the ship manoeuvring in real time;

Step 2: Identify the ships' encounter situation, and calculate the collision risk between ships according to the collision risk model in Section 3.2;

Step 3: According to the autonomous collision avoidance model constructed in Section 3.3, calculate the feasible manoeuvring range that allows all dangerous TSs to be cleared;

Step 4: Calculate the optimization steering angle of ship collision avoidance based on the feasible manoeuvring range;

Step 5: Determine the target course after completing the avoidance and redirection according to step 4, control the ship's steering through the course control system and calculate the corresponding rudder angle;

Step 6: Substitute the rudder angle into the MMG model to calculate the dynamic and static information of OS and the TS at the next moment in the manoeuvre process. After updating the information, go to step 1 until the TS is finally past and clear;

Step 7: Take the completion of the ship avoidance as the initial moment, and search for the time point at which all obstacles can be avoided at one-second intervals, and make it the return time point;

Step 8: Calculate the resume-sailing angle according to the resume-sailing model;

Step 9: Control the ship to resume sailing according to the course control system.

4. Case Study

4.1. Experimental Settings

In this section, to validate the feasibility and effectiveness of the decision support methodology proposed in this paper, five different types of ship encounter scenarios are designed to illustrate the application of the autonomous collision avoidance system in maritime navigation. The scope of the experimental cases includes two-ship encounter scenarios, multi-ship encounter scenarios, the TS maintaining course and speed, and the TS suddenly changing course, etc. In addition, to further verify the effectiveness of the decision support methodology, we recorded the collision avoidance parameter information such as the *DCPA*, *TCPA*, D, and course per unit time step. T_0, T_1 and T_2 are the start time, the middle time and the end time during the collision avoidance process, respectively. The unit of the relative distance is in nautical miles, shown as (nm), the unit of ship speed is knots (kn), the unit of the *DCPA* is meters (m), the unit of the *TCPA* is seconds (s), and the unit of the course is degrees (°). In order to more intuitively display the information such as the motion process and relative distance between ships, a geographical location (Lon. 123°42.8′ E, Lat. 29°24.0′ N) is determined as the origin of the simulation experiment, and an O-X-Y coordinate system with the origin as the centre and nautical miles as the unit is established. The longitudinal and transverse distance from the ship's current position to the coordinate axis is used as the ship's position coordinate in the simulation experiment diagram in this paper. The initial information of the two-ship and multi-ship cases are listed in Tables 1 and 2. Moreover, this paper adopts the classic three-degree-of-freedom MMG; the specific parameters are available in the literature [38]. To verify the precision of the MMG model, the Panama maximum size bulk carrier HUAYANG DREAM is simulated in our study. The simulation results are shown in Appendix A (Table A1, Figures A1 and A2). It can be found that although there are some slight differences between the MMG model and the real ship, the accuracy is generally acceptable. The fuzzy adaptive PID control model is used to control the ship's manoeuvres, and the process of the ship's course/track control system to control the ship's manoeuvres is simulated. Details about fuzzy PID are provided in the literature [39]. In this chapter, the constructed collision avoidance decisions method is connected with the intelligent navigation simulation research platform, and the effectiveness of the collision avoidance decisions method is verified through the simulation platform.

Table 1. Settings of two-ship scenarios.

Case	Ship List	Position	Course (°)	Velocity (kn)	Distance (nm)	Ship Length (m)
Scenario 1	OS	(123°46.519′ E, 29°24.877′ N)	006.5	12	0	225
	TS	(123°46.823′ E, 29°29.763′ N)	186	10	4.8532	180
Scenario 2	OS	(123°46.519′ E, 29°24.877′ N)	040	14	0	225
	TS	(123°48.636′ E, 29°27.213′ N)	040	5	2.9507	180
Scenario 3	OS	(123°46.897′ E, 29°25.019′ N)	022	12	0	225
	TS	(123°49.964′ E, 29°29.231′ N)	235	11	4.9422	180

Table 2. Settings of multi-ship scenarios.

Case	Ship List	Position	Course (°)	Velocity (kn)	Distance (nm)	Ship Length (m)	Course Altering Angle (°)
Scenario 4	OS	(123°46.521′ E, 29°24.222′ N)	022	11	0	225	+16
	TS1	(123°47.420′ E, 29°29.052′ N)	185	10	4.8613	180	+12
	TS2	(123°49.892′ E, 29°27.988′ N)	250	8	4.7404	180	−15
Scenario 5	OS	(123°48.391′ E, 29°26.212′ N)	000	12	0	225	+23
	TS1	(123°49.532′ E, 29°30.150′ N)	225	10	4.0571	180	0
	TS2	(123°45.525′ E, 29°30.042′ N)	135	12	4.5814	180	+12
	TS3	(123°46.854′ E, 29°26.046′ N)	035	10	1.3675	180	0
	TS4	(123°47.925′ E, 29°31.042′ N)	180	10	4.8558	180	0
	TS5	(123°50.315′ E, 29°29.547′ N)	250	12	3.7158	180	−10

Note: "+" means alter course to starboard side, the symbol; "−" means alter course to port side.

4.2. Simulation Scenario 1

In the current status, the course of the OS and TS are 006.5° and 186°, respectively. At this point, according to COLREGs rule 14 and the collision risk model, the two ships are in a head-on situation and there is a collision risk. Every ship should alter their course to the starboard side to avoid the collision. However, there may be situations where the TS keeps her course and speed, and the avoid collision action is taken by the OS solely. Therefore, this scenario is divided into two conditions (situation 1: the TS keeps course and speed; situation 2: the TS alters course to the starboard side) to simulate and verify the collision avoidance decision model.

Figure 6a,b are the simulation results under these two situations when the TS keeps its course and speed and alters its course to the starboard side to avoid collision (the TS alters to starboard by 6°), respectively. For situation 1 and situation 2, according to the decision support methodology constructed in this article, the OS should alter course to the starboard side by 8° and 6°, respectively. The real-time parameter changes are shown in Figure 7. Figure 7a shows the ship's distance curves between the OS and TS in the two situations. The *DCPA* value of the OS and TS first increases and remains stable until the OS starts to resume to the original route, as shown in Figure 7b. Figure 7c gives the curves of the *TCPA*. Figure 7d shows the course change of the OS during the whole collision avoidance process.

4.3. Simulation Scenario 2

In scenario 2, the initial course of the OS and TS are the same, namely 040°. According to COLREGs rule 13, the OS and TS are in an overtaking situation. The OS is a give-way ship and should take charge of performing conflict avoidance actions. According to the calculation result of the autonomous collision avoidance model, the course of the OS should alter 15° to starboard. The results of the ship collision avoidance decisions and manoeuvres are shown in Figure 8a. Figure 9 shows the real-time parameter changes under the crossing encounter situation.

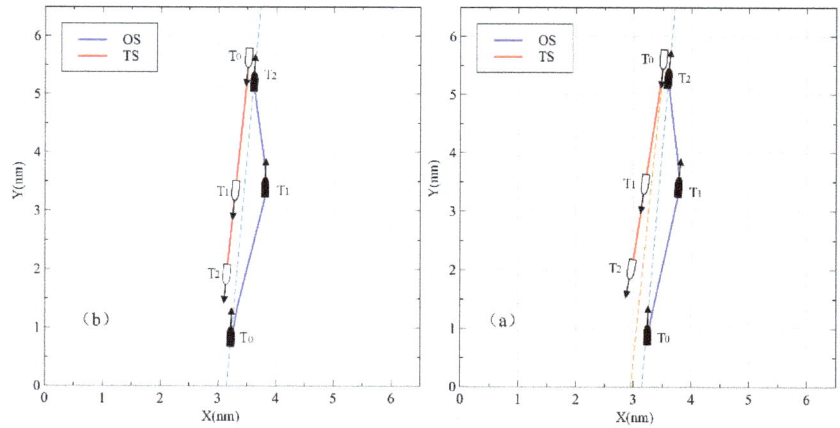

Figure 6. The trajectories of the ships in scenario 1: (**a**) trajectories of the ships in situation 1; (**b**) trajectories of the ships in situation 2.

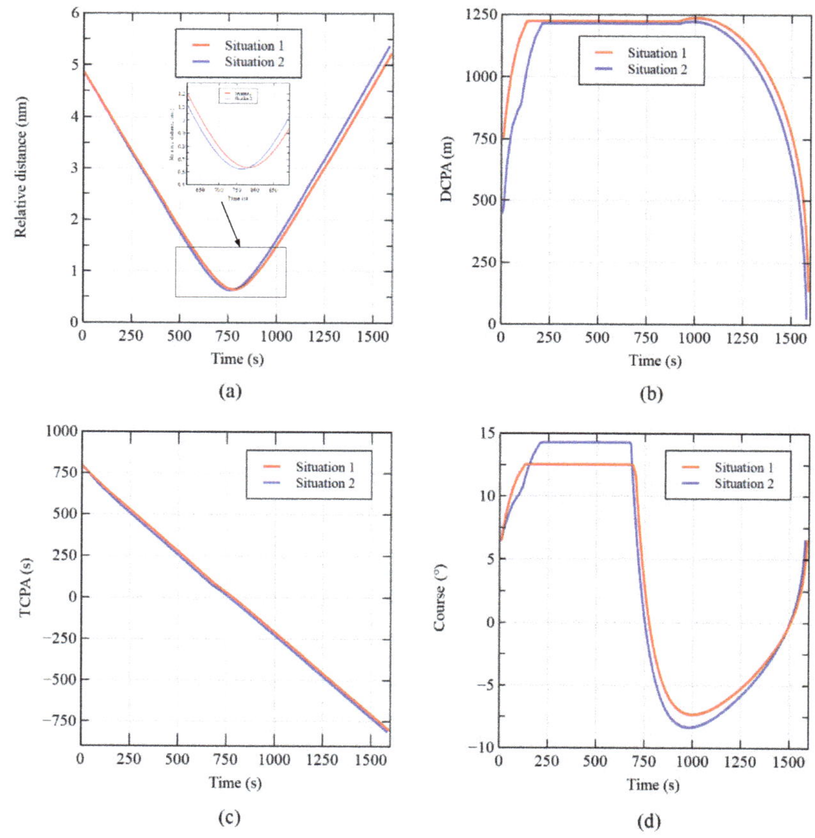

Figure 7. Real-time parameter changes under head-on situation. (**a**) Relative distance between OS and TS; (**b**) *DCPA* between OS and TS; (**c**) *TCPA* between OS and TS; (**d**) the course of OS.

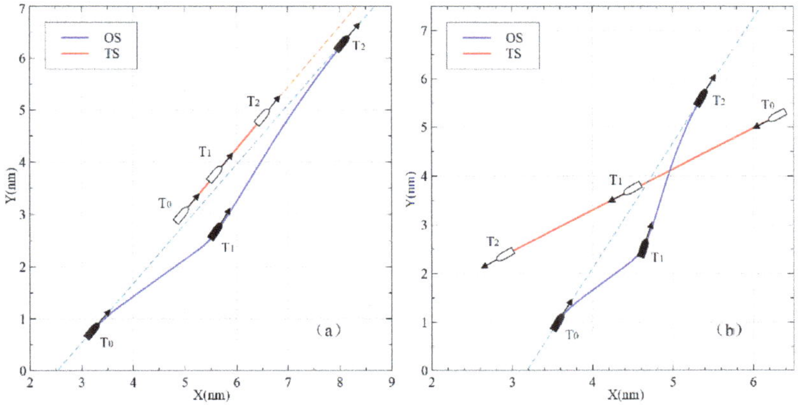

Figure 8. The trajectories of the ships: (**a**) trajectories of the ships in scenario 2; (**b**) trajectories of the ships in scenario 3.

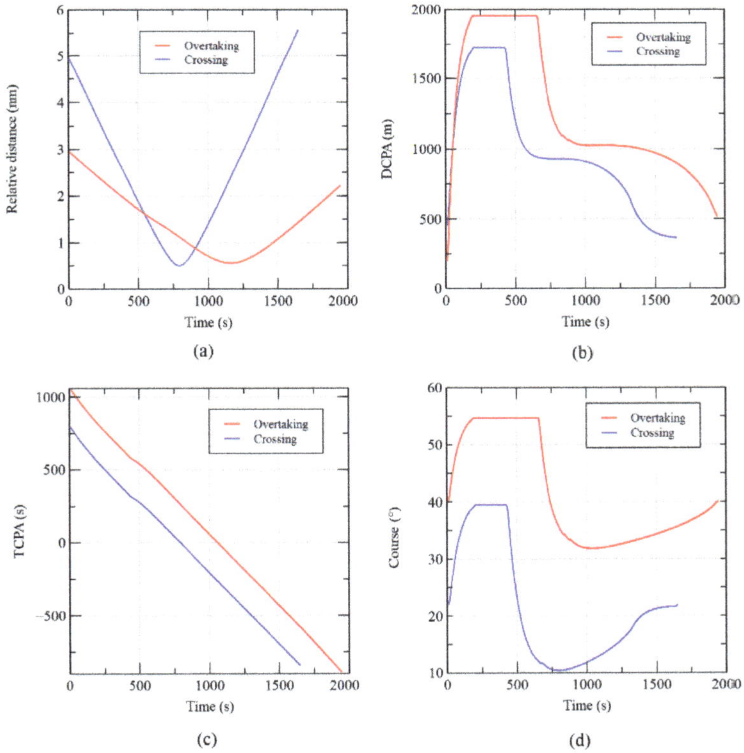

Figure 9. Real-time parameter changes under crossing situation and overtaking situation. (**a**) Relative distance between OS and TS; (**b**) DCPA between OS and TS; (**c**) TCPA between OS and TS; (**d**) the course of OS.

4.4. Simulation Scenario 3

In this simulation scenario, the two ships are in a crossing situation, so the OS should take charge of performing conflict avoidance actions. Figure 8b shows the initial positions and paths of the simulation ships involved in the experiment. According to the autonomous collision avoidance model, the OS makes the decision of altering to starboard by 17°. Due to the influence of ship manoeuvrability, the OS can keep her course stable after course altering is finished, and will keep the new course until 202 s.

Figure 9 shows the real-time parameter changes under the crossing encounter situation and overtaking encounter situation. The relative distance between the ships is shown in Figure 9a. Figure 9b,c show the curve of the *DCPA* and *TCPA* between the OS and TS, respectively. Figure 9d shows the course change of the OS during the whole collision avoidance process.

4.5. Simulation Scenario 4

The scenario considered in this part is a typical three-ship crossing encounter situation. According to the COLREGs rule, the OS, TS1 and TS2 have an obligation to alter their courses to starboard to give way to TS2, TS1 and the OS, respectively. The collision avoidance system recognises that TS1 and TS2 have suddenly altered course to the starboard by 12° and 15°, respectively. According to the autonomous collision avoidance support methodology constructed in this paper, the OS should alter to starboard by 16°.

Figure 10 shows the initial positions of the simulation ships. Figure 11a shows the ship's relative distance curves between TS1, TS2 and the OS, which reaches the lowest point in 839 s and 941 s, respectively, and the minimum distances are 1414 m and 1181 m, respectively. The *DCPA* values of the OS with TS1 and TS2 increase continuously and remain stable until the start of resuming the original route, as shown in Figure 11b. The *TCPA* values between the ships change to negative at 839 s and 942 s in Figure 11c, respectively. Figure 11d shows the course change of the OS during the whole collision avoidance process.

4.6. Simulation Scenario 5

The scenario considered in this part is a typical and more complicated six-ship encounter situation. In this scenario, TS2 and TS3 are located at the far left and form left-crossing encounter situations with the OS. TS4 is located in the left front direction of the OS and forms a head-on situation with the OS. TS1 and TS5 are located in the right front direction of the OS; they all form crossing encounter situations with the OS.

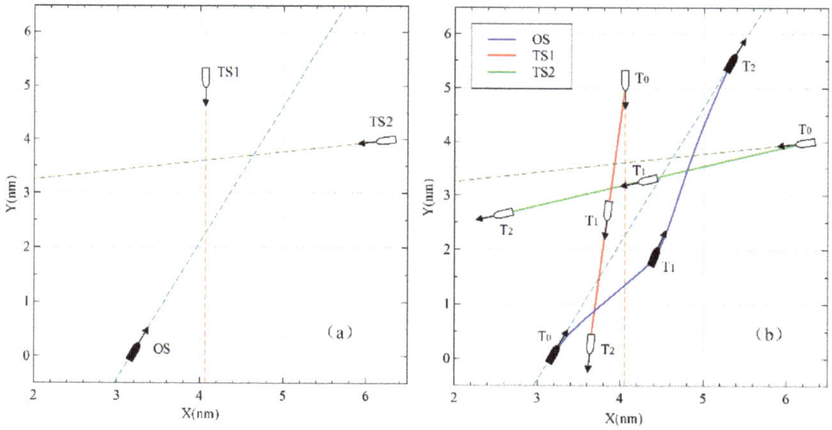

Figure 10. The ships' trajectories under three ships encounter situation: (**a**) initial positions and paths; (**b**) trajectories of three ships.

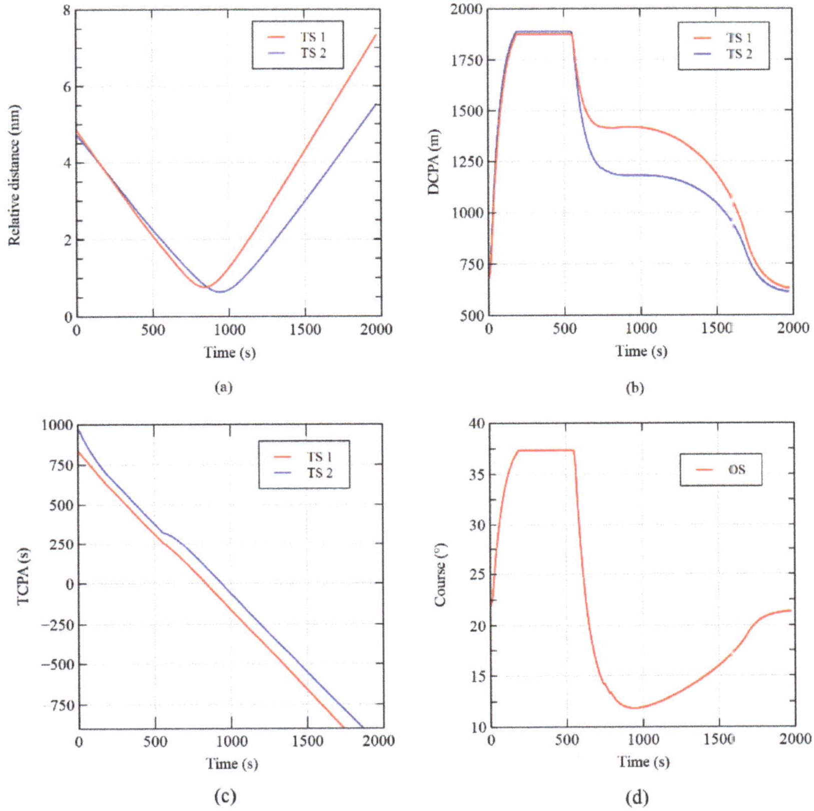

Figure 11. Real-time parameter changes under three-ship encounter situation. (**a**) Relative distance between OS and TSs; (**b**) DCPA between OS and TSs; (**c**) TCPA between OS and TSs; (**d**) the course of OS.

In this scenario, the collision avoidance support methodology recognises that the TS2 is altering 12° to the starboard and TS5 is altering 10° to the port. According to the autonomous collision avoidance system, the OS can only clear all target ships by altering course 23° to starboard. Figure 12 show the initial positions and paths of the experimental ships. The real-time parameter changes under a multi-ship encounter situation are shown in Figure 13. Figure 13a shows the ship's relative distance curve between the OS and the TSs in this scene. It can be seen that the minimum distance between the OS and each TS is larger than the required safe passing distance (grey dotted line). The changes in the *DCPA* curve between the OS and TSs are shown in Figure 13b. The *TCPA* values between the OS and TSs change to negative at 724 s, 757 s, 458 s, 830 s and 622 s in Figure 13c, respectively. Figure 13d shows the course change of the OS during the whole collision avoidance and reversion of course process.

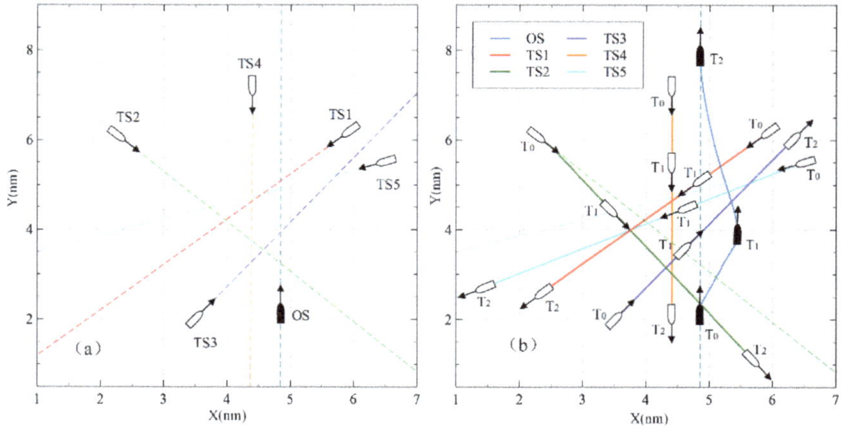

Figure 12. The ships' trajectories under six-ship encounter situation: (**a**) initial positions and paths; (**b**) trajectories of six ships.

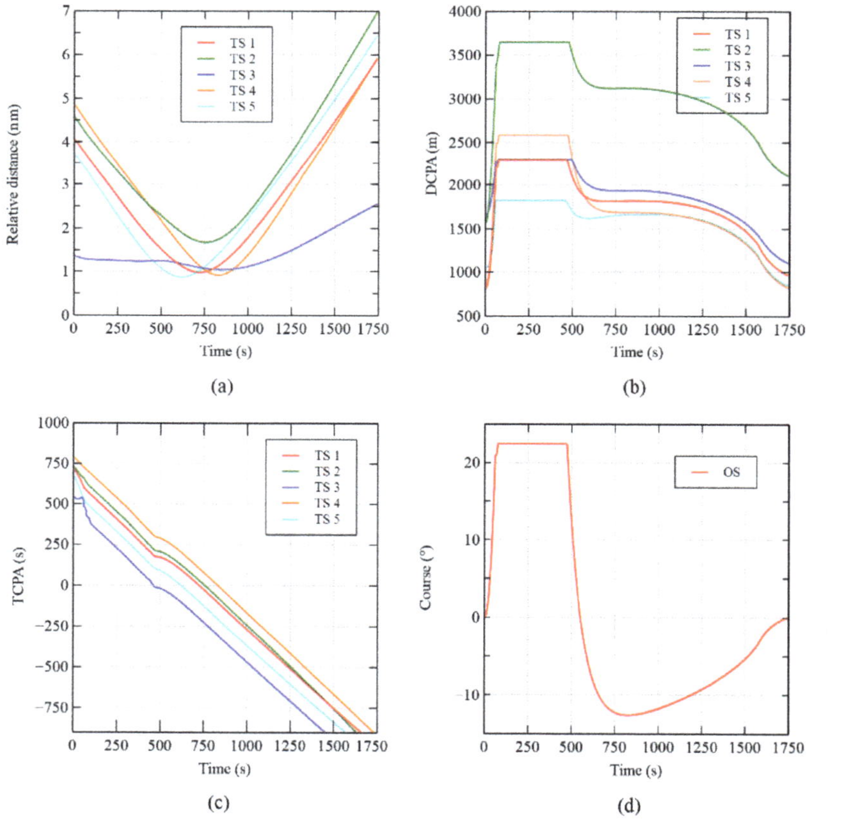

Figure 13. Real-time parameter changes under six-ship encounter situation. (**a**) Relative distance between OS and TSs; (**b**) DCPA between OS and TSs; (**c**) TCPA between OS and TSs; (**d**) the course of OS.

5. Discussions and Analysis

5.1. Results and Discussion

Five different types of ship encounter scenarios are designed to verify the feasibility and effectiveness of the decision support methodology for autonomous collision avoidance in this paper. The results and discussion are as follows:

For Scenario 1, it is divided into two situations (the TS keeps its course and speed; the TS alters its course to the starboard side). By comparing the two situations, it can be seen that in the same scenario, the OS's alteration angle of course is 8° when the TS is keeping speed and course, and when the TS also changes her course for collision avoidance, the OS only alters her course angle by 6°.

Due to the influence of ship manoeuvrability, in these two situations, the OS can only keep her course stable for a short while after changing her course and maintains the new course until 672 s and 685 s, respectively. At 1579 s and 1589 s, the OS completes the resume-sailing process and resumes to the original route in these two scenarios, respectively, with the courses of the OS both 006.5°. As can be seen from Figure 6a, the minimum distances between the TS and OS are almost the same in these two scenarios, which are 1156 m and 1175 m, respectively, and both larger than 899 m. It can be seen that the OS can avoid collision through a smaller redirection angle when the TS also redirects to a certain angle to avoid a collision. Namely, this simulation result is consistent with navigation practice.

For scenarios 2 and 3, the TSs all keep their course and speed; the OS is a give-way vessel. According to the autonomous collision avoidance method, the OS makes a collision avoidance decision of altering to the starboard by 15° and 17°, respectively. By analysing the experimental results, we can see that safe collision avoidance can be achieved well in these two scenarios.

Scenarios 4 and 5 are both multi-ship encounter situations, in which the impacts of the TSs' course changes on the collision avoidance decision during the avoidance process were considered. For example, for scenario 5, *ST*2 alters 12° to the starboard side and TS5 alters 10° to the port side; the other ships keep their speed and course. The collision avoidance method shows that the OS needs to alter to the starboard by 23° to clear all ships. In Figure 13a, it can be seen that the ship's distance curves gradually decrease at first, after reaching the lowest point in 723 s, 756 s, 846 s, 928 s and 619 s, respectively, and then increase gradually. The minimum distances between the OS and TSs are 1810 m, 3123 m, 1932 m, 1689 m and 1623 m, respectively. The minimum relative distance between the OS and all TSs is greater than the safe distance. This means that the collision never occurred. The *TCPA* values between the OS and TSs changes to negative at 724 s, 757 s, 458 s, 830 s and 622 s in Figure 13c, respectively. Figure 13d shows the course change of the OS; the curve can effectively show the course changes of the OS during the whole collision avoidance process. Although the encounter situation in this scenario is more complicated, the decision support methodology can safely avoid all ships through steering to the starboard by 23°, and, finally, realize the resumption to the original route.

In summary, for all scenarios, the minimum relative distances between the OS and TSs during collision avoidance is greater than the safety distance, which are clearly shown in Figures 7a, 9a, 11a and 13a. Figures 7d, 9d, 11d and 13d all show that the OS has good tracking performance for the expected courses, reflecting the manoeuvring characteristic of ship motion control. The simulation results show that the proposed method can be applied to complex scenarios such as two-ship and multi-ship encounters, showing excellent adaptability to and effectiveness in managing complex multi-target situations. In addition, the decision support methodology can take appropriate collision avoidance actions even when the TS changes her course, the collision avoidance manoeuvre of which is effective and reliable.

5.2. Comparison Analysis

In this paper, we propose a decision support methodology of dynamic adaptive autonomous collision avoidance based on ship manoeuvring process deduction for the autonomous ship. This methodology can effectively solve the problem of autonomous collision avoidance under different encounter scenarios. At present, relevant scholars have proposed various methods and techniques for the problem of collision avoidance, such as [4,8,11,18,33,40,41]. However, compared with other research results, this paper has some differences and advantages.

Different from vehicles, ship motion has the characteristics of large inertia, time delay and being non-linear. Therefore, it is very necessary to consider ship manoeuvrability in the decision-making scheme of collision avoidance. Some related studies [11,14,33] ignore ship manoeuvrability, which increases the gap between the collision avoidance algorithms or decision-making systems and practical applications. In addition, most of the researches only focus on the intelligent algorithms of avoidance collision, ignoring the ship manoeuvrability, good seamanship and COLREGs. In this paper, the MMG model and fuzzy adaptive PID method are used to derive the ships' manoeuvre motion process. On this basis, this paper proposes a dynamic adaptive autonomous collision avoidance system based on the second-level update of information, which also takes into account the COLREGs rules, good seamanship and ship manoeuvrability.

In addition, few studies consider the impact of the TS's action uncertainty on the collision avoidance decisions. In this paper, we mainly focus on the change of a ship's course. Most studies assume that the TS is sailing at a constant speed and course, which does not conform to the actual situation of navigation. Even though some algorithms take into account the motion characteristics of the TS, most of them are based on assumptions. In addition, most existing collision avoidance models or systems rarely consider the problem of resuming the original route after collision avoidance. For the safety of ship navigation, the vessel should resume the original route after the collision avoidance action has been successfully completed. In this study, we propose a dynamic adaptive autonomous collision avoidance model based on second-level updating of information, which can solve the problem where other ships do not comply with COLREGs or suddenly take action during the collision avoidance process, and build a resume-sailing model based on a LOS guidance system.

In summary, we present a decision support methodology of dynamic adaptive automatic collision avoidance based on ship manoeuvring process deduction for an autonomous ship. The decision support methodology has intact avoidance manoeuvres, including collision risk detection, collision avoidance manoeuvres and resuming to the original route. It takes full account of various factors, including COLREGs, ship manoeuvrability and good seamanship. Furthermore, this methodology can solve the problem of the autonomous collision avoidance when encountering multiple objects, complex situations and the TS's uncoordinated or temporary actions.

6. Conclusions

In this paper, a decision support methodology of dynamic adaptive autonomous collision avoidance based on ship manoeuvring process deduction for autonomous ships was proposed. The system takes full account of various factors, including COLREGs, ship manoeuvrability and good seamanship. In order to judge the risk of collision between ships in different encounter situations, a new collision risk model is constructed on the basis of the fuzzy set method to synthesize the SCRI and TCRI. The MMG model and fuzzy adaptive PID method are used to derive the ships' manoeuvre motion process. On this basis, the feasible manoeuvring range and optimum steering angle of collision avoidance are calculated according to the deduction of the manoeuvring process and modified VO method. Finally, the dynamic adaptive autonomous collision avoidance model is developed. The feasibility and effectiveness of the decision support methodology proposed in this paper is verified through simulation experiments under five different scenarios.

Although the autonomous collision avoidance system we established is proved to be reasonable, effective and feasible, there are still some deficiencies. For narrow waters or restricted waters, due to the limited manoeuvrability of ships, it is difficult to achieve an effective collision avoidance only by changing course. However, collision avoidance strategies of changing both speed and course are more consistent with navigation practice. Therefore, further work should be focused on collision avoidance strategies that take both course and speed changing into consideration, and carry on simulations and field tests in more complex situations and restricted waters.

Author Contributions: Conceptualization, K.Z. and L.H.; methodology, K.Z.; software, K.Z.; validation, K.Z. and Y.H.; formal analysis, K.Z.; data curation, K.Z.; writing—original draft, K.Z.; writing—review and editing, K.Z., L.H., X.L., J.C., W.H. and X.Z.; visualization, K.Z., Y.H., X.L., W.H. and J.C.; supervision, L.H. and Y.H.; project administration, L.H. and Y.H.; funding acquisition, L.H. All authors have read and agreed to the published version of the manuscript.

Funding: This work is partially supported by the National Key Research and Development Program (Grant number 2019YFB1600603) and the National Natural Science Foundation of China (Grant number 52071249).

Institutional Review Board Statement: Not applicable.

Informed Consent Statement: Not applicable.

Data Availability Statement: Not applicable.

Conflicts of Interest: The authors declare no conflict of interest.

Nomenclature

Symbols	Definitions
V_{OS}	speed of OS
V_{TS}	speed of TS
V_{OT}	relative speed of the OS and TS
R_T	relative distance between the OS and TS
φ_R	relative course
α_T	true relative bearing
v_{x_R}	relative speed of the OS and TS on the X-axes
v_{y_R}	relative speed of the OS and TS on the Y-axes
α_{OT}	relative bearing angle between OS and TS
Δx	relative displacement from TS to OS on the X-axis
u_s	membership function of the fuzzy set SCRI
$(x,y)^t$	position coordinates of the other ship at time t
Δy	relative displacement from TS to OS on the Y-axis
L	ship length
$SDom^t$	sets of location point elements at time t in the OS's domain
u_t	membership function of the fuzzy set TCRI
TCS	time from the current moment to the first time point of a close-quarter situation
D	safe distance between OS and TS
$ConfP(*)$	all the possible positions when a collision happens
$\|\cdot\|$	geographic distance between two vessels
L_i	trajectories considering ship manoeuvrability
$L_i\prime$	trajectories without regard to ship manoeuvrability
C_{range}	feasible manoeuvring range
θ_i	altering angle
θ	the optimization steering angle
ψ_{LOS}	LOS angle
R_{LOS}	radius of the circle

Appendix A

Table A1. HUAYANG DREAM's parameters.

Parameter	Value	Parameter	Value
LOA	225 (m)	Acreage of rudder	56.88 (m^2)
Draft	14.5 (m)	Displacement	90,000 × 10^3 (kg)
C_b	0.8715	Density of sea water	1000 (kg/m^3)
C_p	0.8739	RPM	90 (r/min)
Breadth	32.5 (m)	Propeller pitch	4.738 (m)

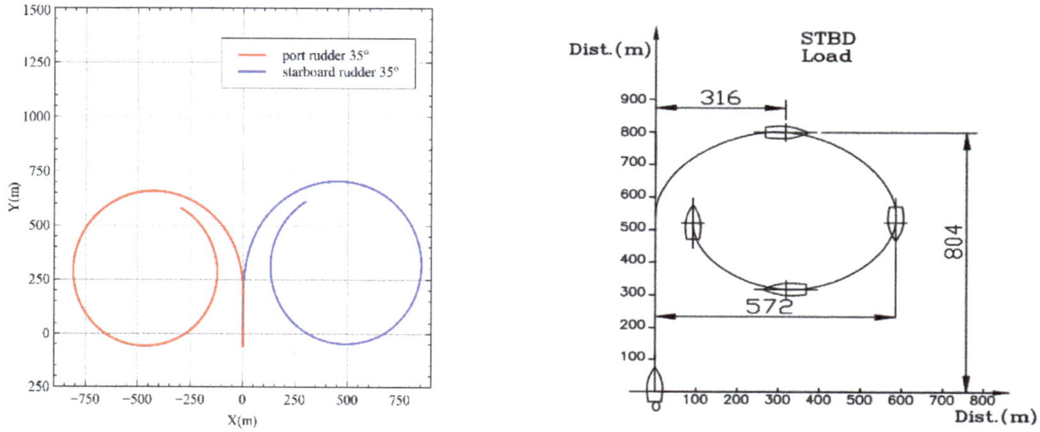

Figure A1. Digital MMG model and real ship model turning cycle.

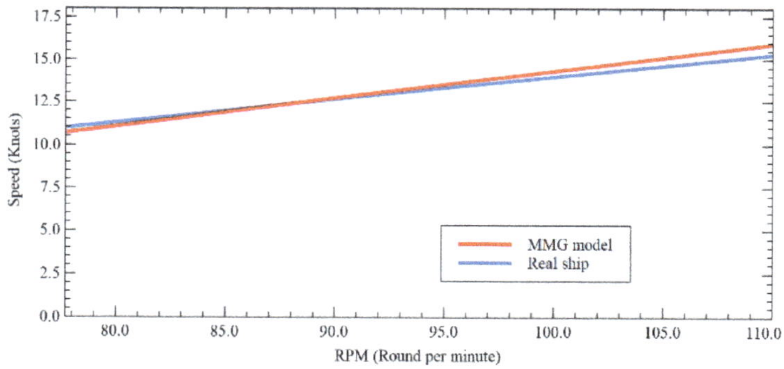

Figure A2. Comparison of speed performance between MMG model and real ship.

References

1. Sawada, R.; Sato, K.; Majima, T. Automatic ship collision avoidance using deep reinforcement learning with LSTM in continuous action spaces. *J. Mar. Sci. Technol.* **2021**, *26*, 509–524. [CrossRef]
2. EMSA. *Preliminary Annual Overview of Marine Casualties and Incidents 2014–2019*; EMSA: Lisbon, Portugal, 2020.
3. Lyu, H.; Yin, Y. Fast path planning for autonomous ships in restricted waters. *Appl. Sci.* **2018**, *8*, 2592. [CrossRef]
4. Lyu, H.; Yin, Y. COLREGS-constrained real-time path planning for autonomous ships using modified artificial potential fields. *J. Navig.* **2019**, *72*, 588–608. [CrossRef]
5. Ahmed, G.; Sheltami, T.; Mahmoud, A.; Yasar, A. IoD swarms collision avoidance via improved particle swarm optimization. *Transp. Res. Part A Policy Pract.* **2020**, *142*, 260–278. [CrossRef]
6. Zaccone, R.; Martelli, M. A collision avoidance algorithm for ship guidance applications. *J. Mar. Eng. Technol.* **2020**, *19*, 62–75. [CrossRef]

7. Yuan, X.; Zhang, D.; Zhang, J.; Zhang, M.; Guedes Soares, C. A novel real-time collision risk awareness method based on velocity obstacle considering uncertainties in ship dynamics. *Ocean Eng.* **2021**, *220*, 108436. [CrossRef]
8. Ni, S.; Liu, Z.; Cai, Y. Ship manoeuvrability-based simulation for ship navigation in collision situations. *J. Mar. Sci. Eng.* **2019**, *7*, 90. [CrossRef]
9. Praczyk, T. Neural anti-collision system for Autonomous Surface Vehicle. *Neurocomputing* **2015**, *149*, 559–572. [CrossRef]
10. Woo, J.; Kim, N. Collision avoidance for an unmanned surface vehicle using deep reinforcement learning. *Ocean Eng.* **2020**, *199*, 107001. [CrossRef]
11. Li, J.; Wang, H.; Guan, Z.; Pan, C. Distributed Multi-Objective Algorithm for Preventing Multi-Ship Collisions at Sea. *J. Navig.* **2020**, *73*, 971–990. [CrossRef]
12. Li, M.; Mou, J.; Chen, L.; Huang, Y.; Chen, P. Comparison between the collision avoidance decision-making in theoretical research and navigation practices. *Ocean Eng.* **2021**, *228*, 108881. [CrossRef]
13. Huang, Y.; Chen, L.; Chen, P.; Negenborn, R.R.; van Gelder, P.H.A.J.M. Ship collision avoidance methods: State-of-the-art. *Saf. Sci.* **2020**, *121*, 451–473. [CrossRef]
14. Li, L.; Wu, D.; Huang, Y.; Yuan, Z.M. A path planning strategy unified with a COLREGS collision avoidance function based on deep reinforcement learning and artificial potential field. *Appl. Ocean Res.* **2021**, *113*, 102759. [CrossRef]
15. Lazarowska, A. A Discrete Artificial Potential Field for Ship Trajectory Planning. *J. Navig.* **2020**, *73*, 233–251. [CrossRef]
16. Zaccone, R. COLREG-compliant optimal path planning for real-time guidance and control of autonomous ships. *J. Mar. Sci. Eng.* **2021**, *9*, 405. [CrossRef]
17. Chiang, H.T.L.; Tapia, L. COLREG-RRT: An RRT-Based COLREGS-Compliant Motion Planner for Surface Vehicle Navigation. *IEEE Robot. Autom. Lett.* **2018**, *3*, 2024–2031. [CrossRef]
18. Huang, Y.; Van Gelder, P.H.A.J.M.; Wen, Y. Velocity obstacle algorithms for collision prevention at sea. *Ocean Eng.* **2018**, *151*, 308–321. [CrossRef]
19. Chen, P.; Huang, Y.; Papadimitriou, E.; Mou, J.; van Gelder, P.H.A.J.M. An improved time discretized non-linear velocity obstacle method for multi-ship encounter detection. *Ocean Eng.* **2020**, *196*, 106718. [CrossRef]
20. Xie, S.; Chu, X.; Zheng, M.; Liu, C. Ship predictive collision avoidance method based on an improved beetle antennae search algorithm. *Ocean Eng.* **2019**, *192*, 106542. [CrossRef]
21. Ahn, J.H.; Rhee, K.P.; You, Y.J. A study on the collision avoidance of a ship using neural networks and fuzzy logic. *Appl. Ocean Res.* **2012**, *37*, 162–173. [CrossRef]
22. Zhao, L.; Roh, M.I.; Lee, S.J. Control method for path following and collision avoidance of autonomous ship based on deep reinforcement learning. *J. Mar. Sci. Technol.* **2019**, *27*, 293–310. [CrossRef]
23. Zhao, L.; Roh, M. COLREGs-compliant multiship collision avoidance based on deep reinforcement learning. *Ocean Eng.* **2019**, *191*, 106436. [CrossRef]
24. Shen, H.; Hashimoto, H.; Terada, D.; Guo, C. Automatic collision avoidance of multiple ships based on deep Q-learning. *Appl. Ocean Res.* **2019**, *86*, 268–288. [CrossRef]
25. Wang, S.; Zhang, Y.; Li, L. A collision avoidance decision-making system for autonomous ship based on modified velocity obstacle method. *Ocean Eng.* **2020**, *215*, 107910. [CrossRef]
26. Zhang, X.; Wang, C.; Jiang, L.; An, L.; Yang, R. Collision-avoidance navigation systems for Maritime Autonomous Surface Ships: A state of the art survey. *Ocean Eng.* **2021**, *235*, 109380. [CrossRef]
27. Hu, Y.; Meng, X.; Zhang, Q.; Park, G.K. A Real-Time Collision Avoidance System for Autonomous Surface Vessel Using Fuzzy Logic. *IEEE Access* **2020**, *8*, 108835–108846. [CrossRef]
28. Zhang, J.; Zhang, D.; Yan, X.; Haugen, S.; Guedes Soares, C. A distributed anti-collision decision support formulation in multi-ship encounter situations under COLREGs. *Ocean Eng.* **2015**, *105*, 336–348. [CrossRef]
29. Mizythras, P.; Pollalis, C.; Boulougouris, E.; Theotokatos, G. A novel decision support methodology for oceangoing vessel collision avoidance. *Ocean Eng.* **2021**, *230*, 109004. [CrossRef]
30. Pietrzykowski, Z.; Wołejsza, P.; Borkowski, P. Decision Support in Collision Situations at Sea. *J. Navig.* **2017**, *70*, 447–464. [CrossRef]
31. Li, M.; Mou, J.; He, Y.; Zhang, X.; Xie, Q.; Chen, P. Dynamic trajectory planning for unmanned ship under multi-object environment. *J. Mar. Sci. Technol.* **2021**, *27*, 173–185. [CrossRef]
32. Wang, X.; Liu, Z.; Cai, Y. The ship maneuverability based collision avoidance dynamic support system in close-quarters situation. *Ocean Eng.* **2017**, *146*, 486–497. [CrossRef]
33. Fiskin, R.; Atik, O.; Kisi, H.; Nasibov, E.; Johansen, T.A. Fuzzy domain and meta-heuristic algorithm-based collision avoidance control for ships: Experimental validation in virtual and real environment. *Ocean Eng.* **2021**, *220*, 108502. [CrossRef]
34. Li, J.; Wang, H.; Zhao, W.; Xue, Y. Ship's Trajectory Planning Based on Improved Multiobjective Algorithm for Collision Avoidance. *J. Adv. Transp.* **2019**, *2019*, 4068783. [CrossRef]
35. He, Y.; Jin, Y.; Huang, L.; Xiong, Y.; Chen, P.; Mou, J. Quantitative analysis of COLREG rules and seamanship for autonomous collision avoidance at open sea. *Ocean Eng.* **2017**, *140*, 281–291. [CrossRef]
36. Fiorini, P.; Shiller, Z. Motion planning in dynamic environments using velocity obstacles. *Int. J. Rob. Res.* **1998**, *17*, 760–772. [CrossRef]

37. Battisti, T.; Muradore, R. A velocity obstacles approach for autonomous landing and teleoperated robots. *Auton. Robots* **2020**, *44*, 217–232. [CrossRef]
38. Yasukawa, H.; Yoshimura, Y. Introduction of MMG standard method for ship maneuvering predictions. *J. Mar. Sci. Technol.* **2015**, *20*, 37–52. [CrossRef]
39. He, Y.; Li, Z.; Mou, J.; Hu, W.; Li, L.; Wang, B. Collision-avoidance path planning for multi-ship encounters considering ship manoeuvrability and COLREGs. *Transp. Saf. Environ.* **2021**, *3*, 103–113. [CrossRef]
40. Xu, X.; Lu, Y.; Liu, X.; Zhang, W. Intelligent collision avoidance algorithms for USVs via deep reinforcement learning under COLREGs. *Ocean Eng.* **2020**, *217*, 107704. [CrossRef]
41. Huang, Y.; Chen, L.; Negenborn, R.R.; Van Gelder, P.H.A.J.M. A ship collision avoidance system for human-machine cooperation during collision avoidance. *Ocean Eng.* **2020**, *217*, 107913. [CrossRef]

Journal of
*Marine Science
and Engineering*

MDPI

Article

Motion Planning for an Unmanned Surface Vehicle with Wind and Current Effects

Shangding Gu [1,2], Chunhui Zhou [1,*], Yuanqiao Wen [3,4], Changshi Xiao [1,3,4] and Alois Knoll [2]

1 School of Navigation, Wuhan University of Technology, Wuhan 430062, China; shangding.gu@tum.de (S.G.); cxiao@whut.edu.cn (C.X.)
2 Department of Informatics, Technical University of Munich, 80333 Munich, Germany; knoll@mytum.de
3 National Engineering Research Center for Water Transport Safety, Wuhan University of Technology, Wuhan 430070, China; yqwen@whut.edu.cn
4 Intelligent Transportation Systems Research Center, Wuhan University of Technology, Wuhan 430062, China
* Correspondence: chunhui@whut.edu.cn

Abstract: Aiming at the problem that unmanned surface vehicle (USV) motion planning is disturbed by effects of wind and current, a USV motion planning method based on regularization-trajectory cells is proposed. First, a USV motion mathematical model is established while considering the influence of wind and current, and the motion trajectory is analyzed. Second, a regularization-trajectory cell library under the influence of wind and current is constructed, and the influence of wind and current on the weight of the search cost is analyzed. Finally, derived from the regularization-trajectory cell and the search algorithm, a motion planning method for a USV that considers wind and current effects is provided. The experimental results indicate that the motion planning is closer to the actual trajectory of a USV in complex environments and that our method is highly practicable.

Keywords: motion planning; unmanned surface vehicle (USV); effects of wind and current; regularization-trajectory cell

Citation: Gu, S.; Zhou, C.; Wen, Y.; Xiao, C.; Knoll, A. Motion Planning for an Unmanned Surface Vehicle with Wind and Current Effects. *J. Mar. Sci. Eng.* **2022**, *10*, 420. https://doi.org/10.3390/jmse10030420

Academic Editor: Dracos Vassalos

Received: 23 February 2022
Accepted: 10 March 2022
Published: 14 March 2022

1. Introduction

The intelligence level of an unmanned surface vehicle (USV) has been improved a lot in recent years; however, it is still challenging to achieve a precise motion planning for USVs in complex environments [1–3]. Due to the particularity of USV navigation environments, it is inevitable that USVs will be affected by the wind, wave, current and other environmental factors in the navigation process. Therefore, it is necessary to consider the impact of environmental factors, where the efficiency and safety of task implementation by the USV need to be taken into account. As far as we know, no previous research has investigated the effects of wind and current on USV motion planning from the perspectives of USV dynamics and cyber-physic systems. In order to fill this gap, we particularly need to figure out how to avoid the adverse influence of wind and current on motion planning for a USV in complex environments, which is one of the key problems that should be solved in the process of USV intelligentization [4].

Although many researchers focus on USV motion and path planning in complex environments [5–13], there are still some problems that need to be solved with respect to the influence of actual wind and current dynamics on USV motion planning. For example, from the perspectives of USV navigation characteristics and cyber-physic systems: How do the wind and current disturb the distance and direction of the USV in the motion planning process? How do the wind and current disturb USV steering in the motion planning process? It is necessary to take these two problems into account during USV navigation. In particular, it is more critical to consider the influence of wind and current on the dynamics of the USV and how the dynamics of the USV changes in the process of motion planning.

The remainder of this paper is organized as follows: an analysis of related works is introduced in Section 2; a major issue and methodology are presented in Section 3; a USV mathematical motion model in terms of wind and current effects is provided in Section 4; the construction of time-varying trajectory cells is presented in Section 5; on the basis of time-varying trajectory cells, regularization-trajectory cells that take into account wind and current effects are also provided in Section 5; a USV motion planning method with respect to wind and current effects is introduced in Section 6; simulation experiments and analyses are introduced in detail in Section 7; the conclusion and outlook of the paper are given in Section 8.

2. Related Works

Path planning, with the effects of wind and current taken into consideration, needs to be optimized from different perspectives such as the distance of the path, the safe path, and the smooth path. Much research has been performed to help solve the USV motion planning problem with the effects of wind and current.

Singh et al. [5] proposed a method on the basis of the A* algorithm [14] to solve the USV path planning problem under the influence of the current. This method mainly projects the planning map to a binary electronic map, in which they set the safety distance between the USV and the obstacle as a certain pixel according to the constraint pixels of the electronic map; it then solves the problem of path planning under the influence of the current. To a certain extent, it provides useful inspirations for solving the path planning problem of the current affecting the USV since they considered the path distance and safety of the planned route.

Ma et al. [6] provided a multi-objective optimization method for the influence of fixed current field and time-varying current field USV path planning. In this method, the multi-objective particle swarm optimization algorithm [15] is constructed. The safety, economy, distance, and smoothness of the path are taken as the objective variables, and then the current function [16,17] is constructed as the environmental variables in order to realize path planning for a USV in the time-varying current field environment. They considered the safety, economy, distance, and smoothness of the planning path.

Thakur et al. [7] proposed the state transition model of GPU to simulate a USV's trajectory planning and then realized this trajectory planning in complex sea conditions. Li et al. [18] researched the dynamics and kinematics model of a USV, simplified it according to the characteristics of a USV trajectory, proposed a three degrees-of-freedom motion model for USV navigation, and then further verified the effectiveness of the designed USV trajectory through simulation. However, Thakur and Li et al. [7,18] ignored the specific dynamic constraints of the USV.

Song et al. [8] analyzed a fast marching algorithm [19,20] to solve the route planning problem of a USV facing a fixed current field, and then constructed a double-layer fast marching algorithm to realize a safe and economic route planning for a USV under the influence of a fixed current field. Song et al. [9] improved the fast marching algorithm based on reference [8] and designed the multi-layer fast marching algorithm in order to consider the time-varying current field route planning of USVs. References [8,9] also considered the safety and distance of the planning path. Oren et al. [10] proposed a velocity obstacle method for path planning in response to wave disturbances while the USV is sailing in complex sea conditions. In their algorithm, the waves disturbing the USV are regarded as moving obstacles, and the velocity obstacle method based on probability prediction [21] is constructed to realize USV path planning. This method provides some reference ideas for solving the navigation problems of a USV in complex sea conditions and has certain practical significance. More specifically, with regard to planning behavior, they considered some constraints such as USV size, and they also considered the safety and distance of the planned path. Based on the Voronoi-Visibility roadmap method and the ant genetic algorithm, Niu et al. [11] proposed the optimal energy path planning method for USVs in the environment of the time-varying current field. They established a Voronoi visibility roadmap and combined it with a genetic algorithm to achieve the optimal energy path in a time-varying environment. Experiments indicate that their method's performance is better than other methods.

Subramani et al. [12] developed a path planning method for strong current environments; they mainly considered the influence of a time-varying environment by establishing partial differential equations and then constructed a navigable area in a time-varying environment by using a probability model. The method mainly considers time-varying optimal path planning under the influence of a strong current environment. However, their method cannot guarantee path planning safety since the probability of navigation points cannot be fully computed in a real-time system. The work most related to our study is that of [13]; in our method, regularization-trajectory cells are developed to take into account the effects of wind and current, and we focus more on trajectory analysis with respect to wind and current effects, which can help in fully considering the USV characteristics.

3. A Major Issue and Methodology

In this section, we will analyse the influence of wind and current on USV motion planning with an experiment, and provide our methodology for USV motion planning considering wind and current effects.

Du et al. [22] developed a trajectory-unit method for USV motion planning that considers the dynamic constraints of the USV. This method works well when USVs are not disturbed by wind and current. However, as shown in Figure 1, the method can not provide good performance in complex environments, it does not consider the impact of wind and current on the dynamics of the USV. If the wind and current disturbances are considered in their method, the planned path will not be spliced. The reasons for this problem are: (1) its search algorithm does not consider environmental factors; (2) its trajectory unit does not consider environmental factors. When the environment changes, the trajectory unit becomes irregular and cannot adapt to complex environments; hence, it cannot effectively make the USV navigate in complex environments.

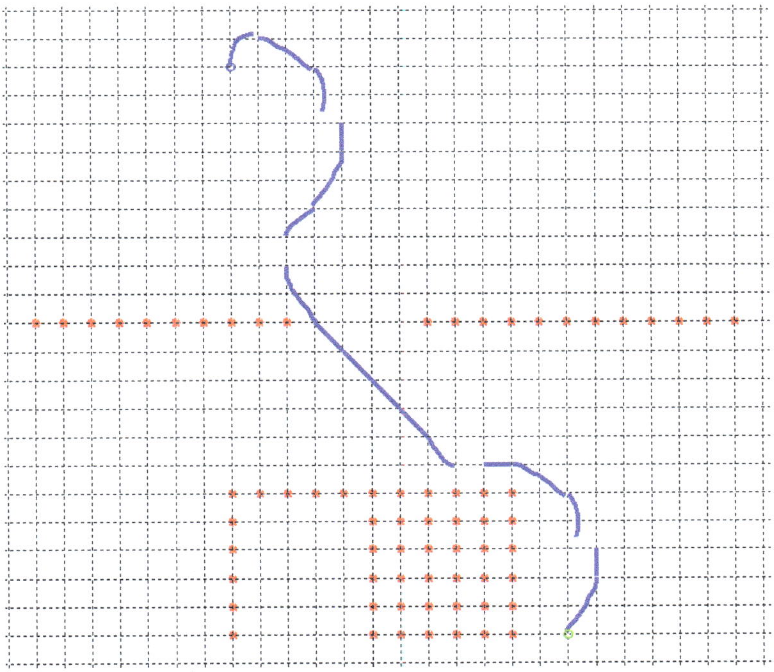

Figure 1. Motion planning algorithm without considering wind and current disturbances.

In order to solve this issue, a USV motion planning method based on the regularization-trajectory cell is proposed (as shown in Figure 2). More specifically, first, a model is built under the influence of wind and current while considering the dynamics of the USV. Second, a regularization-trajectory cell based on the dynamics of the USV is constructed, taking into account the actual dynamics of the USV in the process of determining how to realize path planning. Third, after computing the impact of wind and current on the dynamics of the USV, an effective algorithm for USV motion planning is proposed to achieve an efficient, safe, and economical path search. In the next section, we will introduce a USV mathematical motion model that will play an important role in USV motion planning.

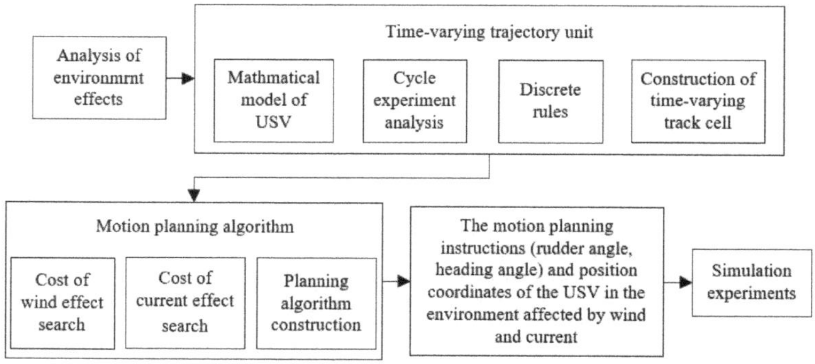

Figure 2. Structure of the motion planning method for USVs based on the regularization-trajectory cell.

4. A USV Mathematical Motion Model

In this section, we establish the mathematical model of a USV that is exposed to wind and current disturbances, and the motion characteristics of the USV under the influence of environmental factors such as wind and current are further analyzed.

In the process of establishing the mathematical model of the USV, it is assumed that mainly the forward (x), traverse (y), and bow (n) of the USV are considered.

Therefore, the mathematical model of the USV here is given by Equation (1):

$$\begin{vmatrix} X \\ Y \\ N \end{vmatrix} = \begin{vmatrix} X_I + X_H + X_P + X_R + X_{wind} \\ Y_I + Y_H + Y_P + Y_R + Y_{wind} \\ N_I + N_H + N_P + N_R + N_{wind} \end{vmatrix} \tag{1}$$

In Equation (1), I, H, P, R, and $wind$ respectively represent the forces (moments) generated by inertia, viscosity, propeller, rudder, and wind. A detailed introduction to the inertial model, viscous model, propeller model, and rudder model of a USV can be found in references [22–25].

4.1. Wind Disturbances

When the wind on the sea surface changes randomly, wind force disturbances are also random. Here, the wind on the sea surface is assumed to be uniform, W_A denotes the absolute wind speed, W_{AD} denotes the direction of the absolute wind speed, W_R denotes the relative wind speed, W_{RD} denotes the direction of the relative wind speed, and V_V denotes the USV speed. The absolute wind speed and direction are calculated in a geodetic coordinate system, and the relative wind speed and direction are calculated in a USV coordinate system [26,27].

The relationship between the absolute wind speed, relative wind speed, and USV speed is as follows:

$$W_R = W_A - V_V \tag{2}$$

We map this relationship to the USV-following coordinate system:

$$\begin{aligned} u_R &= -u - W_A \cos(W_{AD} - \phi) \\ v_R &= -v - W_A \sin(W_{AD} - \phi) \end{aligned} \tag{3}$$

where u_R and v_R are the components on the X and Y axes of the USV-following coordinate system. In the relative coordinate system, when the wind is blowing from the port side of the USV, the relative wind speed W_R and direction W_{RD} are positive, and the calculated relative wind speed W_R and direction W_{RD} are as follows:

$$\begin{aligned} W_{RD} &= \arctan\left(-\tfrac{v_R}{u_R}\right) + \mathrm{sgn}(\pi, v_R) \quad & u_R > 0 \\ W_{RD} &= \arctan\left(-\tfrac{v_R}{u_R}\right) & u_R < 0 \end{aligned} \tag{4}$$

The relative wind speed value obtained from this formula is:

$$W_R^2 = W_A^2 + V_V^2 + 2W_A V_V \cos(W_{AD} - \beta) \tag{5}$$

where β is the drift angle.

4.1.1. Wind Pressure

The wind pressure and wind ballast calculated from the formula above are as follows. The calculated forces and moments acting on the hull wind can be expressed as:

$$\begin{aligned} X_{wind} &= 0.5\rho_a A_f {W_R}^2 C_{wx}(W_{RD}) \\ Y_{wind} &= 0.5\rho_a A_s W_R^2 C_{wy}(W_{RD}) \\ N_{wind} &= 0.5\rho_a A_s L_{OA} W_R^2 C_{wn}(W_{RD}) \end{aligned} \tag{6}$$

In the formula above, ρ_a denotes air density, L_{OA} denotes the length of the USV, A_f denotes the forward projection area of the USV's water part, A_s denotes the side projection area of the USV's water part, $C_{wx}(W_{RD})$ denotes the wind pressure coefficient in the x-axis direction, $C_{wn}(W_{RD})$ is the wind pressure moment coefficient in the z-axis direction, and $C_{wy}(W_{RD})$ denotes the wind pressure coefficient in the y-axis direction.

The forward projection area and the side projection area need to be calculated in detail according to the general layout of the USV; in the absence of a detailed general layout of the USV, it can be roughly calculated according to reference [28].

In addition, we calculate the wind pressure F_{wind} (the coupling of different wind directions is considered in the calculation of wind pressure):

$$F_{wind} = 0.5\rho_a W_R^2 \left(A_s \sin(W_{RD})^2 + A_f \cos(W_{RD})^2\right) C_{wF}(W_{RD}) \tag{7}$$

C_{wF} in formula (7) is the coefficient of the wind pressure resultant force. According to the wind pressure resultant force, the moment in the x-axis, y-axis, and z-axis is decomposed. By combining these formulas, the relationship of the wind pressure coefficients in each direction can be obtained:

$$\begin{aligned} C_{wx}(W_{RD}) &= C_{wF} \cos W_{RF} \left(A_s \sin(W_{RD})^2 + A_f \cos(W_{RD})^2\right) / A_f \\ C_{wy}(W_{RD}) &= C_{wF} \sin W_{RF} \left(A_s \sin(W_{RD})^2 + A_f \cos(W_{RD})^2\right) / A_s \\ C_{wn}(W_{RD}) &= (0.5 - x_F) C_{wF} \sin W_{RF} \left(A_s \sin(W_{RD})^2 + A_f \cos(W_{RD})^2\right) / A_s \end{aligned} \tag{8}$$

where x_F is the position point of the pressure resultant force by wind, and W_{RF} is the angle of the wind pressure resultant force.

4.1.2. Calculation of Wind Pressure and Moment

Generally, these correlation coefficients are obtained from a wind tunnel test. However, due to the limited experimental conditions and the impossibility of wind tunnel testing for every USV, this paper calculates the correlation coefficients according to the approximate calculation formulas given by A. Iwai and H. Kugumiya, and it is estimated for a general cargo ship [28].

$$C_{wF} = 1.325 - 0.05\cos(2W_{RD}) - 0.35\cos(4W_{RD}) - 0.175\cos(6W_{RD}) \tag{9}$$

$$W_{RF} = \left(1 - 0.15\left(1 - \frac{W_{RD}}{90}\right) - 0.80\left(1 - \frac{W_{RD}}{90}\right)^3\right)90 \tag{10}$$

$$x_F = (0.291 + 0.0023W_{RD})L_{PP} \tag{11}$$

where L_{PP} is the length between the perpendiculars of the ship; the wind force disturbances can be obtained according to the correlation coefficients calculated by the formulas above.

4.2. Current Disturbances

In the actual navigation process, it is assumed that the effect of the current on the ship can make the ship's speed and direction deviate [6]. Therefore, the effect of the current will change the x-axis and y-axis speed of the USV. To consider current disturbances, velocity is directly superimposed onto the ship's velocity. As shown in Formula (12), where u denotes transverse velocity after consideration of the current disturbances, v denotes the longitudinal velocity after consideration of the current disturbances, u_{USV} denotes the transverse velocity of the USV, u_C denotes the transverse velocity of the current, v_{USV} denotes the longitudinal velocity of the USV, and v_C denotes the longitudinal velocity of the current.

$$\begin{cases} u = u_{\text{USV}} + u_C \\ v = v_{\text{USV}} + v_C \end{cases} \tag{12}$$

5. Construction of Time-Varying Trajectory Cells

5.1. Analysis of the USV Turning Experiment

Disturbances by wind and current, dynamic constraints, state constraints, and other characteristics of environmental disturbances need to be considered from the perspectives of USV navigation characteristics and cyber-physic systems. Thus, this paper establishes a USV dynamics model under of wind and current disturbances, and based on a force analysis of the USV, we further analyze how the wind and current affect the navigation state of a USV in detail. The simulation experiments are carried out according to our trajectory analysis. The trajectory of the USV motion state without environmental disturbances, the trajectory of the USV motion state affected by the current, the trajectory of the USV motion state affected by the wind, and the trajectory of the USV motion state affected by wind and current at the same time are shown as follows.

As shown in Figure 3a, there are no environmental disturbances in which the angle of the turning rudder is 12.5°. It is obvious that when there are no environmental disturbances, the results of the USV cycle meet actual needs. As shown in Figure 3b, there are current disturbances of 1 m/s in the x-axis direction and 1 m/s in the y-axis direction, and in which the angle of the turning rudder is 12.5°; at this time, the turning path is shifted, and with the current disturbances, the turning trajectory becomes irregular. Figure 3c is the result of the turning experiment of a USV, with a wind speed of 2 m/s and a wind direction of true north, and in which the angle of the turning rudder is 12.5°; in this experiment, the turning experiment trajectory of the USV becomes irregular due to the disturbances of the wind, resulting in an irregular cycle. As shown in Figure 3d, the current velocity is 1 m/s in the x-axis direction and 1 m/s in the y-axis direction. The turning experiment is conducted with disturbances of wind speed of 2 m/s in the true north direction, in which the angle of turning rudder is 12.5°; in this experiment, the turning trajectory of the USV is irregular because of the superposition effect of the current and wind, and the force direction and force of the USV are constantly changing, which makes it challenging to predict the trajectory.

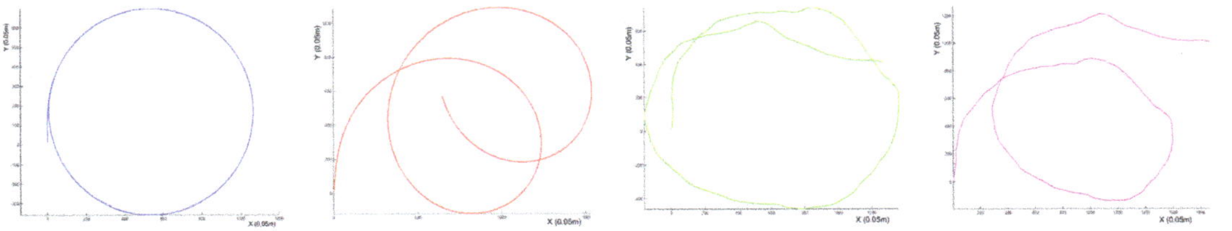

(a) (b) (c) (d)

Figure 3. USV turning experiments. (**a**) Turning experiment of a USV without environmental disturbances. (**b**) Turning experiment of a USV with current disturbances. (**c**) Turning experiment of a USV with wind disturbances. (**d**) Turning experiment of a USV with current and wind disturbances.

5.2. Rules of Time-Varying Trajectories

From these experimental results, when the USV is affected by the wind and current at the same time, the trajectories of the USV show irregular changes. However, the trajectory cell constructed in references [22,23] is a regular trajectory, which is not suitable for the motion planning of a USV with environmental disturbances. Thus, we established a regularization-trajectory cell, which lays the foundation for USV motion planning that takes into consideration the effects of both wind and current.

In this study, the influence of environmental factors is considered, and the rudder angle of the regularization-trajectory cell in this paper needs to be changed at any time through the search algorithm in order to maintain safe USV navigation. Therefore, the rudder angle of the regularization-trajectory cell in this paper is mutable, and it needs to be adjusted according to the environment change. Similar to references [22,23], to facilitate the consideration of the forces and moments on the USV, the following rules need to be considered in the construction of regularization-trajectory cells:

Rule 1: The trajectory cells are divided into 36 categories, and the trajectory distances of each category are equal within a certain error range. Based on this, a regularization-trajectory cell library is constructed.

Because environmental factors such as wind, wave, and current are changing at every moment, the 36 categories of the regularization-trajectory cells are built according to the search direction of the algorithm. There is a large number of regularization-trajectory cells in each category, that is, each category of regularization-trajectory cells can form a sub-regularization-trajectory cell library, which can provide sufficient reachable areas for USV motion planning.

Rule 2: In order to maintain the continuity of the search path, the motion state of the regularization-trajectory cell at the beginning and at the end is kept stable.

Rule 3: In the case of wind and current disturbances in a certain trajectory cell at a certain time, it is necessary to turn the rudder only once in order to continue with navigation, excluding a rudder's return (steering to counteract the disturbances of the wind and current).

To sum up, the three rules specified in this section will lay the foundation for subsequent motion planning, which can better help to realize motion planning while considering environmental influences and USV navigation characteristics.

5.3. Regularization-Trajectory Cells

On the basis of the establishment of trajectory rules, in the grid environment, trajectories need to be constantly adjusted with the rudder angle to make the trajectory meet different navigation requirements based on the reachable points, the trajectory cell heading, and the final state of the trajectory cells. In addition, after generating the regularization-trajectory cells, for the convenience of calculation, the regularization-trajectory cells of the USV are divided into 36 categories, as mentioned above.

5.3.1. Regularization-Trajectory Cell Construction Method

Based on the dynamics model of the USV under the influence of wind and current, the regularization-trajectory cell is constructed on the premise of the regular constraints of the regularization-trajectory cell. First of all, we explore the same rudder angle in different directions, as shown in Figure 4, and search for the reachable points at 0.5° intervals (i.e., change the trajectory cell at 0.5° intervals, and the rudder angle at this time is 15°). It can be seen that the search area of the current point can be covered with full probability by exploring different heading intervals for the same rudder angle; thus, the full probability search can be realized.

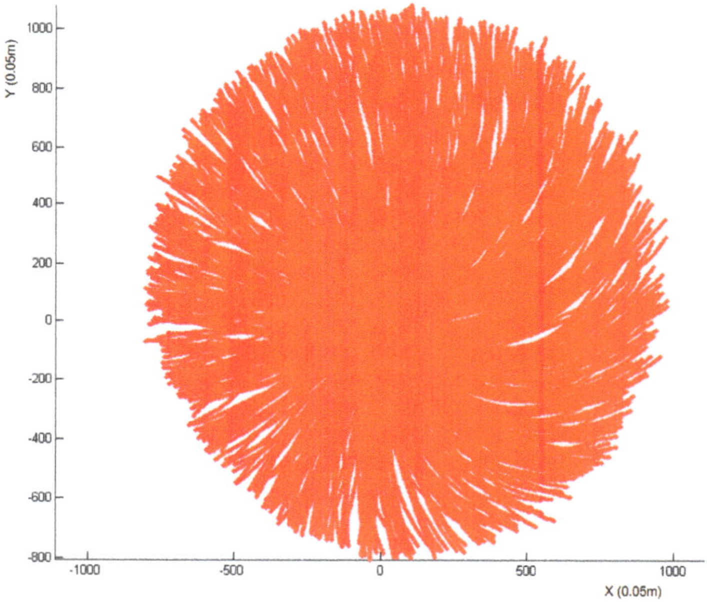

Figure 4. Reachable points of the rudder angle at 15°; the heading is explored at 0.5° intervals.

According to the above-mentioned total probability exploration of the surrounding nodes, when the fixed trajectory cell [29,30] cannot be spliced, the established regularization-trajectory cell can be used (the trajectory cell changes according to the change of environments).

Figure 5 shows the chart for building a regularization-trajectory cell, which comprises of the USV geometry shape and physical characteristics. First, determine the current environmental information, that is, wind and current will have important disturbances on the trajectory cell; second, by building the uncertain trajectory cell library under the influence of wind and current, judge the characteristics and construct the corresponding trajectory cell; finally, resist the environmental disturbances by steering the rudder, that is to say, regularize the trajectory by changing the rudder angle in order to lay a foundation for the trajectory cell to achieve splice.

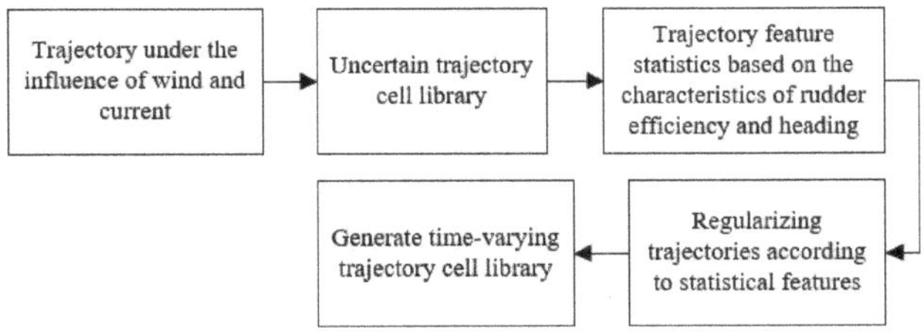

Figure 5. Construction process of a regularization-trajectory cell.

5.3.2. Building a Regularization-Trajectory Cell Library

In this section, based on real-world environments, assume that the wind speed is 2 m/s and the wind direction is 0° (Figure 6 shows a schematic diagram of ship motion under the influence of wind and current). Assume that the current speed in the x-axis direction is 1 m/s and the current speed in the y-axis direction is 1 m/s (the calculation of wind speed and current velocity here is based on the absolute wind speed and current speed, that is, the calculation is carried out in the fixed coordinate system).

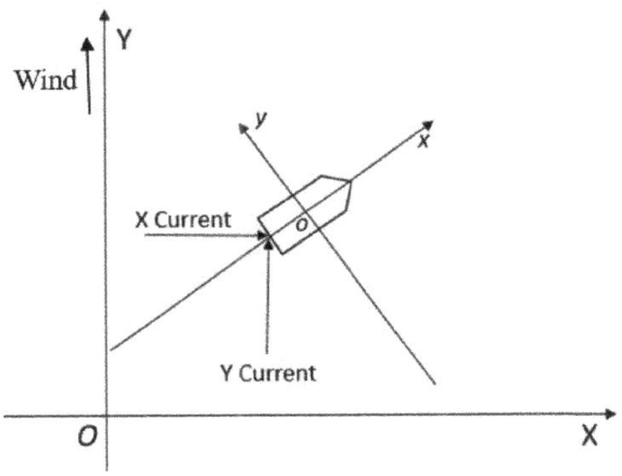

Figure 6. Schematic diagram of ship motion under the influence of wind and current.

Next, the regularization-trajectory cell is constructed on the basis of the above-mentioned method. Figures A1–A3 denote the regularization-trajectory cell library of the above-mentioned environment (Figures A1–A3 can be seen from Appendix A).

6. Motion Planning Method

The trajectory rules and regularization-trajectory cells lay a solid foundation for the following planning method. In this section, based on the regularization-trajectory cell, motion planning for a USV under the influence of wind and current is further realized, which is derived from the A* algorithm.

First, we analyze the effects of wind and current, and then we generate the regularization-trajectory cell according to the influencing factors of wind and current. Second, we splice the trajectory cell under the influence of the environment according to the trajectory rules. We further adjust the path search generation value and the regularization-trajectory cells of the A* algorithm in real time on the basis of the influence of the environments, and we fully consider the motion rules of the USVs under the influence of wind and current. Finally, we construct the motion planning method of the USV under the influence of wind and current on the basis of the search algorithm and the regularization-trajectory cells. Based on this method, not only can the influence of wind and current be considered, but the influence of wave and other complex environments on the path planning of the USV can also be analyzed in more detail in the future.

6.1. Analyze Effects of Wind and Current

6.1.1. Analysis of Wind Effects

In this simulation experiment, the experimental wind speed is under normal conditions (0–8 m/s). According to the experimental results in Figure 7, under the influence of a wind speed of 0–8 m/s after a certain period of navigation, the navigation distance of the USV is almost unchanged, but the navigation direction is changed. From the analysis, we can observe that the change of the navigation direction shows a Gaussian distribution.

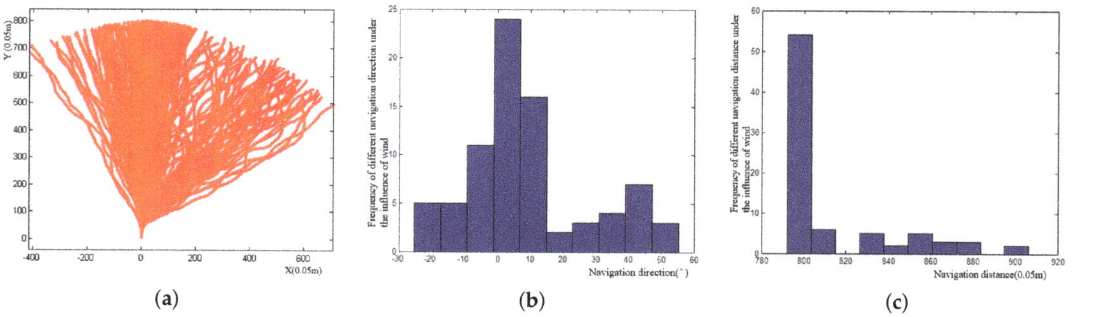

Figure 7. Wind impacts from 0 to 8 m/s. (**a**) Trajectories under wind disturbances from 0 to 8 m/s. (**b**) Heading statistics of wind impacts from 0 to 8 m/s. (**c**) Distance statistics of wind impacts from 0 to 8 m/s.

6.1.2. Analysis of Current Effects

In this section, the influence of the current is directly added to the navigation speed of the USV. Figure 8a,b respectively show the change of the navigation path and the navigation path distance of the USV under the influence of a certain current velocity.

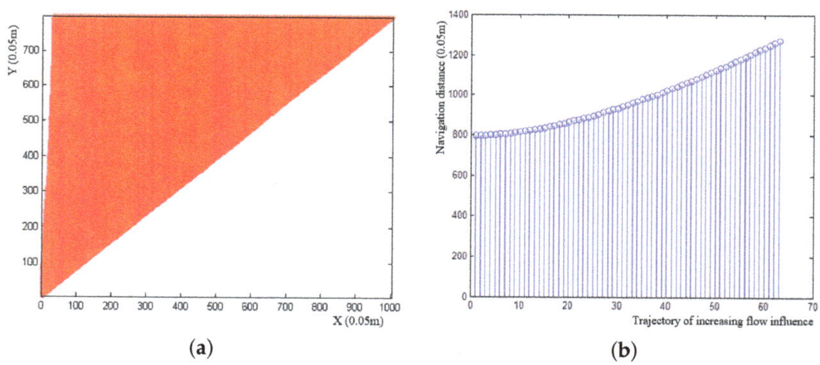

Figure 8. Current impacts at from 0 to 11.3 m/s. (**a**) Trajectory diagram under the current disturbances from 0 to 11.3 m/s. (**b**) Trajectory distance statistics under the current disturbances from 0 to 11.3 m/s.

6.2. Search Algorithm Construction

In the process of constructing the algorithm, the wind and current effects are considered to be part of the path search generation value. Through experimental analysis, the influence of the wind on navigation presents a Gauss distribution, and the effect of the current is directly superimposed onto the speed of the USV.

Figure 9a is a cos (x) figure. The x-axis is the angle between the current direction and the direction of the USV, and the y-axis is the weight of the search cost. According to the simulation results, with the change of the angle between the direction of the current and the course of the USV, the current velocity superimposed onto the velocity of the USV presents a cosine curve change.

Figure 9b is a figure of a sin (x) function from 0 to 180° (similar to other angles) on the basis of current impacts. The x-axis represents the angle between the direction of the current or the wind and the course of the USV, and the y-axis represents the weight of the search cost. When the current or wind is 90° to the heading, the weight value is the largest, that is, the cost of steering becomes higher; this can help a USV to avoid a position where its navigation course will be perpendicular to the direction of the current or the wind in the process of motion planning, and to avoid a situation where it will roll due to the disturbances of the current or the wind.

The A* search Algorithm 1 introduces the evaluation function F(x) when selecting the next exploration node of the current node:

$$\mathrm{F}(\mathrm{x}) = \mathrm{G}(\mathrm{x}) + \mathrm{H}(\mathrm{x}) \tag{13}$$

$\mathrm{F}(\mathrm{x})$ represents the sum of the actual cost $\mathrm{G}(\mathrm{x})$ from the starting point to the current point, and the evaluation cost $\mathrm{H}(\mathrm{x})$ from the current point to the target point (as shown in Formula (13). The weight of $\mathrm{G}(\mathrm{x})$ is expressed as follows: distance cost + steering cost + time cost. The steering cost is proportional to the change of the rudder angle; the change of the USV speed presents the time cost, and the speed of the USV is affected by the wind and current; the distance cost principle is the shortest path principle, where the Euclidean distance is used for the heuristic distance calculation (as shown in Equation (14)).

$$\mathrm{E} = \sqrt[2]{(x_1 - x_0)^2 + (y_1 - y_0)^2} \tag{14}$$

Algorithm 1: Motion planning algorithm of a USV with the effects of wind and current.

Input: map information, starting point, and end point.
Output: USV rudder angles, navigation angles, path points (x, y).
1. Define the open list and close list, add the starting point to the open list and set it as the current point.
2. Move the current point to the close list and explore the F minimum around the current point.
   (1) If a point is already in the close list or cannot be passed, the point is ignored.
   (2) If the point is not in the open list, we add it to the open list and take the current point as the parent node of the point, and then record the F, G, and H values of the point.
   (3) If the point is already in the open list, the path is determined according to the G value, and the point with the lowest G value is the current point.
3. Under the influence of the wind and current environment, start to explore the appropriate trajectory based on the current node and the previous node, and then constantly adjust the rudder angle in the process of exploration.
4. The rudder angle adjustment range requires proper steering.
   (1) If the rudder angle at this time cannot meet the steering requirements in this environment, abandon the current node. Go to step 2 and select the sub-optimal node, and so on, until the node that meets the navigation requirements of the USV is found.
   (2) If the target point has already been added to the open list, then return to the starting point along with the parent node of each cell.
5. End (output planning information).

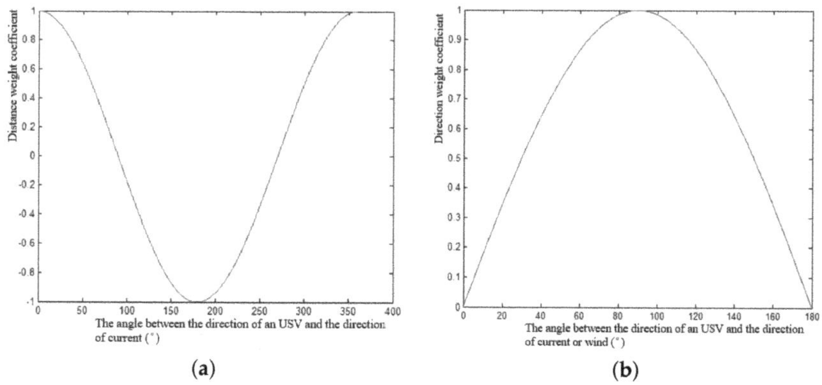

Figure 9. Wind and current impacts on the weight of motion planning. (**a**) Current impacts on the weight of motion planning. (**b**) Wind and current impacts on the planned heading of motion.

7. Simulation Experiments and Analysis

7.1. Experimental Environments and Results

This section will introduce the environments and results of the experiments. Figure 10 shows a schematic diagram of the force on the USV during navigation. Based on the dynamic model above, the corresponding motion model is established, and the corresponding trajectory cell is generated. In the experiment, the speed of the USV is 10 knots, the wind direction is 0°, the wind speed is 2 m/s, the current velocity in the x-axis direction is 1 m/s, and the current velocity in the y-axis direction is 1 m/s.

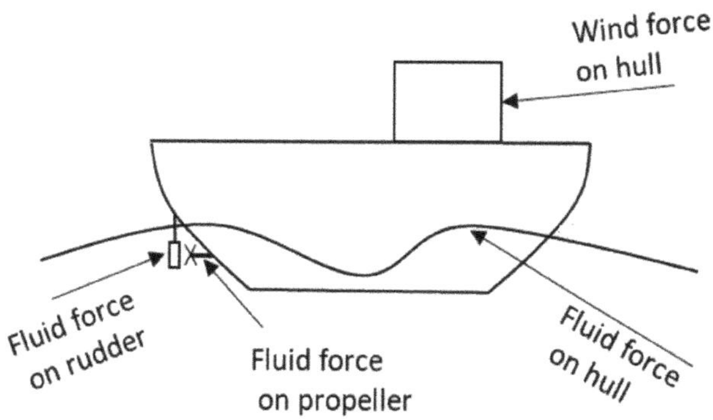

Figure 10. Force diagram of the USV during navigation.

As shown in Figure 11a, the starting point is (5, 24) and the ending point is (20, 10) in the environment without obstacles, and wind and current interfere with the USV motion planning experiment results (in the figure, the blue circle is the starting point, the green circle is the ending point, the green arrow represents current disturbances, and the red arrow represents wind disturbances). As shown in Figure 11b, the starting point is (20, 10) and the ending point is (4, 26). Figure 11c shows in detail the motion planning of the USV in the environment of the starting point (5, 24) and ending point (20, 5), without obstacles but with wind and current disturbances.

The experimental results of USV motion planning with obstacles and wind and current disturbances (the red asterisk is the obstacle) are shown in Figure 12a. The starting point and ending point are (25, 6) and (5, 24), respectively. Note: We consider Figures 11 and 12 as two-dimensional coordinate maps in which the unit of measurement used is meter.

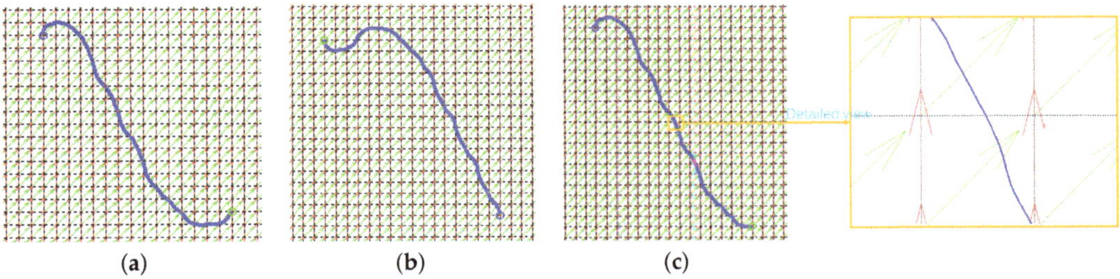

Figure 11. Motion planning for a USV in a free open-sea environment at different starting and ending points, considering wind and current disturbances. (**a**) Motion planning for a USV at the starting point (5, 24) and the ending point (20, 10). (**b**) Motion planning for a USV at the starting point (20, 10) and the ending point (4, 26). (**c**) Local details of USV motion planning at the starting point (5, 24) and the ending point (20, 5).

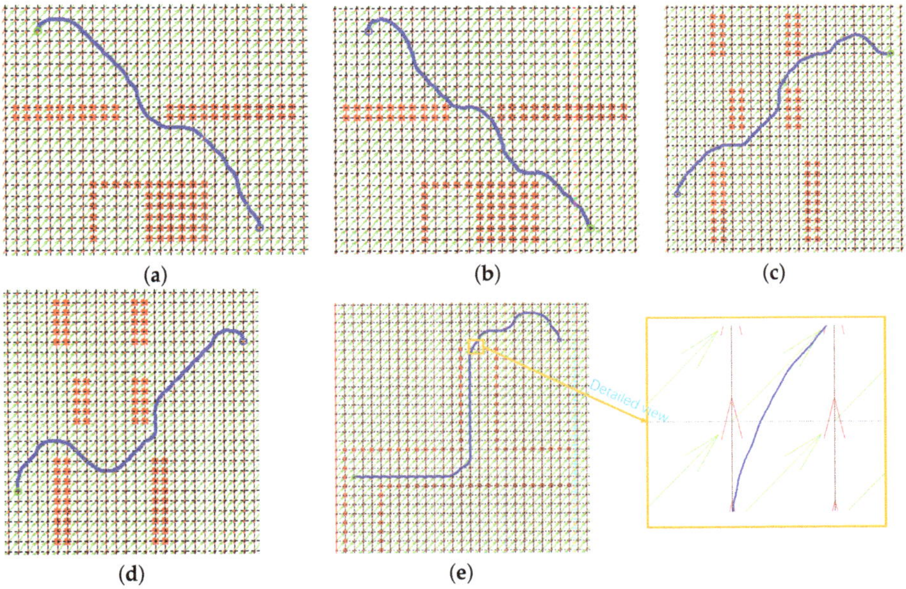

Figure 12. Motion planning for a USV in obstacle environments at different starting and ending points, considering wind and current disturbances. (**a**) Motion planning for a USV at the starting point (25, 6) and the ending point (5, 24). (**b**) Motion planning for a USV at the starting point (5, 24) and the ending point (25, 6). (**c**) Motion planning for a USV at the starting point (2, 9) and the ending point (25, 24). (**d**) Motion planning for a USV at the starting point (25, 24) and the ending point (2, 9). (**e**) Motion planning for a USV at the starting point (25, 24) and the ending point (2, 9).

The experimental results of the USV motion planning in the environment of the starting point (5, 24) and the ending point (25, 6) with obstacles and wind and current disturbances are shown in Figure 12b. As shown in Figure 12c, the starting point is (2, 9) and the ending point is (25, 24). The experimental results of USV motion planning in the environment of obstacles and wind and current disturbances are also shown in Figure 12d. The experimental results of USV motion planning in the environment of the starting point (25, 24) and the ending point (2, 9) with obstacles and wind and current disturbances are shown in Figure 12e, and the detail trajectory is shown in the right of Figure 12e.

7.2. Analysis of Experimental Results

In the experiments, the characteristics of the USV dynamics and the influence of wind and current on the motion of a USV are fully considered. First, the disturbance effects of wind and current are simulated and analyzed, and the relevant characteristic parameters are extracted. Second, according to the influence of the wind and current, the regularization-trajectory cell is constructed to resist wind and current disturbances. Finally, a wind speed of 2 m/s and a current speed in the x-axis and y-axis directions of 1 m/s are simulated. The experiments show that the trajectory of the USV changes irregularly under the environment disturbances, but that the method proposed in this paper makes the trajectory as smooth as possible and achieves a short path planning with safety through the regularization-trajectory cell that allows it to resist the environment disturbances.

From the experiments, the highlights of our method are as follows: (1) On the basis of the regularization-trajectory cell, the smooth rudder command in the trajectory cell is considered; in the process of solving a motion planning problem, the short distance is taken as one of the optimization objectives. (2) Compared with the existing research (such as the method proposed in reference [6]):

- Our method takes into account the force process of the USV in real time by constructing the regularization-trajectory cell and carrying out motion planning according to the changes in force, while the method in reference [6] solves the relevant problems by considering the influence of the current to realize path planning with a multi-objective optimisation method, there is no specific force analysis in the process of the navigation of the USV.
- Our method can realize motion planning under more complex sea conditions by constructing the regularization-trajectory cell, which achieves practical and safe motion planning for a USV.
- In this study, an objective function can be efficiently optimized, which can be easily implemented and may be widely used in the future.

8. Conclusions

In this paper, we tried to solve issues with the disturbances of wind and current on USV motion planning. From the perspectives of USV navigation characteristics, a motion planning method was proposed: first, existing problems were analyzed in detail, and the problem formulation was provided. Second, the USV's dynamics model under the effects of wind and current was established. Third, the regularization-trajectory cell was constructed to provide reachable areas on the basis of a dynamics model. Furthermore, the USV's motion state disturbed by wind and current effects was analyzed in detail. Based on the analysis, the rules and the regularization-trajectory cell were established. Finally, the regularization-trajectory cell, rules, and the A* algorithm were leveraged to construct the motion planning method for a USV under wind and current disturbances. The empirical results indicate the effectiveness of our proposed method that may help to achieve safe and efficient USV motion planning while considering the disturbances of wind and current; this is the key component in future attempts to overcome the influence of more complex environments.

In this study, the influence of waves was not considered, and the influence of waves may be more intense to some extent as compared to the influence of current and wind. Therefore, future research should consider the potential effects of waves more carefully. The problem may also be more complex; we have left this problem for future research.

Author Contributions: Methodology, S.G.; software, S.G.; validation, S.G.; formal analysis, S.G.; investigation, S.G.; resources, S.G.; data curation, S.G.; writing—original draft preparation, S.G.; writing—review and editing, C.Z.; visualization, S.G.; supervision, C.Z. and A.K.; project administration, C.Z., Y.W. and C.X.; funding acquisition, C.Z., Y.W. and C.X. All authors have read and agreed to the published version of the manuscript.

Funding: This work was supported by the National Science Foundation of China (NSFC) through Grant No. 52171349, the National Key R&D Program of China through Grant No. 2018YFC1407405, and the Zhejiang Provincial Science and Technology Program through Grant No. 2021C01010.

Institutional Review Board Statement: Not applicable.

Informed Consent Statement: Not applicable.

Data Availability Statement: The datasets generated during this study are available from the author, who can be reached by contacting gshangd@163.com.

Conflicts of Interest: The authors declare no conflict of interest.

Appendix A. Regularization-TrajectoryCell Library

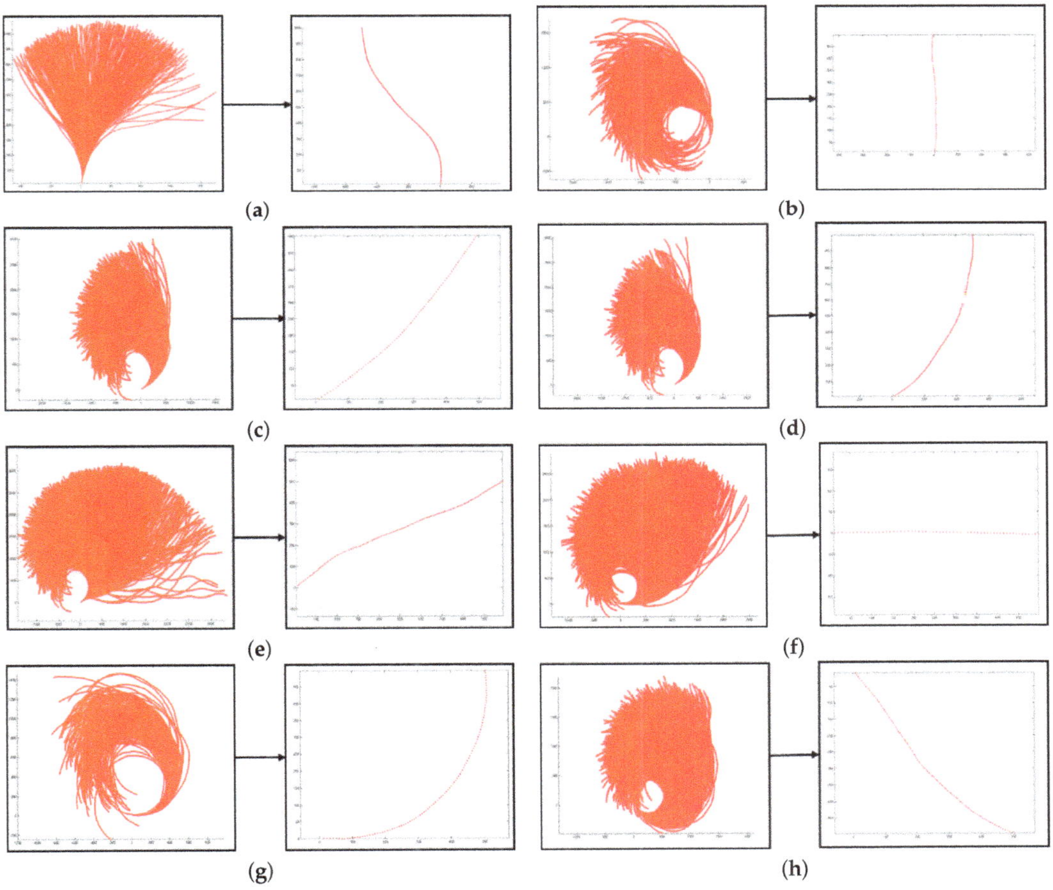

Figure A1. Regularization-trajectory cell library: the first section. (**a**) Regularization-trajectory cells from the initial 0° direction to 45° (rudder angle is 13.6°). (**b**) Time-varying trajectory cell from the initial 0° direction to 0° (rudder angle is −1.92°). (**c**) From the initial 45° direction to the 45° regularization-trajectory cell (rudder angle is −11.52°). (**d**) Time-varying trajectory cell from the initial 45° direction to 0° (rudder angle is −13.66°). (**e**) Time-varying trajectory cell from the initial 45° direction to 90° (rudder angle is 2.58°). (**f**) Time-varying trajectory cell from the initial 90° direction to 90° (rudder angle is −0.66°). (**g**) Time-varying trajectory cell from the initial 90° direction to 0° (rudder angle is −11.52°). (**h**) From the initial 135° direction to the 135° regularization-trajectory cell (rudder angle is −10.74°).

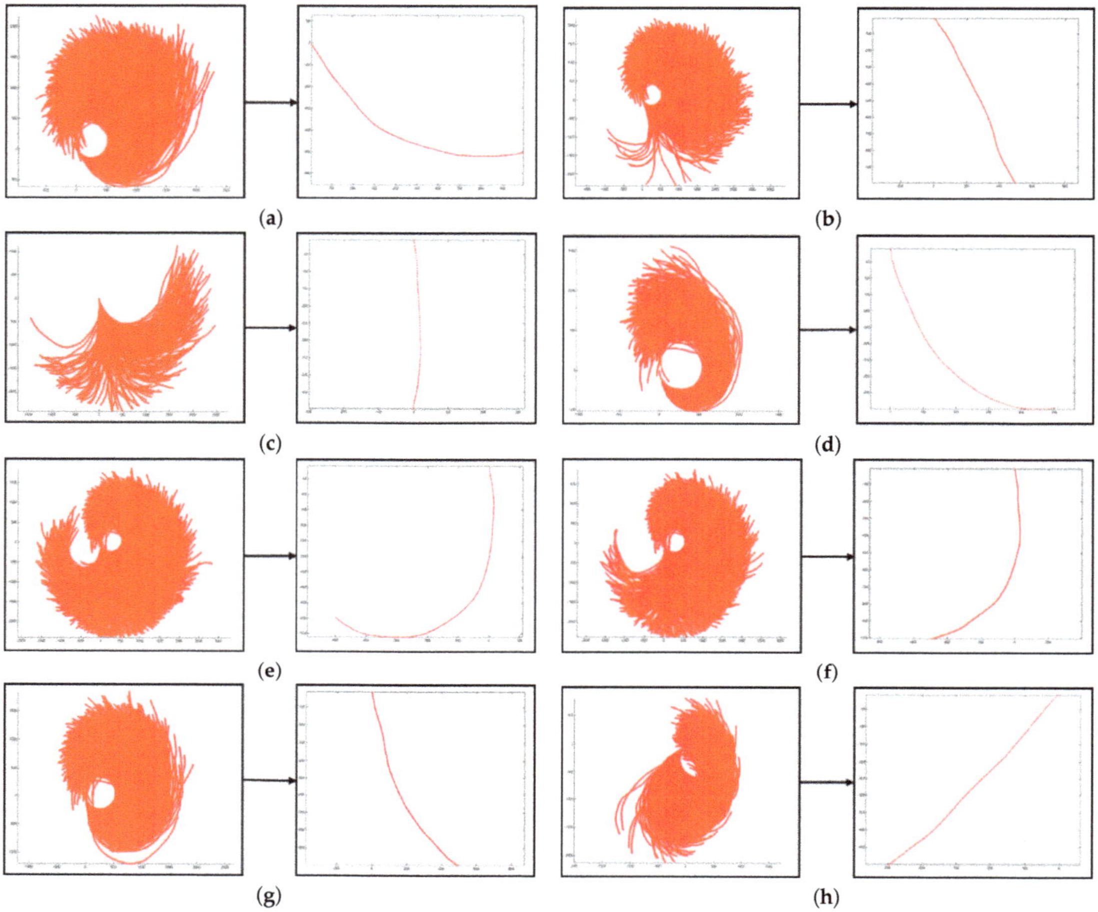

Figure A2. Regularization-trajectory cell library: the second section. (**a**) Time-varying trajectory cell from the initial 135° direction to 90° (rudder angle is −5.32°). (**b**) Time-varying trajectory cell from the initial 135° direction to 180° (rudder angle is 3.8°). (**c**) From the initial 180° direction to the 180° time-varying track cell (rudder angle is −0.78°). (**d**) Time-varying trajectory cell from the initial 180° direction to 90° (rudder angle is −22.78°). (**e**) Time-varying trajectory cell from the initial 180° direction to 270° (rudder angle is 4.33°). (**f**) Time-varying trajectory cell from the initial 180° direction to 225° (rudder angle is −0.05°). (**g**) Time-varying trajectory cell from the initial 180° direction to 135° (rudder angle is −9.61°). (**h**) From the initial 225° direction to the 225° regularization-trajectory cell (rudder angle is −16.65°).

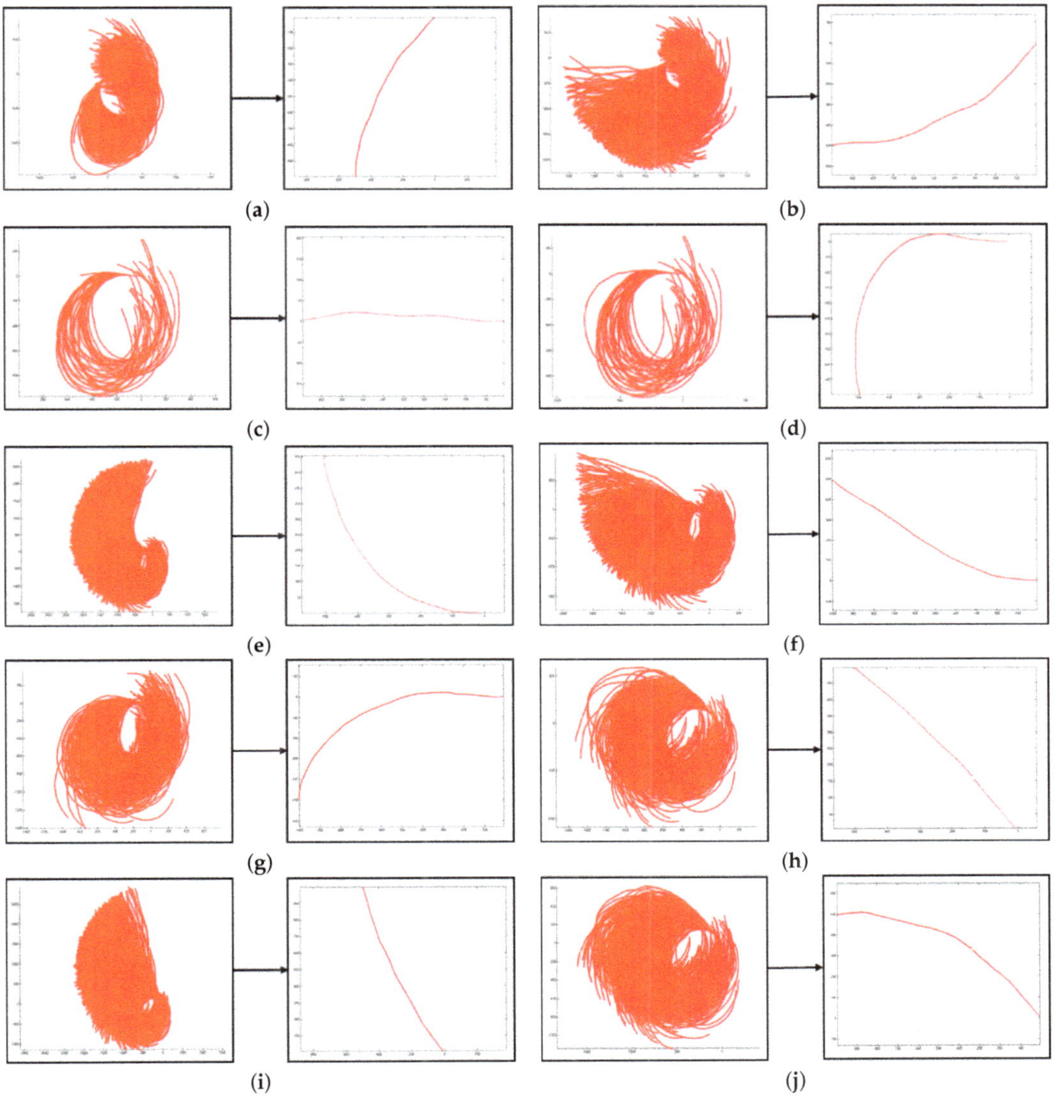

Figure A3. Regularization-trajectory cell library: the third section. (**a**) Time-varying trajectory cell from the initial 225° direction to 180° (rudder angle is −25.79°). (**b**) Time-varying trajectory cell from the initial 225° direction to 270° (rudder angle is −9.74°). (**c**) From the initial 270° direction to the 270° regularization-trajectory cell (rudder angle is −34.68°). (**d**) Time-varying trajectory cell from the initial 270° direction to 180° (rudder angle is −34.64°). (**e**) Time-varying trajectory cell from the initial 270° direction to 0° (rudder angle is 0.84°). (**f**) Time-varying trajectory cell from the initial 270° direction to 315° (rudder angle is −13.68°). (**g**) From the initial 270° direction to the 225° regularization-trajectory cell (rudder angle is −29.17°). (**h**) From the initial 315° direction to the 315° regularization-trajectory cell (rudder angle is −29.96°). (**i**) Time-varying trajectory cell from the initial 315° direction to 0° (rudder angle is −8.25°). (**j**) Time-varying trajectory cell from the initial 315° direction to 270° (rudder angle is −27.19°).

References

1. Liu, Y.; Bucknall, R. Path planning algorithm for unmanned surface vehicle formations in a practical maritime environment. *Ocean Eng.* **2015**, *97*, 126–144. [CrossRef]
2. Gu, S.; Zhou, C.; Wen, Y.; Xiao, C.; Du, Z.; Huang, L. Path Search of Unmanned Surface Vehicle Based on Topological Location. *Navig. China* **2019**, *42*, 52–58.
3. Zhou, C.; Huang, H.; Gu, S.; Chen, R.; Wen, Y.; Gan, L. Design and Implementation of Virtual Warning Buoy System for Over-Water Construction. *Navig. China* **2020**, *43*, 65–69.
4. Zhou, C.; Gu, S.; Wen, Y.; Du, Z.; Xiao, C.; Huang, L.; Zhu, M. The review unmanned surface vehicle path planning: Based on multi-modality constraint. *Ocean Eng.* **2020**, *200*, 107043. [CrossRef]
5. Singh, Y.; Sharma, S.; Sutton, R.; Hatton, D.; Khan, A. A constrained A* approach towards optimal path planning for an unmanned surface vehicle in a maritime environment containing dynamic obstacles and ocean currents. *Ocean Eng.* **2018**, *169*, 187–201. [CrossRef]
6. Ma, Y.; Hu, M.; Yan, X. Multi-objective path planning for unmanned surface vehicle with currents effects. *ISA Trans.* **2018**, *75*, 137–156. [CrossRef] [PubMed]
7. Thakur, A.; Svec, P.; Gupta, S.K. GPU based generation of state transition models using simulations for unmanned surface vehicle trajectory planning. *Robot. Auton. Syst.* **2012**, *60*, 1457–1471. [CrossRef]
8. Song, R.; Liu, W.; Liu, Y.; Bucknall, R. A two-layered fast marching path planning algorithm for an unmanned surface vehicle operating in a dynamic environment. In Proceedings of the OCEANS 2015-Genova, Genova, Italy, 18–21 May 2015; pp. 1–8.
9. Song, R.; Liu, Y.; Bucknall, R. A multi-layered fast marching method for unmanned surface vehicle path planning in a time-variant maritime environment. *Ocean Eng.* **2017**, *129*, 301–317. [CrossRef]
10. Gal, O. Unified approach of unmanned surface vehicle navigation in presence of waves. *J. Robot.* **2011**, *2011*, 703959. [CrossRef]
11. Niu, H.; Ji, Z.; Savvaris, A.; Tsourdos, A. Energy efficient path planning for Unmanned Surface Vehicle in spatially-temporally variant environment. *Ocean Eng.* **2020**, *196*, 106766. [CrossRef]
12. Subramani, D.N.; Wei, Q.J.; Lermusiaux, P.F. Stochastic time-optimal path-planning in uncertain, strong, and dynamic flows. *Comput. Methods Appl. Mech. Eng.* **2018**, *333*, 218–237. [CrossRef]
13. Wu, M.; Zhang, A.; Gao, M.; Zhang, J. Ship Motion Planning for MASS Based on a Multi-Objective Optimization HA* Algorithm in Complex Navigation Conditions. *J. Mar. Sci. Eng.* **2021**, *9*, 1126. [CrossRef]
14. Hart, P.E.; Nilsson, N.J.; Raphael, B. A formal basis for the heuristic determination of minimum cost paths. *IEEE Trans. Syst. Sci. Cybern.* **1968**, *4*, 100–107. [CrossRef]
15. Goh, C.K.; Tan, K.C.; Liu, D.; Chiam, S.C. A competitive and cooperative co-evolutionary approach to multi-objective particle swarm optimization algorithm design. *Eur. J. Oper. Res.* **2010**, *202*, 42–54. [CrossRef]
16. Alvarez, A.; Caiti, A.; Onken, R. Evolutionary path planning for autonomous underwater vehicles in a variable ocean. *IEEE J. Ocean. Eng.* **2004**, *29*, 418–429. [CrossRef]
17. Eichhorn, M. Optimal routing strategies for autonomous underwater vehicles in time-varying environment. *Robot. Auton. Syst.* **2015**, *67*, 33–43. [CrossRef]
18. Wang, L. Trajectory Planning and Path Following Techniques of Unmanned Surface Vehicle. Ph.D. Thesis, Harbin Engineering University, Harbin, China, 2016.
19. Sethian, J.A. *Level Set Methods and Fast Marching Methods: Evolving Interfaces in Computational Geometry, Fluid Mechanics, Computer Vision, and Materials Science*; Cambridge University Press: Cambridge, UK, 1999; Volume 3.
20. Tsitsiklis, J.N. Efficient algorithms for globally optimal trajectories. *IEEE Trans. Autom. Control* **1995**, *40*, 1528–1538. [CrossRef]
21. Poonganam, S.N.J.; Gopalakrishnan, B.; Avula, V.S.S.B.K.; Singh, A.K.; Krishna, K.M.; Manocha, D. Reactive navigation under non-parametric uncertainty through hilbert space embedding of probabilistic velocity obstacles. *IEEE Robot. Autom. Lett.* **2020**, *5*, 2690–2697. [CrossRef]
22. Du, Z.; Wen, Y.; Xiao, C.; Zhang, F.; Huang, L.; Zhou, C. Motion planning for unmanned surface vehicle based on trajectory unit. *Ocean Eng.* **2018**, *151*, 46–56. [CrossRef]
23. Du, Z.; Wen, Y.; Xiao, C.; Huang, L.; Zhou, C.; Zhang, F. Trajectory-cell based method for the unmanned surface vehicle motion planning. *Appl. Ocean. Res.* **2019**, *86*, 207–221. [CrossRef]
24. Ogawa, O. MMG report I: Mathematical model of control movement. *Jpn. Shipbuild. Soc.* **1977**.
25. Zhu, M.; Xiao, C.; Gu, S.; Du, Z.; Wen, Y. A Circle Grid-based Approach for Obstacle Avoidance Motion Planning of Unmanned Surface Vehicles. *arXiv* **2022**, arXiv:2202.04494.
26. Tian, C. Numerical Simulation of Ship's Maneuvering Motion in the Wind, Wave and Current. Master's Thesis, Wuhan University of Technology, Wuhan, China, 2003.
27. Shi, C. Study on Simulation Mathematical Model of Ship Maneuvering Motion in Wind and Wave. Master's Thesis, Harbin Engineering University, Harbin, China 2011.
28. Jia, X.; Yang, Y. *Mathematical Model of Ship Motion-Mechanism Modeling and Identification Modeling*; Publishing House of Dalian Maritime University: Dalian, China, 1999.

29. Zhou, C.; Gu, S.; Wen, Y.; Du, Z.; Xiao, C.; Huang, L.; Zhu, M. Motion planning for an unmanned surface vehicle based on topological position maps. *Ocean Eng.* **2020**, *198*, 106798. [CrossRef]
30. Gu, S.; Zhou, C.; Wen, Y.; Zhong, X.; Zhu, M.; Xiao, C.; Du, Z. A motion planning method for unmanned surface vehicle in restricted waters. *Proc. Inst. Mech. Eng. Part M J. Eng. Marit. Environ.* **2020**, *234*, 332–345. [CrossRef]

Journal of
Marine Science and Engineering

MDPI

Review

Review of Ship Behavior Characteristics in Mixed Waterborne Traffic

Yingjie Tang [1,2], Junmin Mou [1,2], Linying Chen [1,2] and Yang Zhou [3,*]

1 Hubei Key Laboratory of Inland Shipping Technology, Wuhan University of Technology, Wuhan 430063, China; yingjie_tang@whut.edu.cn (Y.T.); moujm@whut.edu.cn (J.M.); linyingchen@whut.edu.cn (L.C.)
2 School of Navigation, Wuhan University of Technology, Wuhan 430063, China
3 Faculty of Civil Engineering and Geosciences, Delft University of Technology, 2628 CN Delft, The Netherlands
* Correspondence: y.zhou-5@tudelft.nl

Abstract: Through the continuous development of intellectualization, considering the lifecycle of ships, the future of a waterborne traffic system is bound to be a mixed scenario where intelligent ships of different autonomy levels co-exist, i.e., mixed waterborne traffic. According to the three modules of ships' perception, decision-making, and execution, the roles of humans and machines under different autonomy levels are analyzed. This paper analyzes and summarizes the intelligent algorithms related to the three modules proposed in the last five years. Starting from the characteristics of the algorithms, the behavior characteristics of ships with different autonomous levels are analyzed. The results show that in terms of information perception, relying on the information perception techniques and risk analysis methods, the ship situation can be judged, and the collision risk is evaluated. The risk can be expressed in two forms, being graphical and numerical. The graphical images intuitively present the risk level, while the numerical results are easier to apply into the control link of ships. In the future, it could be considered to establish a risk perception system with digital and visual integration, which will be more efficient and accurate in risk identification. With respect to intelligent decision-making, currently, unmanned ships mostly use intelligent algorithms to make decisions and tend to achieve both safe and efficient collision avoidance goals in a high-complexity manner. Finally, regarding execution, the advanced power control devices could improve the ship's maneuverability, and the motion control algorithms help to achieve the real-time control of the ship's motion state, so as to further improve the speed and accuracy of ship motion control. With the upgrading of the autonomy level, the ship's behavior develops in a safer, more efficient, and more environment-friendly manner.

Keywords: mixed waterborne traffic; ship behavior; ship autonomy; information perception; intelligent decision-making; execution

Citation: Tang, Y.; Mou, J.; Chen, L.; Zhou, Y. Review of Ship Behavior Characteristics in Mixed Waterborne Traffic. *J. Mar. Sci. Eng.* **2022**, *10*, 139. https://doi.org/10.3390/jmse10020139

Academic Editor: Marco Cococcioni

Received: 30 December 2021
Accepted: 19 January 2022
Published: 20 January 2022

1. Introduction

With the continuous development and breakthrough of modern technologies in information, communication, sensors, and artificial intelligence, intellectualization has become an important development direction for waterborne transportation systems. Intelligent shipping is an emerging modern shipping industry state led by the deep integration of traditional shipping elements with modern technologies. From a global perspective, various countries are actively promoting research on intelligent vessels, with plans to have unmanned autonomous vessels in operation within the next decade. In June 2017, the International Maritime Organization (IMO) clarified that the Maritime Autonomous Surface Ships (MASS) is defined as a ship which, to a varying degree, can operate independently of human interaction [1]. In 2018, the fully autonomous ferry Falco, developed by Rolls-Royce and Finnish FinFerries, realized an automated voyage with 80 passengers on board. In September 2019, Nippon Yusen Kaisha (NYK) completed the world's first sea trial of a "manned autonomous ship", performing various tests under the latest IMO

Interim Guidelines for MASS trials. In December 2019, the maiden voyage of "Tendon 0" was taken in Zhuhai Dong'ao Island, which was China's first self-developed cargo ship with an independent navigation function [2]. Horizon 2020 [3], the biggest EU Research and Innovation program, has supported a range of intelligent shipping and intelligent ship-related projects, including the AUTOSHIP project [4], the NOVIMAR project [5], the MOSES project [6], and the AEGIS project [7]. The autonomous ships are also considered for the Arctic development of the Northern Sea Route [8].

However, the intellectualization of the waterborne traffic system cannot be quickly achieved. In 2020, the global commercial shipping fleet reached 99,800 ships of 100 gross tonnage and above [9]. The intellectualization of existing vessels is unlikely to be completed in a short amount of time—it requires a rather long process. According to the prediction by the World Maritime University, the proportion of remotely monitored autonomous ships would reach 15% by around 2040. In addition, IMO has divided the autonomy level of MASS into four classes, and ships can switch among different levels as required [1].

Level one: Ships with automated processes and decision support. The crew on board operates and controls the ship, while some operations can be automated and at times be unsupervised, but with crew on board ready to take control.

Level two: Remotely controlled ships with crew on board. The ship is controlled and operated from another location. The crew is available on board to take control of the shipboard system and functions.

Level three: Remotely controlled ships without crew on board. The ship is controlled and operated from another location, without crew on board.

Level four: Fully autonomous ships. The operating system of the ship is capable of making decisions and determining actions by itself.

Therefore, regardless of the current development stage of the intelligence of a waterborne traffic system, its future is bound to be a mixed system where intelligent ships of different autonomy levels co-exist, i.e., a mixed traffic scenario.

In the field of road traffic, mixed traffic is usually defined as the phenomenon of mixing different types of transport, which mostly involves motor vehicles, bicycles, and pedestrians interfering with each other [10]. With the development of autonomous driving technology, the hybrid traffic of manned and unmanned vehicles has gradually attracted research interest, such as the analysis of macroscopic hybrid traffic flow characteristics and the development of microscopic driving behavior models [11]. This paper focuses on microscopic ship behavior in mixed waterborne traffic scenarios, where ship behavior is usually defined as the mode and pattern of action of a particular class of ships. In existing studies, ship behavior generally refers to behavior such as path, speed, and course, etc. A solution framework to support situation awareness in a mixed environment is considered by Perera et al. [12]. The review on ship behavior established a generic behavior identification model, but which did not consider the behavioral differences of different ship types under specific external environmental constraints [13]. A behavioral clustering-based classification for harbor ships using Automatic Identification System (AIS) data is proposed in a subsequent study [14]. In the further research, the impacts of external factors on ship behavior considering changes in ship size is investigated [15].

So far, few researchers have investigated the ship behavior under mixed traffic scenarios. It is significant to extract the behavioral characteristics of various types of autonomous ships in mixed waterborne traffic scenarios and master the methods and rules of ship navigation and collision avoidance, which would enhance the navigation safety, reduce the navigation conflicts, improve the waterway capacity, and upgrade the level of waterborne traffic supervision.

This paper analyzes the roles of man and machine on different autonomy levels based on the three modules of "perception–decision–execution" and proposes a classification method for the ship autonomy level. The relevant intelligent algorithms on the three main modules in the last five years are analyzed and summarized. Starting from the algorithm features, the behavioral characteristics of ships with different autonomy levels adopting the corresponding algorithms are further analyzed.

2. Research Method

In this section, the research method is introduced. A framework of intelligent ships illustrating the three modules is presented, based on which a classification of the ship autonomy level is proposed.

2.1. Basic Framework of Intelligent Ships

The intelligent ship is a huge and complex system, which involves theories and technologies in multiple fields, such as ship design and manufacturing, sensor techniques, intelligent decision-making, maritime communication, information fusion, etc. Generally, an intelligent ship system consists of three subsystems, namely an information perception system, decision-making system, and execution system, respectively. These three modules interact with each other, and the stability and reliability of any module will affect the final task execution efficiency and the ship safety. The inner relationship among subsystems of an intelligent ship system is shown in Figure 1.

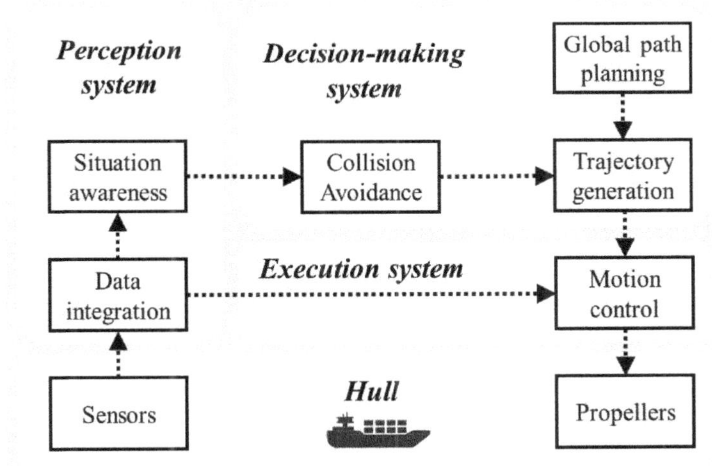

Figure 1. Illustration of subsystems within an intelligent ship system.

The perception system collects the information provided by various sensors and obtains the ship's situation awareness through information fusion techniques. Based on the ship's perceived situation, considering the requirements of global path planning and collision avoidance operation, the decision-making system generates the planned trajectory and transmits its instructions to the execution system. Finally, according to the instructions received and the fusion of sensor information, the execution system drives the main engine, propeller, rudder, and other devices. Afterwards, the execution results are fed back to the perception system through sensors to update the ship situation awareness information for next-step decision-making.

2.2. Autonomy Hierarchy

So far, several international authorities have announced their approaches of autonomous ship classification, which includes Lloyd's register of shipping (LR). On 8 July 2016, LR declared to divide the autonomy level of ships into seven levels, from AL0 to AL6. It was the first time classifying the ship autonomy level, and this classification method has also been widely adopted by the academic community. The classification standard for autonomous ships issued by LR is presented in Table 1 [16]. This classification standard ranges from the ordinary ship without any independent function (AL0), to the human–machine joint decision-making system (AL1 and AL2), then to the autonomously navigated ship with onshore operator monitoring (AL3–AL5), and finally to the fully autonomous ship without manual operation (AL6).

Table 1. The ship autonomy classification standard by LR.

Level	Description
AL0	No automation function, manual navigation of a ship.
AL1	On-ship decision support system, data available to crew.
AL2	Off-ship decision support system, shore monitoring.
AL3	Semi-autonomous ship with active human-in-the-loop where crew can intervene.
AL4	Human-in-the-loop, the ship operates autonomously with human supervision.
AL5	Fully autonomous ship with means of human control.
AL6	Fully autonomous ship without the need for any human intervention.

Afterwards, the Norwegian Forum for Autonomous Ships (NFAS), Danish Maritime Authority (DMA), Marine Autonomous Systems Regulatory Working Group (MASRWG), IMO, and other institutions successively proposed their classification standards of ship autonomy.

Based on the classification criteria of ship autonomy by the institutions, this paper proposes an autonomy classification method on the basis of the three modules of intelligent ships, "perception–decision–execution". The level of ship autonomy is determined by analyzing and comparing the involvement of humans and machines in each module. This paper classifies the ship autonomy into five levels: manual operation level, machine-aided level, remote-control level, onshore supervision level, and fully autonomous level, which are listed in Table 2 in detail.

Table 2. Classification of ship autonomy.

	Perception	Decision	Execution
Manual operation	Manual	Manual	Manual
Machine-aided	Human-in-the-loop (On- and off-board data)	Human-in-the-loop (On- and off-board data)	Manual
Remote-control	Human supervision (Broad level)	Human supervision (Broad level)	Human supervision (Broad level)
Onshore supervision	Rarely supervised	Rarely supervised	Rarely supervised
Fully autonomous	Unsupervised	Unsupervised	Unsupervised

The correspondence between the proposed classification method and the methods used by other institutions is presented in Table 3. Manual operation level refers to the ordinary ships whose perception, decision-making, and execution are completed manually, corresponding to AL0 by LR, AAB by NFAS, and AL0 by MASRWG. Compared to the manually operated ships, machine-aided ships have improved the participation of machines in the perception system to realize the automatic acquisition of information at both ends of the ship and shore. This way, the crew can make navigation decisions using such information. From remote-control ships to fully autonomous ships, there are no crew on board. As the autonomy level increases, the human intervention on the ship control gradually diminishes. The machines would gradually replace the role of humans to control the ship operation. Eventually, the full autonomy of the ship is achieved, corresponding to the highest level defined by other institutions.

Table 3. Correspondence table for ship autonomy criteria.

	LR	NFAS	IMO/DNA	MASRWG
Manual operation	AL0	AAB		AL0
Machine-aided	AL1, AL2		M	
Remote-control	AL4		RU	AL2, AL3
Onshore supervision	AL5	PUS		AL4
Fully autonomous	AL6	CUS	A	AL5

The manual operation level refers to the ships not equipped with any navigational aids, which have been gradually phased out from the market. Currently, the majority of operating ships are at the machine-aided stage, which aims at safer navigation via the assistance of machines. The research and development of unmanned ships are mostly at the remote-control stage, yet limited to the small-size ships. In November 2020, the British company SEA-KIT remotely commanded the Unmanned Surface Vessel (USV) "Maxlimer" through three satellites to complete a 22-day exploration mission [17]. The Yara Birkeland, the world's first zero-emission, fully automated container ship developed by the Yara Group in cooperation with the Kongsberg Group, started its maiden voyage in November 2021 [18]. This ship is expected to come into commercial operation carrying humans from 2022. Starting with a two-year technical test period, the ultimate goal of the ship is to achieve fully autonomous navigation.

3. Characteristics of Intelligent Algorithms

To research the behavior of autonomous ships, this section describes the three modules of perception, decision-making, and execution of a ship in detail. Besides the explanation of the module-related intelligent algorithms, a set of evaluation criteria are introduced for each type of algorithm to compare the characteristics. The possible limitations are discussed at the end of the introduction of each part.

3.1. The Perception System

3.1.1. Information Perception Technology

The information perception system includes perception sensors and information fusion techniques. It obtains the information about the ship's own state and the environment that the ship is in. By identifying the ship's situation, the perception system provides a data basis for collision avoidance and path planning, which is expected to lead to safer and more reliable navigation. The information on the ship's own state includes the position, speed, course, and other navigational information, as well as the information on the operational state of the equipment systems and the state of the carried cargo. The environmental information refers to the meteorological and hydrological conditions, and the status of surrounding vessels and obstacles. Depending on the type of equipped sensors, the intelligent ship information perception technologies can be distinguished in the following four categories.

1. Radar-based information perception technology

Radar is an essential sensor in the maritime field. The perception is mostly in the form of processing of radar images and signals. This category of equipment includes the ultrasonic radar, millimeter-wave radar, continuous-wave radar, etc. Radar has high resolution and accuracy, and can be used in all-weather and wide-area detection. However, there are also certain limitations: the detection of radar is susceptible to weather conditions, and the radar images often suffer from problems such as noise interference, uneven brightness, target loss, etc. [19].

2. Lidar-based information perception technology

Light Detection and Ranging (Lidar) works in a similar way to radar. It is mainly used for purposes of target detection and obstacle avoidance. Achtert et al. combined Doppler radar with a steady-motion platform to measure the atmospheric wind profiles [20]. The results show that the data coverage of this method is comparable to that of land-based measurements. However, this method provides a more detailed and higher temporal resolution view of atmospheric boundary variability, compared to the radiosonde measurements. Compared to radar, Lidar has a higher range accuracy and stronger anti-interference capability, but the detection range is smaller.

3. AIS-based information perception technology

AIS, as an important tool for the perception of waterborne traffic information, provides a complementary source of data to radar, since it has no dead zone. However, this method only applies to vessels equipped with AIS transponders. Thus, AIS cannot be taken as the only form of information perception for intelligent ships. Prasad et al. utilized AIS data and multi-sensor information to augment the data from weather sensors, which can be applicable for the control and navigation of ships in foggy weather or other restricted-visibility conditions [21]. Zhou et al. developed a regression model using AIS data to quantify the impacts of wind and current on ship behavior without the input of specific ship maneuvering details [15]. Regarding the data quality, the current AIS data mostly contain errors and data loss, which may lead to wrong or at least incomplete information acquisition.

4. Vision-based information perception technology

Vision sensors acquire the image information of the surrounding environment by machine vision and process the captured images to achieve environmental perception. Wang et al. proposed a framework for automatic detection and localization of USV real-time targets based on binocular vision [22]. It extracts and matches the features within a target area determined by a deep convolution network. Then, the target is localized using the calibrated binocular camera parameters in the triangulation measurement principle. The experimental results proved the delivery of both accurate detection and high-precision positioning results in real-time applications. Currently, vision sensors would be the cutting-edge method to perceive the surrounding information. However, to achieve a full overview of the surroundings, a sophisticated sensor system is needed, which possibly implies a high risk of machine failure.

3.1.2. Risk Perception Method

Intelligent ships rely on sensors to obtain basic information, apply information fusion techniques to judge the ship's situation, and estimate the collision risk. There have been a number of studies on the risk analysis or assessment of ship collision. Some researchers tend to indicate the risk as a numerical index, such as the Minimum Safety Passing Distance (MSPD) [23,24] and the Collision Risk Index (CRI) [25–29], which is deemed as a numerical form of risk. Whereas, the other way to reflect risk is in two-dimensional graphics, such as Ship Domain (SD) [30–33], dangerous region (DR) [34–36], and action lines (AL) [37,38].

The MSPD method usually provides a deterministic result of a collision event in the given scenario, i.e., occurrence or non-occurrence. The premise is that when both the own ship (OS) and the target ship (TS) keep their course and speed, if the Distance at the Closest Point of Approach (DCPA) is smaller than the MSPD, a collision occurs. On the contrary, OS can safely pass TS. This method has been widely adopted for manned and unmanned ships [23,24]. In addition, the MSPD is also an essential risk indicator in CRI calculation [39].

The assessment of ship collision is influenced by multiple factors, such as ship speed, course, distance to TS, speed ratio, and meteorological and hydrological conditions, etc. Besides, the presence of sensor errors also leads to the uncertainty of the collision process. The CRI measurement provides an exact value of the threat level, which is an intuitive indicator of the collision risk. The current main CRI measurement methods include the DCPA and the Time to the Closest Point of Approach (TCPA) weighting methods [25,26], fuzzy logic algorithms [27], and neural networks [28]. When using the weighting method to measure CRI, the different dimensions of DCPA and TCPA are usually ignored, which makes the calculation result inaccurate. In addition, in the multi-ship situation, it is impossible to objectively reflect the threat level of each ship. Fuzzy logic methods are quite subjective when calculating CRI, and can only be applied in certain specific scenarios. Neural network algorithms require a substantial storage of expert experience and knowledge in advance and plenty of sample learning. Thus, this algorithm cannot satisfactorily fulfill the instantaneity requirement of CRI.

SD is a graphic depiction of the ship collision risk, which is usually a group of areas around OS to visualize the risk. When TS enters or is about to enter the area, a collision alert is triggered. Szlapczynski et al. defined two SD-based safety parameters: Degree of Domain Violation (DDV) and Time to Domain Violation (TDV) [30]. The results show that the accuracy of DDV/TDV is higher than that of the DCPA/TCPA. Qiao et al. developed a quadratic ship domain model considering the uncertainty of ship position and proposed a method to calculate the spatial collision risk, which had been improved in further applications [31]. Some researchers assess the collision risk by developing new SD models. Bakdi et al. developed an adaptive SD model for risk identification through a spatial risk function based on the type of encounter situation and collision hazard [32]. The results showed competitive advantages in terms of intuitiveness and computational efficiency. Aiming at the shortage of including single factors in SD in previous studies, Guan et al. established an SD model based on fuzzy logic considering multiple variables [33]. The obtained results support the decision-making of collision avoidance and early prediction of collision risk. However, the calculation of this method is complex, and not suitable for risk identification in the case of multi-ship encounters.

The DR is designed to collect a set of OS's speed or course that leads to a conflict with TS and display this set of speed or course to the Officer on Watch (OOW) in a graphical form. Velocity Obstacle (VO) is a typical algorithm in this category. It is capable of seeking out the optimal collision-free solution in two-ship and multi-ship encounter scenarios [34], considering ship dynamics [35] and maneuverability [36].

The method of AL focuses on identifying a line of action around OS in geographic space, which indicates the final timing of OS to complete collision avoidance through a series of actions. AL is usually obtained by simulations. Szlapczynski et al. determined AL by a series of simulations of various types of ship encounters under different conditions using a hydrodynamic model of ship movement [37]. Namgung et al. established an adaptive neuro fuzzy inference system to judge the CRI of the optimal position and timing [38]. The system ensures that the OOW has sufficient time to make decisions and take the necessary actions of collision avoidance.

Based on the manifestations of the risk perception methods presented above, the models referred by literature research are summarized in Table 4.

Table 4. Overview of ship risk perception methods and the application of corresponding models.

Category	Abbreviation of Method	Description	Model Applications in the Literature
Numerical	MSPD	Provide deterministic results of collisions by comparing the distance between ships and the size of MSPD.	[23,24]
	CRI	Assign weights to influencing factors to calculate the probability of a collision occurring.	[25–29]
Graphical	SD	Deemed to be a risk of collision when TS enters or will enter SD.	[30–33]
	DR	Collect the set of speeds or courses that cause the OS to conflict with the TS, then displays the set in the image to the OOW.	[34–36]
	AL	The final time for OS to take evasive action to avoid a collision	[37,38]

Among the risk perception methods, the number of models adopting CRI and SD is the highest. Only a few models apply the method of MSPD, DR, and AL. MSPD can be deemed as a basis for the calculation of CRI, while DR and AL can provide decision-making support for ship collision avoidance on the basis of SD display. In the future, more models adopting DR and AL can be developed for collision risk perception purposes.

When the collision risk is indicated in a numerical form, it can be intuitively compared. A higher value indicates higher risk, and vice versa. For intelligent ships, risks in the numerical form are easier to involve in the system control link, and provide an informative basis for intelligent decision-making. In the multi-ship encounter situation, the graphic expression of risk allows to intuitively divide TSs into several groups defined by graphical indicator, SD, DR, or AL. However, in such a form, the risks of TSs in each group still cannot be further compared. The graphic form is indeed more intuitive for the operator, which can be integrated into the map to support the OOW in obtaining an overview of the surrounding situation. To make full use of the advantages of both methods, a risk perception platform with integrated collision risk digitalization and visualization can be considered in the future. It is expected to more efficiently and accurately perceive the risks.

In current navigation practice, the collision risk can be assessed with the assistance of some techniques and systems on board. However, for manned ships, the OOW mostly tends to judge the risk by good seamanship and situational awareness, considering the perceived information from the assistance system, instead of directly adopting the indicated risk result. Thus, the risk perception result still largely depends on the experience, knowledge, and skills of the crew. However, it is difficult for OOWs to maintain good situational awareness and precautions when simultaneously monitoring multiple ships. For intelligent ships, the collision risk can be monitored in real-time via reliable and timely identification of obstacles by sensors and information fusion techniques.

Most of the risk perception models are based on AIS data. Generally, the uncertainty of trajectory data is not considered. Besides, there could be some special circumstances when the AIS equipment is off, or the signal transmission fails. In such a situation, AIS data are no longer available, let alone the data accuracy. These facts lead to the error between the theoretical research and the reality. In the future, it is necessary to introduce parameters of uncertainty when developing risk perception models to consider the corresponding impacts and integrate alternative methods without AIS data as mandatory input.

3.2. The Decision-Making System

An intelligent decision-making system involves various techniques, such as path planning, risk conflict detection, intelligent collision avoidance, energy efficiency management, etc. It can continuously generate smooth and feasible optimal paths based on the perception system information, mission requirements, and environmental status. Afterwards, the system sends the decision command to the execution system. In the system, path planning and intelligent collision avoidance are the core techniques in the field of unmanned ships. Path planning aims to find out the collision-free paths on the map considering static obstacles, while intelligent collision avoidance focuses on avoiding collisions with dynamic or unknown obstacles. Six groups of ship collision avoidance techniques for decision-making are identified, which are introduced as follows [40].

Rule-based (RB) methods use a set of pre-defined rules to guide collision avoidance. One approach is to incorporate the International Regulations for Collision Avoidance at Sea (COLREGs) and good seamanship into the rule system. The system is able to propose rule-compatible operations for OS in various scenarios, which are usually based on fuzzy logic [41–43] and neural network algorithms [42]. Since it is impossible to enumerate all rules for all scenarios, the method can only provide collision avoidance guidance for specific scenarios.

Virtual vector (VV) methods obtain the ship's motion by generating a virtual vector field, in which the Artificial Potential Field (APF) method is the specific algorithm. Lazarowska introduced a method to solve the planning of a safe path, using a discrete APF and path optimization algorithm to calculate an optimized collision-free trajectory. The solution is conflict-free, but may not be optimal [44].

The methods of discretization of solutions with collision check (DSCC) discretize the solution space of collision avoidance and eliminate the dangerous solution via collision detection. Then, the collision-free solution is selected from the rest. The typical algorithms include the Dynamic-Window (DW) algorithm, the discrete input optimization (DIO) algorithm, etc. Serigstad et al. proposed a Hybrid Dynamic-Window (HDW) algorithm as a reactive collision avoidance method to improve the trajectory planning when approaching an obstacle [45].

Continuous solutions with collision constraints (CSCC) methods formulate collisions as constraints and find collision-free solutions in a continuous solution space. The approach uses polygons or circles to represent obstacles for collision detection and then applies certain algorithms to calculate a set of control inputs that lead to collisions, such as the VO algorithm [34,46], Vision Cone (VC) [47], etc. Accordingly, the optimal collision avoidance solution can be acquired. Another approach is to use collision detection as a constraint in the optimization to obtain a collision-free solution. A typical algorithm is the Model Predictive Control (MPC)-based collision avoidance [48], which provides a solution at minimum cost.

Re-planning (RP) methods transform the collision avoidance problem to a path planning problem by searching for the collision-free paths in the free configuration space. Two groups of algorithms are found. One group relies on graph searching methods, such as the Fast Marching Method (FMM) [49] and Particle Swarm Optimization (PSO) [50]. The other directly uses evolutionary algorithms to find paths, such as Ant Colony Optimization (ACO) [51], etc. The obtained path does not directly depend on the graphical map.

The presented algorithms found in the literature are usually adopted in a combined way in maritime practice to perform collision avoidance, which is a hybrid of algorithms (HA). Chen et al. proposed a Time-varying Collision Risk-based Fast Marching Square (TCR-FM2) algorithm that combines FM and VO for path planning of autonomous vessels [52].

Table 5 lists the classification of the collision avoidance methods together with brief descriptions.

Table 5. Classification of intelligent collision avoidance methods during decision-making.

Abbreviation of Methods	Description
RB	Adopt pre-defined rules to guide collision avoidance.
VV	Determine the ship's motion by generating a virtual vector field.
DSCC	Search the discrete solution space and find a collision-free solution or an optimal solution.
CSCC	Formulate collisions as constraints and find the optimal solution in continuous space.
RP	Transform the collision avoidance problem to a path planning one and search for a collision-free path in the free configuration space.
HA	Combine some of the above-mentioned methods.

Looking into the collision avoidance decision models in the literature, two types of behavioral features are extracted: decision-making category and decision-making preference. The decision-making is classified into three categories by a decrease of the randomness: large operation, small operation, and trajectory. Large operation refers to the operation of only steering to portside or starboard as the decision, which involves much randomness and probably leads to low accuracy and efficiency of ship collision avoidance. Small operation changes the specific rudder angle or other operations with less randomness as the decision, which improves the accuracy and efficiency compared to the large operation. The trajectory is a series of operations including course alteration and speed change in the form of a path to realize the optimal collision avoidance result. The decision preference is divided into three types: course alteration, speed change, and a combination of both, which reflects the collision avoidance operations by an increase of complexity. The combination mode refers to the operation of continuous course alterations and speed changes within a period of time. According to the intelligent decision algorithm categorization, the decision-making category, and the type of decision-making preference, the behavioral characteristics of the intelligent decision-making process of the referred models in the literature are summarized in Table 6.

Table 6. Overview of the behavioral characteristics of the intelligent decision-making process in the models in the literature.

Model	Decision-Making Algorithm	Decision-Making Category			Decision-Making Preference		
		Large Operation	Small Operation	Trajectory	Course Alteration	Speed Change	Combination
[27]	RB		√				√
[40]	CSCC		√				√
[41]	RB	√			√		
[42]	RB	√			√		
[43]	RB		√		√		
[44]	VV		√		√		
[45]	DSCC			√			√
[46]	CSCC		√				√
[47]	CSCC			√			√
[48]	CSCC			√			√
[49]	RP			√			√
[50]	RP		√				√
[51]	RP			√			√
[52]	HA			√			√
[53]	DSCC		√				√
[54]	DSCC		√				√
[55]	DSCC	√			√		
[56]	HA	√					√
[57]	DSCC		√		√		
[58]	RP			√			√
[59]	HA			√			√
[60]	HA			√			√
[61]	RP			√			√

Abbreviations: RB: rule-based; VV: virtual vector; DSCC: discretization of solutions with collision check; CSCC: continuous solutions with collision constraints; RP: re-planning methods; HA: hybrid of algorithms.

The behavioral characteristics of the decision-making methods are summarized as follows. The RB method mainly adopts the operation of course alteration to fulfill collision avoidance, and seldom concerns trajectory planning. It is probably due to a lack of quantitative analysis of the predefined rules in the RB method. The VV method performs the collision avoidance operation mainly by determining the magnitude of altering course. The CSCC method adopts the operations with less randomness as the decision and always combines the operation of course alteration and speed change to realize the precise execution of the collision avoidance action. However, the DSCC method has a wider range of decision types and preferences. The RP method always guides the collision avoidance operation via planned trajectories. The HA method also tends to take a reasonable collision avoidance trajectory as the decision.

Most intelligent decision-making algorithms adopt operations with small randomness, such as a specific course change or planned trajectory, as collision avoidance decisions. In terms of decision-making preference, the operations with high complexity are preferred for collision avoidance, being the combined manner. As can be seen from Table 6, no model solely relies on speed change to accomplish collision avoidance, which is probably because of the large inertia of the ship. However, it does not mean that the option with only speed change can never work. In special circumstances when course alteration is restricted, speed change can be an alternative.

Currently, the decision-making of collision avoidance for the manned ships mostly relies on the empirical knowledge. From the perspective of navigation practice, the obtained decision scheme should satisfy the four principles of "early, large, wide, clear" required by COLREGs, but with the efficiency often ignored. There is a lack of standardized machine-language-based framework rule interpretation. Thus, for specific encounter situations, COLREGs and other navigational guidance cannot provide quantitative support for the judgment. However, the application of decision-making algorithms could be a solution to these problems. Improving the existing algorithms and considering more constraints to reach more efficient and reasonable decision-making capabilities would be an important research and development direction in the future.

3.3. The Execution System

The core objective of the execution system is to generate the appropriate control variables at the proper timing in cooperation with the information perception system and the decision-making system to achieve the expected execution result within the capability of the ship actuator. The main function of this system is to perform the operational motion control of the intelligent ship. The main task is to ensure that, during the sailing along the optimal planned path, two subgoals can be achieved. On the one hand, the intelligent ship can avoid collision with the surrounding static and dynamic obstacles. On the other hand, the instances of course alteration and speed change can be as few as possible.

The execution system includes two parts: the power control unit and the motion control algorithm.

3.3.1. Power Control Unit

The power control unit consists of a propulsion unit and a course control unit. The common ship propulsion modes include diesel propulsion, water jet propulsion, electric propulsion, etc. The course control is usually fulfilled by rudder and steering engine equipment.

The marine diesel engine has been widely used in the field of ship propulsion, since it is economical, easy to start, and has high thermal efficiency. The ship propulsion system is composed of the main engine, transmission device, shaft system, propeller, etc. The operating principle is to transmit the power from the main engine to the propeller via the transmission device and shaft system, which completes the mission of propelling the ship. However, this system still has some shortcomings, e.g., complex structure, large space occupation by the shaft system, low transmission efficiency, loud noise, frequent equipment failure, etc. These disadvantages drive the attention toward more advanced water jet propulsion systems and electric propulsion systems.

Water jet propulsion uses the water pump as the ship propeller, which propels the ship via the reactive force of the water jet from the pump. This method has good maneuverability and flexibility, strong adaptability to different working conditions, and low resistance. Thus, it has been more widely applied in high-speed and high-performance ships. However, there still exists some disadvantages, such as low propulsion efficiency at low speed, easy inhalation of debris at the water inlet, complicated maintenance, etc.

The electric propulsion system consists of the prime motor, electric generator, electromotor, control equipment, and propeller. The prime motor converts mechanical energy to electric energy, which is transmitted to the propeller via the electromotor. The electric energy is converted to thrust energy to propel the ship. Pod propulsion and shaftless rim-driven thruster are the two cutting-edge propulsion methods, both of which have the advantages of flexible arrangement, rapid response, convenient maneuvering, adaptive flexibility, and high reliability. They will be optimal alternatives for future intelligent ocean ships.

The currently operating ships generally adopt a combination of follow-up and automatic rudder as the steering engine, which can be freely switched between themselves. The follow-up rudder controls the ship via the helmsman's steering, while the automatic rudder realizes the purpose of keeping stable or altering course according to the instructions from the automatic navigation system. In this way, the workload of manual steering can be reduced, and the accuracy of the path control can be guaranteed.

3.3.2. Motion Control Algorithm

The ship motion control algorithm is the core technique of the execution system. By integrating factors such as navigation, weather, route information, and ship maneuverability, the algorithm can derive the course or rudder angle command as required. Accordingly, the ship can be controlled to sail along the planned path. In this way, it is expected to reduce the workload of the crew, save energy consumption, and improve navigation safety. PID control is widely used in engineering practices because of its simple structure and few adjustment parameters. However, the traditional PID algorithm cannot handle the uncertainty problem well. To compensate for the disturbances caused by environmental factors, control algorithms, such as adaptive PID [62], adaptive MPC [63–66], and adaptive dynamic surface control [67,68], have been gradually applied to ship control. With the development of modern control theory, intelligent algorithms, such as sliding mode control (SMC) [69], backstepping [70–75], fuzzy control [76], and neural network [77–80], are also widely adopted in the field of ship control.

This section analyzes the characteristics of ship behavior in the scenarios with the control algorithm applied, including the error convergence rate and path control accuracy. The error convergence rate refers to the time when the ship's tracking error converges to zero or an acceptable range, and can be further categorized into straight-line tracking and curve tracking depending on the tracking trajectory. The path control accuracy refers to the error between the ship's trajectory and the planned path. It is used to quantitatively evaluate the performance of path tracking, including tracking error, yaw error, and Integral Absolute Error (IAE). Table 7 summarizes the characteristics of the ship behavior using the control algorithm.

Table 7. Classification of intelligent collision avoidance methods during decision-making.

Model	Error Convergence Rate	Path Control Accuracy		
		Tracking Error	IAE	Yaw Error
[62]	15 s	T: 0.032 m		
[63]	35.5 s	T: 0.495 m		
[64]	L: 29.52 s; C: 167.25 s			
[65]	L: 9 s; C: 50 s			
[66]		T: 0.7603 m		
[67]	163 s		T: 81.7	
[69]			T: 1.1×10^3	
[70]	200 s	0		
[71]	48 s	0		
[72]	20 s	0		
[73]		H: 0.4805 m; V: 0.4784 m		
[74]	20 s	arbitrarily small error		
[75]	10 s			
[76]				2.27°
[77]		T: <1 m		
[78]			H: 1.11×10^3; V: 0.93×10^3	
[80]				0.39°

Abbreviation: L: line; C: curve; H: horizontal direction; V: vertical direction; T: total value without direction specification.

In the simulation experiments, most control models can keep the tracking error within 1 m or even zero, which achieves good path control results. However, the quantification criteria of control accuracy are different. A unified quantitative indicator of the path control accuracy can be defined to further evaluate the model performance.

The problem of the error convergence time of control algorithms is not addressed in some studies, while instantaneity is the essential requirement of the control algorithm.

Thus, it is necessary to develop the ship motion control model under the premise of real-time control.

The present manned ships mostly rely on manual operation of the engine and rudder to realize ship motion control. There is always a time delay from the order by OOW to the execution of equipment. For intelligent ships, certain improvements are expected in execution time and execution accuracy due to the improvement of execution equipment and the application of control algorithms.

4. Ship Behavior on Different Autonomy Levels

In this section, the behaviors of autonomous ships at the different autonomy levels described in Section 2 will be individually evaluated using the algorithm characteristics introduced in Section 3. The ship behaviors at different autonomy levels during collision avoidance have been qualitatively illustrated for an intuitive comparison.

4.1. Manual Operation Level

In the process of information perception, decision-making, and execution, the quality of navigation of ships at the manual operation level completely depend on the crew's expertise and skills, which makes it difficult to be guaranteed.

In terms of information perception, the crew mainly obtains environmental information via their visual and auditory senses, which has low reliability due to the limited observation range and the high possibility of interference by external factors. As for collision avoidance decisions, the crew refers to COLREGs and local rules and applies their experience and expertise for collision avoidance operation. However, COLREGs only define basic behavior principles in specific encounter scenarios without detailed quantitative instructions to support situational assessment, which can hardly be directly applied in decision-making. Moreover, the pursuit of efficiency and economy is neglected as well. Regarding the execution, the huge intrinsic inertia and the random interference by the environmental factors make the ship motion control difficult. For ships at the manual operational level, the execution is usually poorly performed in terms of accuracy and stability control.

Figure 2 illustrates the operation of a manual operation level ship in a collision avoidance scenario. When approaching an obstacle or an oncoming ship and starting the actions, there is a time delay mainly composed of the reaction and decision time of the OOW and the response time of the execution equipment. To reach the purpose of safe navigation, the ship usually follows the safety principle in such a scenario, adopting a large turning angle. To use as little steering effort as possible, the ship normally sails in a straight line and resumes to the original course after being "past and clear", as suggested by COLREGs.

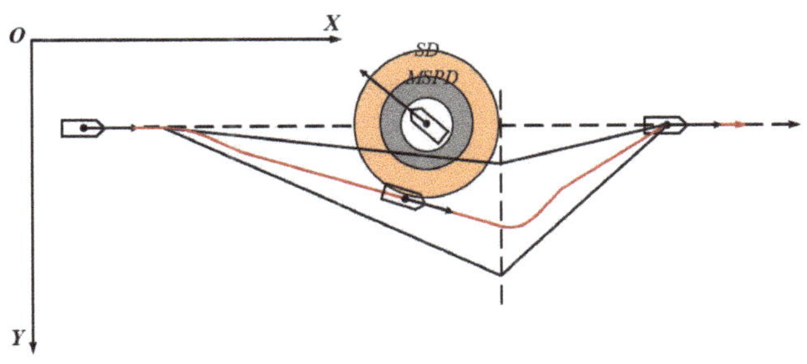

Figure 2. Illustration of the collision avoidance process of a ship at the manual operation level. The red line represents the collision avoidance path decided by the OOW, while the area between the two black lines shows the distribution range of the positions of ships.

4.2. Machine-Aided Level

Ships at the machine-aided level realize the automation of some specific functions with the assistance of machines, such as automatic course tracking. However, crews are still needed on board to complete the full sailing operations. The vast majority of ships currently in operation are at this autonomy level.

Based on the perception systems of the previous level, the machine aids such as AIS and Radar are added to have a broader monitoring range and higher information perception efficiency. For example, the application of maritime radar and AIS can acquire information about the surrounding environment nearly in real time. The ships at the machine-aided level complement the manual operation level in decision-making by providing other applicable alternatives of operations. On the aspect of execution, some navigation operations are also automated. For instance, the automatic rudder is used to keep the course stable, which reduces the workload of manual steering.

Despite the machine assistance, ships at this level are essentially manually operated and unable to respond to changes of environment and situation as quickly as high-autonomy ships. Time delays due to human factors still exist, as well as for the equipment execution. Secondly, there are also deficiencies in the trajectory control accuracy and energy efficiency management. The real-time control of speed and course is difficult to be realized, too. However, in comparison with ships of manual operation, the ships at the machine-aided level have improvements in time delay (shorter time delay) and track control accuracy (smaller distribution range), which enhances the navigation safety, as shown in Figure 3.

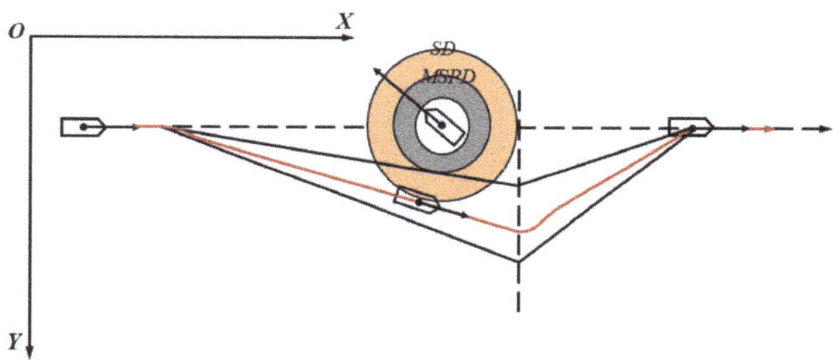

Figure 3. Illustration of the collision avoidance process of a ship at the machine-aided level.

4.3. Remote-Control Level

Before the maturity of fully autonomous navigation technology, the autonomous ships cannot meet the maneuvering requirements of navigation safety under the circumstances of complicated waters, bad weather, or equipment damage. Thus, a remote-control center is necessary to ensure the operation safety by remotely switching to manual control. In normal circumstances, the ships at this level can operate autonomously under the monitoring of the control center. Currently, the research on unmanned ships is mainly at this stage.

The remote-control ship returns the perceived environmental information and ship status to the control center. By applying the information fusion techniques using multi-source sensor data and the display technique of Mixed Reality (MR), the visualized remote interaction with intelligent ships can be realized.

The remote intelligent decision-making system applies the intelligent decision-making algorithms to realize the intelligent design and real-time optimization for the ship's navigation, which makes the optimal decision of the next-step action for the autonomous ship. Such a decision mostly considers the requirements of both safety and efficiency. The decision-making process is carried out under the monitoring of the control center, which requires the validity and completeness of information transmission between the remote-control center and the ship. In most waters, the ship–shore communication relies on satellite communication, which has the defects of vulnerability to interference by weather conditions, small communication bandwidth, and a long communication time delay. Besides, the decision-making time and quality of the person in charge in the control center is negatively affected due to the quality of the received information. Thus, when the autonomous ship sails in the above-mentioned extreme situations, the uncertainty of navigation will increase. Considering all influencing factors of humans and machines, higher uncertainty would lead to higher risks of navigation safety.

Ships at this level are usually equipped with a multi-directional power propulsion system, applying a motion control algorithm in cooperation with the dynamic positioning technique. It effectively improves the accuracy of ship motion control and maneuverability, which also enables the intelligent control of course and trajectory in open waters. Compared with the current operating ships, it shows obvious advantages in control accuracy and collision avoidance efficiency. A similar case for a ship at this level is illustrated in Figure 4.

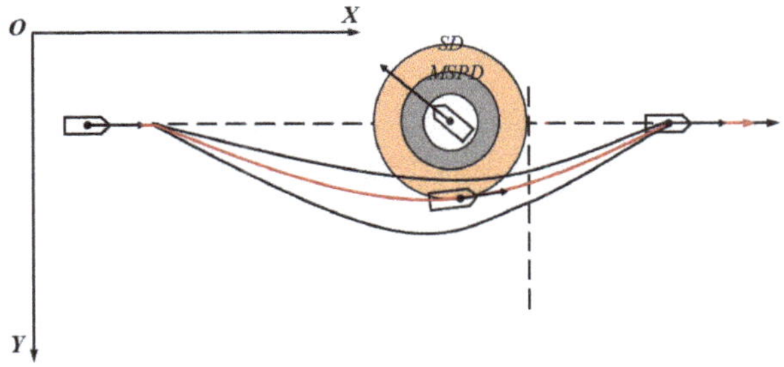

Figure 4. Illustration of the collision avoidance process of a ship at the remote-control level.

4.4. Onshore Supervision Level

In comparison with the ships at the previous level, the ships under onshore supervision have stronger autonomous control and scenario adaptability. It allows effective collision avoidance control in waters with complex traffic conditions and during berthing and unberthing phases, which means all-water collision avoidance and autonomous berthing and unberthing. The main function of the control center is transformed to supervise the operation of the ship, but still with the right to intervene in the high-impact decisions.

The ship at this level sails autonomously under the supervision of the onshore control center. Under the premise of energy efficiency management, it is capable of autonomously making decisions in scenarios such as dense traffic, bad weather, berthing and unberthing, navigation optimization, etc. It uses the power control unit and motion control algorithms to effectively release the manual workload and improve the working environment, which fundamentally reduces the influence of human factors on the navigation safety of ships. The illustrated path for the onshore supervision ship is presented in Figure 5.

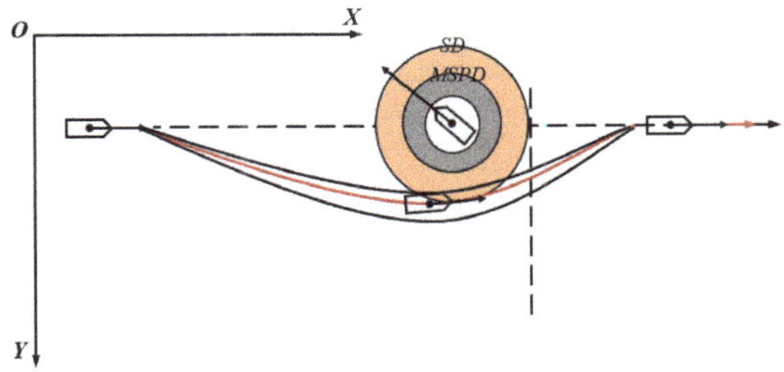

Figure 5. Illustration of the collision avoidance process of a ship at the onshore supervision or fully autonomous level.

4.5. Fully Autonomous Level

The fully autonomous ship no longer requires the continual supervision of the control center. The intervention only occurs in cases of emergency or at the request of the system. Its behavior performance is almost the same as the ship at the onshore supervision level, as shown in Figure 5.

4.6. Discussion

With the continuous development of the ship's intellectualization, intelligent ships at different autonomy levels will be put into operation in succession. However, in the current research on ship behavior, the differences among ships are only distinguished by manned and unmanned ships. The autonomy differences among intelligent ships are seldom considered. Research on ship behavior at different autonomy levels can be conducted in the future.

The research on ship behavior usually adopts simulation results based on collected or given data, generally irrespective of the possible existence of external data sources. This leads to errors when applying intelligent algorithms. In the future, it is necessary to consider the uncertainty of parameters in the study of ship behavior under the undefined influences.

5. Conclusions

With the continuous development and breakthrough of modern technologies in information, communication, sensors, and artificial intelligence, various countries have been actively promoting the research of intelligent ships. The ships are continuously developing towards intellectualization, to eventually become unmanned. However, the intellectualization of the whole waterborne traffic system cannot be quickly achieved, and is bound to be a hybrid system with intelligent ships at different autonomy levels co-existing. Therefore, it is important to investigate the ship's behavior under the mixed traffic scenario and master the methods and principles of ship navigation and collision avoidance to improve the navigation safety, reduce the navigation conflicts, improve the waterway capacity, and upgrade the maritime traffic supervision level.

This paper identified the three modules of the intelligent ship system, i.e., information perception, decision-making, and execution control. For each module, the currently adopted methods were analyzed. Based on the proposed classification of ship autonomy level, the behavioral characteristics of various types of autonomous ships were compared and discussed. For information perception, risk can be judged in a combined form of numerical indicators and visual graphics by applying data fusion techniques using multi-source ship sensor data. It enables the ships to evaluate their own situations in complex surroundings. In terms of decision-making, the researchers proposed intelligent algorithms to provide the optimal collision avoidance decision considering the ship's motion characteristics under the premise of energy efficiency management. Regarding execution, the studies adopted new power control unit and motion control algorithms to optimize the results of ship behavior control. It can be seen that as the level of ship autonomy upgrades, human functions in the three modules of perception, decision-making, and execution are gradually replaced by machines. The interaction between humans and intelligent systems is becoming more and more frequent.

Promoting autonomous shipping is not just a matter of continuing the existing research on manned and unmanned ships (machine-aided and full autonomy) but filling the blanks (remote-control and onshore supervision) between them. Specifically, rendering unmanned ships more operator-friendly and exploring more functions of existing manned ships. The future research could focus on: (1) studying the transition phase from manned to unmanned ships, (2) considering the effects of uncertainty on the models, (3) integrating realistic environment simulations, and (4) adding safety verification. Achieving safe and efficient navigation of ships in future mixed waterborne traffic scenarios would be the ultimate goal.

Author Contributions: Conceptualization, Y.T., L.C., and Y.Z.; methodology, L.C. and J.M.; investigation, Y.T.; writing—original draft preparation, Y.T.; writing—review and editing, Y.Z.; visualization, Y.T. and Y.Z.; supervision, L.C. and Y.Z.; project administration, J.M.; funding acquisition, J.M. and L.C. All authors have read and agreed to the published version of the manuscript.

Funding: This research was funded by the National Natural Science Foundation of China, Grant No. 52001242. The APC was funded by Delft University of Technology Library.

Institutional Review Board Statement: Not applicable.

Informed Consent Statement: Not applicable.

Conflicts of Interest: The authors declare no conflict of interest.

References

1. International Maritime Organization. *Report of the Maritime Safety Committee on Its Ninety-Eighth Session;* IMO: London, UK, 2017.
2. China's First Unmanned Cargo Ship Makes Maiden Test. Available online: https://www.navyrecognition.com/index.php/news/defence-news/2019/december/7811-china-s-first-unmanned-cargo-ship-makes-maiden-test.html (accessed on 10 November 2021).
3. Horizon 2020. Available online: https://ec.europa.eu/programmes/horizon2020/en/home (accessed on 10 November 2021).
4. AUTOSHIP. Available online: https://www.autoship-project.eu/ (accessed on 10 November 2021).
5. NOVIMAR. Available online: https://www.novimar.eu/ (accessed on 10 November 2021).
6. MOSES. Available online: https://moses-h2020.eu/about-moses/concept-innovations/ (accessed on 10 November 2021).
7. AEGIS. Available online: https://www.aegis-project.org/ (accessed on 10 November 2021).
8. Lee, S.W.; Jo, J.S.; Kim, S.W. Leveraging the 4th Industrial Revolution technology for sustainable development of the Northern Sea Route (NSR)—The case study of autonomous vessel. *Sustainability* **2021**, *13*, 8211. [CrossRef]
9. United Nations. *Review of Maritime Transport 2021;* United Nations: New York, NY, USA, 2021.
10. Liu, M.; Xiong, S. A refined and dynamic cellular automaton model for pedestrian–vehicle mixed traffic flow. *Int. J. Mod. Phys. C* **2015**, *27*, 1650053. [CrossRef]
11. Zhu, W.X.; Zhang, H.M. Analysis of mixed traffic flow with human-driving and autonomous cars based on car-following model. *Phys. A-Stat. Mech. Its Appl.* **2018**, *496*, 274–285. [CrossRef]
12. Perera, L.P.; Murray, B. Situation Awareness of Autonomous Ship Navigation in a Mixed Environment Under Advanced Ship Predictor. In Proceedings of the ASME 2019 38th International Conference on Ocean, Offshore and Arctic Engineering, Scotland, UK, 9–14 June 2019.
13. Zhou, Y.; Daamen, W.; Vellinga, T.; Hoogendoorn, S. Review of maritime traffic models from vessel behavior modeling perspective. *Transp. Res. Part C Emerg. Technol.* **2019**, *105*, 323–345. [CrossRef]
14. Zhou, Y.; Daamen, W.; Vellinga, T.; Hoogendoorn, S.P. Ship classification based on ship behavior clustering from AIS data. *Ocean Eng.* **2019**, *175*, 176–187. [CrossRef]
15. Zhou, Y.; Daamen, W.; Vellinga, T.; Hoogendoorn, S.P. Impacts of wind and current on ship behavior in ports and waterways: A quantitative analysis based on AIS data. *Ocean Eng.* **2020**, *213*, 107774. [CrossRef]
16. Lloyd's Register. *Cyber-Enabled Ships Shipright Procedure-Autonomous Ships;* Lloyd's Register: London, UK, 2016.
17. Sea-Kit's Uncrewed Surface Vessel Ends 22-Day Offshore Mission. Available online: https://www.oedigital.com/news/481024-sea-kit-s-uncrewed-surface-vessel-ends-22-day-offshore-mission (accessed on 10 November 2021).
18. Yara Birkeland. Available online: https://www.yara.com/news-and-media/press-kits/yara-birkeland-press-kit/ (accessed on 10 November 2021).
19. Ji, X.; Zhuang, J.; Su, Y. Marine Radar Target Detection for USV. *Adv. Mater. Res.* **2014**, *1006*, 863–869. [CrossRef]
20. Achtert, P.; Brooks, I.M.; Brooks, B.J.; Moat, B.I.; Prytherch, J.; Persson, P.O.G.; Tjernström, M. Measurement of wind profiles by motion-stabilised ship-borne Doppler lidar. *Atmos. Meas. Tech.* **2015**, *8*, 4993–5007. [CrossRef]
21. Prasad, D.K.; Prasath, C.K.; Rajan, D.; Rachmawati, L.; Rajabally, E.; Quek, C. Maritime situational awareness using adaptive multi-sensor management under hazy conditions. *arXiv* **2017**, arXiv:1702.00754.
22. Wang, Y.; Peng, M.; Liu, Z.; Wan, W.; Di, K.; Hu, C.; Liu, L.; Lv, T.; Yang, C. Binocular Visual Environment Perception Technology for Unmanned Surface Vehicle. *Int. Arch. Photogramm. Remote Sens. Spat. Inf. Sci.* **2020**, *XLIII-B2-2020*, 1297–1302. [CrossRef]
23. Gao, M.; Shi, G.-Y. Ship collision avoidance anthropomorphic decision-making for structured learning based on AIS with Seq-CGAN. *Ocean Eng.* **2020**, *217*, 107922. [CrossRef]
24. Li, Y.; Zheng, J. Real-time collision avoidance planning for unmanned surface vessels based on field theory. *ISA Trans.* **2020**, *106*, 233–242. [CrossRef] [PubMed]
25. Liu, H.; Sun, Y.; Li, B. Evaluation Modeling Establishment for the Risk Degree of Ship Collision. In *Application of Intelligent Systems in Multi-modal Information Analytics;* Springer: Cham, Switzerland, 2021; pp. 54–63.
26. Yoo, Y.; Lee, J.-S. Evaluation of ship collision risk assessments using environmental stress and collision risk models. *Ocean Eng.* **2019**, *191*, 106527. [CrossRef]

27. Hu, Y.; Zhang, A.; Tian, W.; Zhang, J.; Hou, Z. Multi-Ship Collision Avoidance Decision-Making Based on Collision Risk Index. *J. Mar. Sci. Eng.* **2020**, *8*, 640. [CrossRef]
28. Zheng, K.; Chen, Y.; Jiang, Y.; Qiao, S. A SVM based ship collision risk assessment algorithm. *Ocean Eng.* **2020**, *202*, 107062. [CrossRef]
29. Yu, H.; Fang, Z.; Murray, A.T.; Peng, G. A Direction-Constrained Space-Time Prism-Based Approach for Quantifying Possible Multi-Ship Collision Risks. *IEEE Trans. Intell. Transp. Syst.* **2021**, *22*, 131–141. [CrossRef]
30. Szlapczynski, R.; Szlapczynska, J. An analysis of domain-based ship collision risk parameters. *Ocean Eng.* **2016**, *126*, 47–56. [CrossRef]
31. Qiao, Z.; Zhang, Y.; Wang, S. A Collision Risk Identification Method for Autonomous Ships Based on Field Theory. *IEEE Access* **2021**, *9*, 30539–30550. [CrossRef]
32. Bakdi, A.; Glad, I.; Vanem, E.; Engelhardtsen, Ø. AIS-Based Multiple Vessel Collision and Grounding Risk Identification based on Adaptive Safety Domain. *Materials* **2019**, *8*, 5. [CrossRef]
33. Guan, Z.; Wang, Y.; Zhou, Z.; Wang, H. Research on Early Warning of Ship Danger Based on Composition Fuzzy Inference. *J. Mar. Sci. Eng.* **2020**, *8*, 1002. [CrossRef]
34. Yuan, X.; Zhang, D.; Zhang, J.; Zhang, M.; Guedes Soares, C. A novel real-time collision risk awareness method based on velocity obstacle considering uncertainties in ship dynamics. *Ocean Eng.* **2021**, *220*, 108436. [CrossRef]
35. Huang, Y.; Chen, L.; van Gelder, P.H.A.J.M. Generalized velocity obstacle algorithm for preventing ship collisions at sea. *Ocean Eng.* **2019**, *173*, 142–156. [CrossRef]
36. Huang, Y.; van Gelder, P. Time-Varying Risk Measurement for Ship Collision Prevention. *Risk Anal. Off. Publ. Soc. Risk Anal.* **2020**, *40*, 24–42. [CrossRef]
37. Szlapczynski, R.; Krata, P.; Szlapczynska, J. Ship domain applied to determining distances for collision avoidance manoeuvres in give-way situations. *Ocean Eng.* **2018**, *165*, 43–54. [CrossRef]
38. Namgung, H.; Kim, J.-S. Collision Risk Inference System for Maritime Autonomous Surface Ships Using COLREGs Rules Compliant Collision Avoidance. *IEEE Access* **2021**, *9*, 7823–7835. [CrossRef]
39. Lee, J.-S.; Kwon, Y.; Choi, J.-S. Basic Research for the Development of Collision Risk Model of Passing Vessels at an Anchorage (Safety Domain). *J. Korean Soc. Mar. Environ. Saf.* **2021**, *27*, 67–73. [CrossRef]
40. Huang, Y.; Chen, L.; Chen, P.; Negenborn, R.R.; van Gelder, P.H.A.J.M. Ship collision avoidance methods: State-of-the-art. *Saf. Sci.* **2020**, *121*, 451–473. [CrossRef]
41. Xu, X.; Lu, Y.; Liu, X.; Zhang, W. Intelligent collision avoidance algorithms for USVs via deep reinforcement learning under COLREGs. *Ocean Eng.* **2020**, *217*, 107704. [CrossRef]
42. Meyer, E.; Heiberg, A.; Rasheed, A.; San, O. COLREG-Compliant Collision Avoidance for Unmanned Surface Vehicle Using Deep Reinforcement Learning. *IEEE Access* **2020**, *8*, 165344–165364. [CrossRef]
43. Nasibov, E.; Fkn, R.; Yardimci, M.O. A Fuzzy Ship Domain-Based Method for Collision Avoidance at Sea. In Proceedings of the IEEE International Conference on Computer Science and Engineering (UBMK'19), Samsun, Turkey, 11–15 September 2019.
44. Lazarowska, A. A Discrete Artificial Potential Field for Ship Trajectory Planning. *J. Navig.* **2020**, *73*, 233–251. [CrossRef]
45. Serigstad, E.; Eriksen, B.-O.H.; Breivik, M. Hybrid Collision Avoidance for Autonomous Surface Vehicles. *IFAC-Pap.* **2018**, *51*, 1–7. [CrossRef]
46. Chen, P.; Huang, Y.; Papadimitriou, E.; Mou, J.; van Gelder, P.H.A.J.M. An improved time discretized non-linear velocity obstacle method for multi-ship encounter detection. *Ocean Eng.* **2020**, *196*, 106718. [CrossRef]
47. Fan, Y.; Sun, X.; Wang, G. An autonomous dynamic collision avoidance control method for unmanned surface vehicle in unknown ocean environment. *Int. J. Adv. Robot. Syst.* **2019**, *16*, 1729881419831581. [CrossRef]
48. Chen, L.; Hopman, H.; Negenborn, R.R. Distributed model predictive control for vessel train formations of cooperative multi-vessel systems. *Transp. Res. Part C Emerg. Technol.* **2018**, *92*, 101–118. [CrossRef]
49. Song, R.; Liu, Y.; Bucknall, R. A multi-layered fast marching method for unmanned surface vehicle path planning in a time-variant maritime environment. *Ocean Eng.* **2017**, *129*, 301–317. [CrossRef]
50. Guo, X.; Ji, M.; Zhao, Z.; Wen, D.; Zhang, W. Global path planning and multi-objective path control for unmanned surface vehicle based on modified particle swarm optimization (PSO) algorithm. *Ocean Eng.* **2020**, *216*, 107693. [CrossRef]
51. Lazarowska, A. Ship's Trajectory Planning for Collision Avoidance at Sea Based on Ant Colony Optimisation. *J. Navig.* **2015**, *68*, 291–307. [CrossRef]
52. Chen, P.; Huang, Y.; Papadimitriou, E.; Mou, J.; van Gelder, P. Global path planning for autonomous ship: A hybrid approach of Fast Marching Square and velocity obstacles methods. *Ocean Eng.* **2020**, *214*, 107793. [CrossRef]
53. Wang, T.; Wu, Q.; Zhang, J.; Wu, B.; Wang, Y. Autonomous decision-making scheme for multi-ship collision avoidance with iterative observation and inference. *Ocean Eng.* **2020**, *197*, 106873. [CrossRef]
54. Woo, J.; Kim, N. Collision avoidance for an unmanned surface vehicle using deep reinforcement learning. *Ocean Eng.* **2020**, *199*, 107001. [CrossRef]
55. Lazarowska, A. A discrete planning approach in collision avoidance for smart ships. *Procedia Comput. Sci.* **2020**, *176*, 380–389. [CrossRef]
56. Eriksen, B.; Bitar, G.; Breivik, M.; Lekkas, A.M. Hybrid Collision Avoidance for ASVs Compliant with COLREGs Rules 8 and 13–17. *Front. Robot. AI* **2019**, *7*, 11. [CrossRef]

57. Li, S.; Liu, J.; Negenborn, R.R. Distributed coordination for collision avoidance of multiple ships considering ship maneuverability. *Ocean Eng.* **2019**, *181*, 212–226. [CrossRef]
58. Singh, Y.; Sharma, S.; Sutton, R.; Hatton, D. Towards use of Dijkstra Algorithm for Optimal Navigation of an Unmanned Surface Vehicle in a Real-Time Marine Environment with results from Artificial Potential Field TransNav Int. *J. Mar. Navig. Saf. Sea Transp.* **2018**, *12*. [CrossRef]
59. Lyu, H.; Yin, Y. Fast Path Planning for Autonomous Ships in Restricted Waters. *Appl. Sci.* **2018**, *8*, 2592. [CrossRef]
60. Wang, Y.; Yu, X.; Liang, X.; Li, B. A COLREGs-based obstacle avoidance approach for unmanned surface vehicles. *Ocean Eng.* **2018**, *169*, 110–124. [CrossRef]
61. Candeloro, M.; Lekkas, A.M.; Sørensen, A.J. A Voronoi-diagram-based dynamic path-planning system for underactuated marine vessels. *Control Eng. Pract.* **2017**, *61*, 41–54. [CrossRef]
62. Liu, Z.; Zhang, Y.; Yuan, C.; Luo, J. Adaptive Path Following Control of Unmanned Surface Vehicles Considering Environmental Disturbances and System Constraints. *IEEE Trans. Syst. Man Cybern. Syst.* **2021**, *51*, 339–353. [CrossRef]
63. Liu, C.; Zheng, H.; Negenborn, R.; Chu, X.; Xie, S. Adaptive predictive path following control based on least squares support vector machines for underactuated autonomous vessels. *Asian J. Control* **2021**, *23*, 432–448. [CrossRef]
64. Liu, C.; Li, C.; Li, W. Computationally efficient MPC for path following of underactuated marine vessels using projection neural network. *Neural Comput. Appl.* **2020**, *32*, 7455–7464. [CrossRef]
65. Liu, C.; Wang, D.; Zhang, Y.; Meng, X. Model predictive control for path following and roll stabilization of marine vessels based on neurodynamic optimization. *Ocean Eng.* **2020**, *217*, 107524. [CrossRef]
66. Liang, H.; Li, H.; Xu, D. Nonlinear Model Predictive Trajectory Tracking Control of Underactuated Marine Vehicles: Theory and Experiment. *IEEE Trans. Ind. Electron.* **2021**, *68*, 4238–4248. [CrossRef]
67. Du, J.; Hu, X.; Sun, Y. Adaptive Robust Nonlinear Control Design for Course Tracking of Ships Subject to External Disturbances and Input Saturation. *IEEE Trans. Syst. Man Cybern. Syst.* **2020**, *50*, 193–202. [CrossRef]
68. Peng, Z.; Jiang, Y.; Wang, J. Event-Triggered Dynamic Surface Control of an Underactuated Autonomous Surface Vehicle for Target Enclosing. *IEEE Trans. Ind. Electron.* **2021**, *68*, 3402–3412. [CrossRef]
69. Weng, Y.; Wang, N.; Guedes Soares, C. Data-driven sideslip observer-based adaptive sliding-mode path-following control of underactuated marine vessels. *Ocean Eng.* **2020**, *197*, 106910. [CrossRef]
70. Yang, Z.; Lili, D. Adaptive back-stepping control on container ships for path following. *J. Syst. Eng. Electron.* **2020**, *31*, 780–790. [CrossRef]
71. Li, M.; Li, T.; Gao, X.; Shan, Q.; Chen, C.L.P.; Xiao, Y. Adaptive NN event-triggered control for path following of underactuated vessels with finite-time convergence. *Neurocomputing* **2020**, *379*, 203–213. [CrossRef]
72. Li, J.-H. Path tracking of underactuated ships with general form of dynamics. *Int. J. Control* **2016**, *89*, 506–517. [CrossRef]
73. Kurtoglu, D.; Bidikli, B.; Tatlicioglu, E.; Zergeroglu, E. Periodic disturbance estimation based adaptive robust control of marine vehicles. *Ocean Eng.* **2021**, *219*, 108351. [CrossRef]
74. Gao, T.; Huang, J.; Zhou, Y.; Song, Y.-D. Robust adaptive tracking control of an underactuated ship with guaranteed transient performance. *Int. J. Syst. Sci.* **2017**, *48*, 272–279. [CrossRef]
75. Dong, Z.; Wan, L.; Li, Y.; Liu, T.; Zhang, G. Trajectory tracking control of underactuated USV based on modified backstepping approach. *Int. J. Nav. Archit. Ocean Eng.* **2015**, *7*, 817–832. [CrossRef]
76. Xu, S.; Wang, X.; Yang, J.; Wang, L. A Fuzzy Rule Based PID Controller for Dynamic Positioning of Vessels in Variable Environmental Disturbances. *J. Mar. Sci. Technol.* **2020**, *25*, 914–924. [CrossRef]
77. Zhu, G.; Du, J.; Kao, Y. Robust adaptive neural trajectory tracking control of surface vessels under input and output constraints. *J. Frankl. Inst.* **2020**, *357*, 8591–8610. [CrossRef]
78. Zhang, G.; Yao, M.; Xu, J.; Zhang, W. Robust neural event-triggered control for dynamic positioning ships with actuator faults. *Ocean Eng.* **2020**, *207*, 107292. [CrossRef]
79. Wang, Y.; Wang, Y.; Nguyen, H.D. Experimental Study of Intelligent Autopilot for Surface Vessels Based on Neural Network Optimised PID Controller. In Proceedings of the 2019 Chinese Control and Decision Conference (CCDC), Nanchang, China, 3–5 June 2019; pp. 27–34.
80. Wang, Y.; Chai, S.; Nguyen, H.D. Unscented Kalman Filter trained neural network control design for ship autopilot with experimental and numerical approaches. *Appl. Ocean Res.* **2019**, *85*, 162–172. [CrossRef]

MDPI
St. Alban-Anlage 66
4052 Basel
Switzerland
Tel. +41 61 683 77 34
Fax +41 61 302 89 18
www.mdpi.com

Journal of Marine Science and Engineering Editorial Office
E-mail: jmse@mdpi.com
www.mdpi.com/journal/jmse

www.ingramcontent.com/pod-product-compliance
Lightning Source LLC
Chambersburg PA
CBHW061846210926
51850CB00006B/982
* 9 7 8 3 0 3 6 5 7 4 4 3 1 *